About IFPRI

The International Food Policy Research Institute (IFPRI), established in 1975, provides research-based policy solutions to sustainably reduce poverty and end hunger and malnutrition. The Institute conducts research, communicates results, optimizes partnerships, and builds capacity to ensure sustainable food production, promote healthy food systems, improve markets and trade, transform agriculture, build resilience, and strengthen institutions and governance. Gender is considered in all of the Institute's work. IFPRI collaborates with partners around the world, including development implementers, public institutions, the private sector, and farmers' organizations.

About ASTI

Extensive empirical evidence demonstrates that agricultural research and development (R&D) investments have greatly contributed to economic growth, agricultural development, and poverty reduction in developing regions. Numerous regional and subregional initiatives emphasize the importance of agricultural R&D to achieving the productivity growth urgently needed to feed expanding populations; reduce poverty levels; and address new challenges, such as those imposed by climate change. Agricultural Science and Technology Indicators (ASTI), led by the International Food Policy Research Institute (IFPRI) and operating within the portfolio of the CGIAR Research Program on Policies, Institutions, and Markets (PIM), contributes to this agenda by collecting, analyzing, and publishing quantitative and qualitative information and trends on funding sources, spending levels and allocations, human resource capacities, and institutional developments in agricultural research in low- and middle-income countries. Working with a large network of country-level collaborators, ASTI conducts primary surveys to collect data from government, higher education, nonprofit, and (when possible) private agricultural R&D agencies in around 80 developing countries worldwide.

Agricultural Research in Africa

Investing in Future Harvests

Agricultural Research in Africa

Investing in Future Harvests

Edited by John Lynam, Nienke Beintema, Johannes Roseboom, and Ousmane Badiane

A Peer-Reviewed Publication

International Food Policy Research Institute
Washington, DC

International Food Policy Research Institute
2033 K Street, NW
Washington, DC 20006-1002, USA
Telephone: 202-862-5600

DOI: http://dx.doi.org/10.2499/9780896292123

Library of Congress Cataloging-in-Publication Data

Names: Lynam, John K., editor. | Beintema, Nienke M., editor. | Roseboom, Johannes, editor. | Badiane, Ousmane, editor.
Title: Agricultural research in Africa : investing in future harvests / edited by John Lynam, Nienke Beintema, Johannes Roseboom, and Ousmane Badiane.
Description: Washington, D.C. : International Food Policy Research Institute, [2016] | Includes index.
Identifiers: LCCN 2016006138 | ISBN 9780896292123
Subjects: LCSH: Agriculture--Research--Africa.
Classification: LCC S542.A35 A37 2016 | DDC 338.1072096—dc23 LC record available at http://lccn.loc.gov/2016006138

Cover design: Anne C. Kerns/Anne Likes Red, Inc.
Project manager: Patricia Fowlkes/IFPRI
Book layout: Laurel Muller/Cohographics

Dedicated to Carl K. Eicher (1920–2014)

Carl was a pioneer in advocating for agricultural development in Africa. He devoted more than 50 years of his life to educating students, building institutions, and helping to develop African agricultural research capacity.

Contents

Tables, Figures, and Boxes

Tables

Figures

Boxes

Abbreviations and Acronyms

AATF	African Agricultural Technology Foundation
ABC	Brazilian Cooperation Agency
ACBF	African Capacity Building Foundation
ACCI	African Centre for Crop Improvement (University of Kwazulu-Natal, South Africa)
AERC	African Economic Research Consortium
AETS	Agricultural Expenditure Tracking Survey
AFSI	L'Aquila Food Security Initiative
AgGDP	agricultural gross domestic product
AGORA	Access to Global Online Research in Agriculture
AGRA	Alliance for a Green Revolution in Africa
AHI	African Highlands Initiative
AIS	agricultural innovation systems
ANAFE	African Network for Agriculture, Agroforestry and Natural Resources Education
APHRC	African Population and Health Research Center
APPSA	Agricultural Productivity Program for Southern Africa
AR4D	agricultural research for development
ARC	Agricultural Research Council (South Africa)
ARCN	Agricultural Research Council of Nigeria
ARS	Agricultural Research Service (USDA)

ASARECA	Association for Strengthening Agricultural Research in Eastern and Central Africa
ASTI	Agricultural Science and Technology Indicators
ATA	Agricultural Transformation Agency (Ethiopia)
AU	African Union
AUC	African Union Commission
AWARD	African Women in Agricultural Research and Development
BASIC	Building African Scientific and Institutional Capacity (network)
BMGF	Bill & Melinda Gates Foundation
BSc	bachelor of science
BecA	Biosciences eastern and central Africa
CAADP	Comprehensive Africa Agriculture Development Programme
CABI	Centre for Agriculture and Biosciences International
CARGS	competitive agricultural research grant scheme
CARMPoLEA	Center for Agricultural Research Management and Policy Learning for Eastern Africa
CARTA	Consortium for Advanced Research Training in Africa
CBC	classical biological control
CCAFS	Climate Change, Agriculture, and Food Security
CCARDESA	Centre for Coordination of Agricultural Research and Development for Southern Africa
CFA	Communauté Financière Africaine
CIAT	International Center for Tropical Agriculture
CIDA	Canadian International Development Agency
CIFOR	Center for International Forestry Research
CIMMYT	International Maize and Wheat Improvement Center
CIP	International Potato Center
CMAAE	Collaborative Master of Science in Agricultural and Applied Economics
CNRA	National Center for Agricultural Research (Côte d'Ivoire)

COMESA	Common Market for Eastern and Southern Africa
CORAF/ WECARD	West and Central African Council for Agricultural Research and Development
COST	European Cooperation in Science and Technology
CRP(s)	CGIAR Research Program(s)
CRSPs	collaborative research support programs
CSIR	Council for Scientific and Industrial Research (Ghana)
DAC	Development Assistance Committee (OECD)
DALYs	disability-adjusted life years
DFID	Department for International Development (UK)
DRD	Department of Research and Development (Tanzania)
DREAM	Dynamic Research Evaluation for Management model
DTMA	Drought Tolerant Maize for Africa
EAAPP	East Africa Agricultural Productivity Program
EACI	Education for African Crop Improvement
ECOWAS	Economic Community of West African States
EDRI	Ethiopian Development Research Institute
EDULINK	Educational Linkage
Embrapa	Brazilian Agricultural Research Corporation
epIA	ex post impact assessment
ERA-NET	European Research Area Network
ERC	European Research Council
ERS	Economic Research Service (USDA)
ESSP	Ethiopian Strategy Support Program
ETPs	European Technology Platforms
EU	European Union
FAAP	Framework for African Agricultural Productivity
FAC	Future Agricultures Consortium
FANR	Food, Agriculture and Natural Resources Directorate (SADC)
FAO	Food and Agriculture Organization of the United Nations

FARA	Forum for Agricultural Research in Africa
FFS	farmer field school
FIRCA	Interprofessional Fund for Agricultural Research and Advisory Services (Côte d'Ivoire)
FOFIFA	National Center for Applied Research and Rural Development (Madagascar)
FTE(s)	full-time equivalent(s)
FTF	Feed the Future (USAID)
G8	Group of Eight
G20	Group of Twenty
GAFSP	Global Agriculture and Food Security Program
GDP	gross domestic product
GFAR	Global Forum on Agricultural Research
GIS	geographic information system
GM	genetically modified
GSARS	Global Strategy to Improve Agriculture and Rural Statistics
IAC	InterAcademy Council
IAR4D	Integrated Agricultural Research for Development
ICARDA	International Center for Agricultural Research in the Dry Areas
ICRAF	World Agroforestry Centre
ICRISAT	International Crops Research Institute for the Semi-Arid Tropics
ICTs	information and communications technologies
IDO	intermediate development outcome
IDRC	Canada's International Development Research Centre
IFPRI	International Food Policy Research Institute
IITA	International Institute of Tropical Agriculture
ILCA	International Livestock Center for Africa
ILRAD	International Laboratory for Research on Animal Diseases

ILRI	International Livestock Research Institute
INNOVTE	Innovation for Agricultural Training and Education
INIA	National Agricultural Research Institute (Uruguay)
INRM	integrated natural resource management
IPR	intellectual property rights
IRAG	Guinean Agricultural Research Institute
IRR	internal rate of return
IRRI	International Rice Research Institute
ISNAR	International Service for National Agricultural Research
ISPC	International Science and Partnership Council
ISRA	Senegalese Agricultural Research Institute
KAPAP	Kenya Agricultural Productivity and Agribusiness Program
KALRO	Kenya Agricultural & Livestock Research Organization
KARI	Kenya Agricultural Research Institute
KSC	Kenya Seed Company
KTDA	Kenya Tea Development Agency
M&E	monitoring and evaluation
MAFAP	Monitoring African Food and Agricultural Policies
MDG(s)	Millennium Development Goal(s)
MDTF	Multi-Donor Trust Fund
MOU	memorandum of understanding
MSc	master of science
NAADS	National Agricultural Advisory Services (Uganda)
NAIP(s)	national agricultural investment plan(s)
NARI(s)	national agricultural research institute(s)
NARO	National Agricultural Research Organisation (Uganda)
NARO(s)	national agricultural research organization(s)
NARS(s)	national agricultural research system(s)
NEPAD	New Partnership for Africa's Development

NGO(s)	nongovernmental organization(s)
NPV	net present value
NRA	nominal rate of assistance
NRI	Natural Resources Institute (UK)
NRM	natural resource management
ODA	official development assistance
OECD	Organisation for Economic Co-operation and Development
OECD-CRS	Organisation for Economic Co-operation and Development Creditor Reporting System
OECD-DAC	Organisation for Economic Co-operation and Development Development Assistance Committee
OED	Operations Evaluation Department (World Bank)
OER	open educational resources
OFSP	orange-fleshed sweet potato
PAEs	public agricultural expenditures
PAEPARD	Platform for African–European Partnership on Agricultural Research for Development
PCP	Pearl Capital Partners
PEARL	Partnership to Enhance Agriculture in Rwanda through Linkages
PhD	doctor of philosophy
PIM	CGIAR Research Program on Policies, Institutions, and Markets
PIPA	participatory impact pathway analysis
PPP	purchasing power parity (exchange rates)
R&D	research and development
RAC	Reaching Agents of Change
RAIP(s)	regional agricultural investment plan(s)
RBM	results-based management
RCTs	randomized controlled trials
RECs	Regional Economic Communities

RELC(s)	research–extension linkage committee(s)
ReSAKSS	Regional Strategic Analysis and Knowledge Support System
RIU	Research into Use (DFID program)
RNRRS	Renewable Natural Resources Research Strategy
RUFORUM	Regional Universities Forum for Capacity Building in Agriculture
S&T	science and technology
S3A	Science Agenda for Agriculture in Africa
SACCAR	Southern African Centre for Cooperation in Agricultural Research
SADC	Southern African Development Community
SCARDA	Strengthening the Capacity for Agricultural Research and Development in Africa
SDG(s)	Sustainable Development Goal(s)
SMART	specific, measurable, attainable, relevant, and time-bound or trackable
SPAAR	Special Program for African Agricultural Research
SPEED	Statistics on Public Expenditure for Economic Development
SPFS	Special Program for Food Security
SPHI	Sweetpotato for Profit and Health Initiative
SPIA	Standing Panel on Impact Assessment (CGIAR)
SPREAD	Sustaining Partnerships to Enhance Rural Enterprise and Agribusiness Development
SRO(s)	subregional organization(s)
SSA	Africa south of the Sahara
SSA-CP	Sub-Saharan Africa Challenge Programme
T&V	training and visit
TEEAL	Essential Electronic Agriculture Library
TFP	total factor productivity
TOC	theory of change
TRFK	Tea Research Foundation of Kenya

UNESCO	United Nations Educational, Scientific and Cultural Organization
UNSD	United Nations Statistics Division
USAID	United States Agency for International Development
USDA	United States Department of Agriculture
WACCI	West Africa Centre for Crop Improvement (University of Ghana)
WARDA	West Africa Rice Development Association
WWII	World War II

Foreword

Despite unprecedented agricultural productivity growth in recent years, Africa south of the Sahara still lags far behind the rest of the developing world, and its success has been mainly driven by increased use of natural resources, including land. Moreover, growth has been unequal, such that many countries—especially the smaller ones—continue to face serious challenges and are becoming more dependent on food imports, not less. Africa desperately needs innovative methods to increase agricultural production efficiently, while ensuring environmental sustainability. This level of technical change will only occur with greater and more consistent investment in agricultural research and development (R&D)—a decades-long challenge that is ongoing.

Agricultural Research in Africa: Investing in Future Harvests takes a comprehensive look at the evolution, current status, and future goals of agricultural R&D in the region, offering analyses of the complex underlying issues that make Africa's development challenges unique, as well as insights into how they might be overcome. The authors focus on these issues in terms of human, institutional, and financial resources, as well as the effectiveness of R&D and its impact evaluation. In addition to highlighting the need to develop rural innovation capacity and increase the efficiency and effectiveness of R&D, the book makes the case for greater investment in African agricultural R&D as a crucial prerequisite to raising agricultural productivity and competitiveness, increasing rural incomes and food security, reducing poverty and food-import dependence, and halting environmental degradation.

Agricultural Research in Africa: Investing in Future Harvests is intended as a contribution to the creation of lasting support for agricultural research in the region, and ultimately to fulfilling the long-held aspiration of seeing Africa reach its great potential for equitable economic growth and development. On behalf of IFPRI, I thank the many experts on agricultural R&D and related policy in Africa who contributed to this book.

Shenggen Fan
Director General, IFPRI

Preface

Africa south of the Sahara (SSA) has seen unprecedented economic growth since the turn of the millennium, and in recent years poverty rates have steadily declined; yet, while this recent growth has improved rural livelihoods in many of the region's countries, especially the larger ones, numerous countries continue to face an array of serious challenges. Agriculture is the economic mainstay of many of the countries in SSA, providing a significant source of employment and staple food needs. Rapid population growth, rising and volatile food prices, increased agricultural imports, health and nutritional issues, and the adverse impacts of climate change—among numerous other issues—necessitate an acceleration of agricultural productivity without delay. Research and development (R&D) has returned as a priority for donors and policy and decisionmakers. The heads of state at the 2012 G20 meeting in Mexico, for example, highlighted the importance of R&D in promoting agricultural productivity and food security. The key role of R&D in increasing food protection while protecting natural resources was also stressed in the UN post-2015 development agenda and has also gained more attention through the Comprehensive Africa Agricultural Development Programme (CAADP). More recently, the Science Agenda for Agriculture in Africa (S3A) was adopted at the 2014 African Heads of State Summit, necessitating the development of a continentwide implementation plan.

Genesis of the Book and ASTI's Role

Quantitative information is fundamental to the understanding of the contribution of agricultural science and technology (S&T) to agricultural growth. Indicators derived from such information allow the performance, inputs,

and outcomes of agricultural S&T systems to be measured, monitored, and benchmarked. These indicators assist S&T stakeholders in formulating policy, setting priorities, and undertaking strategic planning, monitoring, and evaluation; they are also fundamental to measuring progress in the ongoing implementation of CAADP and of the new Science Agenda.
Agricultural Science and Technology Indicators (ASTI), led by the International Food Policy Research Institute (IFPRI) and operating within the portfolio of the CGIAR Research Program on Policies, Institutions, and Markets (PIM), is generally recognized as the authoritative source of information on the structure, financing, and capacity of agricultural S&T in low- and middle-income countries. Lessons learned over time have prompted the development of a number of approaches to enhance the dissemination and use of ASTI's outputs, including forming strong partnerships, tailoring information to different stakeholders' needs, and creating a set of interactive data tools available through ASTI's website. ASTI's data collection largely began as a series of ad hoc activities, mainly focusing on updating out-of-date datasets. The first of three grants in 2008 from the Bill & Melinda Gates Foundation, supplemented by funding from other donors, enabled ASTI to be transformed into a sustainable, decentralized system of frequent data compilation and analysis. This included the institutionalizion of a network of national and regional focal points to facilitate more frequent data gathering, synthesis, and analysis and to enhance local ownership of the data to stimulate their use for the purposes of country-level advocacy and analysis. As of 2015, ASTI began initiating a new round of data collection, which will include new output indicators and additional human resource information.

ASTI and the Forum for Agricultural Research in Africa (FARA) convened the conference, "Agricultural R&D: Investing in Africa's Future—Analyzing Trends, Challenges, and Opportunities" in December 2011 to define a roadmap for revitalizing agricultural research in the region structured around four overarching themes: (1) sustainable financing of agricultural research; (2) training the next generation of agricultural scientists; (3) effectively evaluating the performance of research institutes and systems; and (4) efficient organization of national agricultural research activities supported by regional and international initiatives. The conference papers and deliberations primarily focused on the current state of agricultural R&D in SSA.[1] After the con-

1 The commissioned conference papers, which were published as an ASTI/IFPRI–FARA Working Paper series, and a brief synopsis, "Reflections on the Conference," are available at www.asti.cgiar.org/2011conf.

ference it was determined that important issues pointing to future solutions needed further attention, so the plan unfolded to expand and revise some of the conference papers, to commission new ones both to fill gaps and provide a more forward-looking perspective, and to publish these in book form as a timely input into Africa's emerging development agenda.

Overview of ASTI's Data and Methodology

The analysis in this volume is based on comprehensive datasets derived from primary surveys conducted by ASTI in the region during 2001–2004, 2009–2010, and 2012–2013, along with subsequent country- and regional-level documents published during 2001–2014. ASTI datasets are collected and processed using internationally accepted definitions and statistical procedures for compiling R&D statistics developed by the Organisation for Economic Co-operation and Development (OECD) and the United Nations Educational, Scientific and Cultural Organization (UNESCO). Key procedures are noted below (more detailed information is available at www.asti.cgiar.org/methodology).

1. ASTI defines agricultural research to include activities related to crops, livestock, forestry, fisheries, natural resources, and the socioeconomic aspects of primary agricultural production. On-farm storage and processing of agricultural products are also included, but research relating to off-farm postharvest or food processing activities is excluded.

2. ASTI facilitates cross-country comparisons, all financial data are converted to 2005 PPP (purchasing power parity) prices using the World Bank's World Development Indicators; PPPs measure the relative purchasing power of currencies across countries by eliminating national differences in pricing levels for a wide range of goods and services.

3. Human resource and financial data are calculated in full-time equivalents (FTEs), which take into account the proportion of time scientists/faculty members spend on research as opposed to managerial, administrative, teaching, or other nonresearch-related activities; for example, four scientists estimated to spend 25 percent of their time on research would individually represent 0.25 FTEs and collectively be counted as 1 FTE.

4. Agricultural R&D is defined to include government, higher education, and nonprofit agencies involved in agricultural R&D, irrespective of the

source of funding; hence, research outsourced by the private sector to public agencies is included, but research directly conducted by the private sector is excluded because private firms seldom share their financial and other data.

5. ASTI measures financial and human resources on a "performer" basis, meaning the entity undertaking the research, not the entity or entities funding it. Note that agricultural research entities can be counted as individual faculties, departments, centers or institutes with a university (such as Department of Forestry and a Department of Agriculture) or numerous research centers within a single council, as opposed to just the university or council.

Structure of the Book

Part One provides an overview of the evolution, current status, and future goals of African agricultural R&D. In Chapter 1, Ousmane Badiane and Julia Collins review the current status of agricultural productivity in African countries against the goals outlined in national-level strategy documents, investment plans, and compacts under CAADP. The authors present a discussion of the diverse challenges countries face in achieving these ambitious goals, thereby providing necessary context on the role and importance of national agricultural research systems (NARSs) and technical change. In Chapter 2, Johannes Roseboom and Kathleen Flaherty detail the evolution and current status of agricultural R&D in SSA regarding its organizational and institutional structure, briefly addressing the policy context and highlighting key institutional changes at national, regional, and international levels. The discussion addresses various fundamental design issues and the various structural linkages, such as across research entities and related functions. In Chapter 3, Keith Fuglie and Nicholas Rada present empirical evidence on the returns to agricultural research in SSA, offering some implications for agricultural science policy. The authors look at the correlation between enhanced productivity and the adoption of new technologies developed by international agricultural research centers and NARSs, as well as how payoffs to agricultural R&D investments are affected by a country's size and the size of its agricultural economy.

Part Two focuses on financial investments. In Chapter 4, Gert-Jan Stads assesses long-term spending trends in agricultural R&D, highlighting differences across countries, offering insight into funding sources and mechanisms, and explaining why spending has increased in some countries and fallen in others. Stads quantifies and assesses volatility in agricultural R&D spending in and across countries, identifies the main drivers of volatility and measures needed to cushion funding shocks, and assesses past resource allocation by national governments. In Chapter 5, Samuel Benin, Linden McBride, and Tewodaj Mogues present an overview of the evidence of returns to agricultural R&D spending compared with other agricultural, as well as nonagricultural, investments. The authors make a strong case for the importance of agricultural R&D, even in the presence of competing needs, suggesting that the high returns are themselves evidence of underinvestment. The authors then go on to analyze why—despite the evidence—agricultural research is still not a high enough political priority and what steps can be taken to change this. In Chapter 6, Prahbu Pingali, David Spielman, and Fatima Zaidi discuss global economic shifts since the turn of the millennium that have prompted changes in public and private donor commitments to agriculture and agricultural R&D. New donors, donor strategies, and funding streams have spurred rapid changes in the way agricultural R&D is structured globally. These changes are having both intended and unintended consequences for national strategies designed to drive agricultural productivity, food security, and economic growth. The authors examine these changing trends and explore their likely impacts. In Chapter 7, Carl Pray, Derek Byerlee, and Latha Nagarajan explore private-sector potential for increasing investment in agricultural R&D in SSA, particularly based on its limitations to date compared with Asia and Latin America. R&D targeting agribusiness and innovation is likely to grow rapidly in the region in response to numerous factors, and if this trend is to continue and be taken advantage of, African governments will need to build and maintain support for agribusiness research and innovation contributed by the private sector.

Part Three addresses human resource development and tertiary or university-level education. In Chapter 8, Nienke Beintema and Howard Elliott assess long-term staffing trends, education levels, turnover, and future human resource requirements by the region's NARSs and discuss key human resource challenges many SSA countries are facing. The authors highlight a number of important current capacity-building initiatives at the regional levels. The chapter concludes with an overview of some of the human resource strategies that have been adopted in some countries that could be replicated in others. In Chapter 9, Moses Osiru, Paul Nampala, and Adipala Ekwamu discuss SSA's higher-education sector in the context of agricultural R&D for development, which has evolved significantly since the 1990s in response to significant investment. The rapidly expanding university sector has brought new challenges for the region's agricultural faculties, especially in meeting sharply rising demand for higher education and new and better skill levels. The authors also analyze the potential for agricultural faculties to upgrade their postgraduate programs and meet the expanded staffing needs of both faculties and NARIs, as well as from the private sector and civil society. In Chapter 10, Joyce Lewinger Moock identifies different models of strategic networks currently making progress toward the advancing university-based training and research, thereby enhancing agricultural productivity. The types of networks featured are critical mechanisms for building the next generation of innovation-savvy agricultural scientists.

Part Four explores measuring and improving the effectiveness of agricultural R&D. In Chapter 11, George Norton and Jeffrey Alwang review past impact assessment of agricultural research in SSA, emphasizing ex post assessments, but also presenting lessons for ex ante analysis and priority-setting. The authors briefly describe and critique methods used to evaluate agricultural research, then go on to summarize and categorize empirical evidence on the benefits of agricultural research in the region by type of research, and finally specify lessons on the role of impact assessment for agricultural research in SSA. In Chapter 12, Howard Elliott and John Lynam review evaluation methods applied in agricultural development projects in SSA generally, together with recent experience in developing performance monitoring systems in agricultural research bodies such as the CGIAR Consortium. Developing a results framework for agricultural research offers particular challenges given the complexity of farming systems, time lags in the attainment of impact, and technology adoption's dependence on contextual factors such as markets and institutions.

Part Five looks into rationalizing and aligning institutional structures. In Chapter 13, John Lynam, Joseph Methu, and Michael Waithaka assess the potential for designing and managing agricultural research within an agricultural innovation systems framework. Such systems are increasingly providing a framework for donor investment, focusing on improving innovation capacity and the linkages among actors in the system, and ensuring the application of research knowledge within rural innovation. Nevertheless, such systems have been most successfully applied in more developed market economies. In Chapter 14, Johannes Roseboom presents an overview of current thinking on agricultural technology spillovers, as well as recent attempts to quantify technology spillover potential in agriculture globally, across Africa's subregions, and among member countries of the region's respective subregional organizations (SROs). He then goes on to assess how this spillover potential is likely to affect regional collaboration in agricultural research, and what the desired impact should be. In Chapter 15, Harold Roy-Macauley, Anne-Marie Izac, and Frank Rijsberman review the CGIAR Consortium's involvement in agricultural R&D in SSA over the past 30 years. They analyze the Consortium's capacity to respond to the region's development challenges and discuss recent CGIAR reform in the context of aligning with regional priorities and ensuring that public goods make a positive contribution to development challenges at the national level. In addition, the authors analyze the role and importance of the SROs and the associated implications for research funding, as well as critically assessing the opportunities and challenges CGIAR faces in forging stronger partnerships with the NARSs, FARA, the SROs, and other regional organizations.

Finally, in Part Six, Chapter 16 presents the editors' synthesis, highlighting key arguments and findings, and delineating policy issues and the way forward.

John Lynam, Nienke Beintema,
Johannes Roseboom, and Ousmane Badiane

Acknowledgments

The editors and authors of this manuscript would first like to thank the Bill & Melinda Gates Foundation and the CGIAR Program on Policies, Institutions, and Markets (PIM), whose ongoing financial support of Agricultural Science and Technology Indicators (ASTI) made this work possible. The editors would also like to thank the original contributors to and participants of the 2011 conference that ultimately gave rise to the book: "Agricultural R&D: Investing in Africa's Future—Analyzing Trends, Challenges, and Opportunities," convened by ASTI and the Forum for Agricultural Research in Africa (FARA).

In particular, the editors and authors are indebted to Mary Jane Banks, a consulting writer/editor, whose considerable expertise in technical editing—including a unique aptitude for restructuring, condensing, and focusing material—substantially improved the manuscript, increased its accessibility to a broad audience, and freed its editors to focus on issues of content rather than form.

The editors offer sincere thanks to Gershon Feder and Corinne Garber, chair and secretary, respectively, of the Publications Review Committee of the International Food Policy Research Institute (IFPRI), who guided the book through the review process. Gratitude is also offered to numerous reviewers who provided valuable feedback at various stages of the book's production, including during IFPRI's official review process. The editors would also like to thank Andrea Pedolsky, the former head of IFPRI's Publications Unit, who provided key guidance throughout the development of the book, and Patricia Fowlkes, the IFPRI project manager assigned to the manuscript, who ably coordinated its final production, and whose patience, attention to detail, and good humor were greatly appreciated.

Finally, the editors would like to thank the book's authors for their ongoing contributions and commitment to the book throughout its extended evolution.

PART 1

Overview of Evolution, Current Status, and Future Goals

Chapter 1

AGRICULTURAL GROWTH AND PRODUCTIVITY IN AFRICA: RECENT TRENDS AND FUTURE OUTLOOK

Ousmane Badiane and Julia Collins

As in any other part of the world, agriculture in Africa is dependent on constraints and limitations imposed by the natural resource base. Foremost among these constraints are those related to land availability, land fertility, and access to water. Traditionally, African countries have faced fewer land constraints because of their comparatively low population density, but many rural areas may be facing rapidly rising population density (Jayne, Chamberlin, and Muyanga 2012). Water constraints similarly limit production in many areas. Almost all agricultural production in Africa is rainfed. Barely 4 percent of the cultivated area in Africa south of the Sahara (SSA) was equipped for irrigation in 2005, compared with 18 percent globally (Svendsen, Ewing, and Msangi 2009). Rainfall, which varies dramatically across Africa, often limits agricultural production during dry seasons and droughts (Xie et al. 2014). Soil degradation is another constraint that tends to worsen with increasing population density and associated shortened fallow periods in many areas. Country-level estimates of yearly productivity losses resulting from land degradation, summarized in Bojo (1996), ranged from less than 1 percent of agricultural gross domestic product (AgGDP) to more than 15 percent.

For many decades, African countries have lagged behind other developing regions in terms of economic performance in general and agricultural performance in particular. It is now widely accepted that African economies are undergoing a remarkable recovery. Although the recent acceleration in growth is welcome and impressive, the gap is still significant, and enormous work remains to be done to sustain the recovery and build on recent progress toward reducing poverty. African countries are showing greater willingness and intention to increase agricultural investments, but most countries still appear to be underinvesting in agricultural research and development (R&D), hindering their ability to generate the technological innovations needed for agricultural productivity growth. Future progress by African countries in spurring agricultural growth and reducing poverty

requires adequate strategies to promote technical change and raise factor productivity—especially labor productivity—among the rural poor. The key is a policy and institutional framework that supports agricultural innovation and competitiveness by helping African
countries successfully address the resource limitations and constraints they are facing.

This chapter reviews the performance of African countries in terms of agricultural growth, productivity, and poverty reduction. After several decades of economic and agricultural stagnation, and even decline, agricultural growth and productivity have increased impressively in the past one to two decades. This chapter analyzes the magnitude and nature of the changes that have taken place in order to draw lessons for future policies and programs focusing on technological innovation. It then assesses current efforts by African countries to sustain the recent recovery under the Comprehensive Africa Agriculture Development Programme (CAADP) of the New Partnership for Africa's Development (NEPAD). A large number of countries have put ambitious national agricultural investment plans in place under CAADP that would significantly increase agricultural growth and reduce poverty, if the stated goals can be achieved. Key questions raised include to what degree countries are making the necessary investments to foster agricultural productivity growth, what gaps remain to be filled, and what the implications are for innovation policies and programs. The discussion in this chapter provides a backdrop for subsequent chapters in this volume that examine various aspects of the science and technology (S&T) agenda in Africa, including the effectiveness and impact of national agricultural research systems (NARSs), the role and reach of investments in human and institutional resources to promote innovation and raise productivity, and goals and targets designed to guide future strategies.

Agricultural Growth: From Decline to Recovery

African countries entered the postcolonial era with strong overall economic growth performance. On average, gross domestic product (GDP) grew at 4.7 percent per year from 1960 to 1970, surpassing South Asia's rate of 4.0 percent.[1] Growth performance began to deteriorate rapidly in the following decade, reaching crisis proportions in the 1980s. Average GDP growth in African countries fell from 3.4 percent per year in the 1970s to 1.4 percent per

1 Growth rates refer to SSA and were calculated from World Bank (2014b).

year in the 1980s. Per capita GDP grew at 2.1 percent per year during the 1960s, compared with 1.7 percent in South Asia, but slowed thereafter. By the 1980s, Africa entered a period of negative per capita GDP growth: –1.4 percent per year in the 1980s and –0.4 percent per year in the 1990s.

The pace of agricultural growth followed the same declining trends. During the 1970s, the average rate of agricultural growth was a mere 2.5 percent per year, representing a low point in Africa's post-independence performance. Growth increased to 3.2 percent per year in the 1990s, accelerating in the following decades to reverse the negative trends of the 1970s and 1980s. In the new millennium, growth increased to reach 4.6 percent per year between 2002 and 2010. During the food and financial crises that devastated the global economy in 2008–2009, Africa managed to maintain relatively healthy positive growth. Agricultural growth continued to accelerate into the current decade at an average rate of 5.1 percent—nearly twice the rate of population growth of 2.7 percent. Over the past 15 years, African countries have experienced the longest period of sustained economic and agricultural growth since independence. More strikingly, growth has not only accelerated, but also spread broadly across all the major subregions (Figure 1.1).

The aggregated GDP and agricultural growth trends discussed above mask significant subregional and national differences that, notably, are the levels at which actual policies and programs are designed and implemented (Figure 1.2). For most countries, both GDP and agricultural growth generally move in the same direction, the exception being mineral-rich countries like Guinea and Zambia. This fact is not new, but is also not likely to change in the foreseeable future. However, as seen in the ensuing discussion, it has significant implications for future growth and development strategies, and particularly for the agricultural S&T agenda. A number of countries are still facing considerable growth challenges (Figure 1.2). With the exception of West Africa, the majority of countries in all the other subregions exhibit agricultural growth rates that are still markedly below the 6 percent target set under CAADP.

Consequently, while the current recovery is encouraging, and growth is moving in the right direction, significant efforts are still needed in many countries to sustain and further accelerate the recovery process. This becomes even clearer when looking beyond the absolute rates of growth to trends in per capita agricultural production over the past five decades. The rapid growth in the past 15 years has at best allowed countries to make up for the lost decades of the 1970s and 1980s (Figure 1.3). With the notable exception of West Africa, average regional per capita production is still well below the levels of the 1960s.

FIGURE 1.1 Agricultural GDP growth recovery among African countries, 1990–2000 and 2000–2013

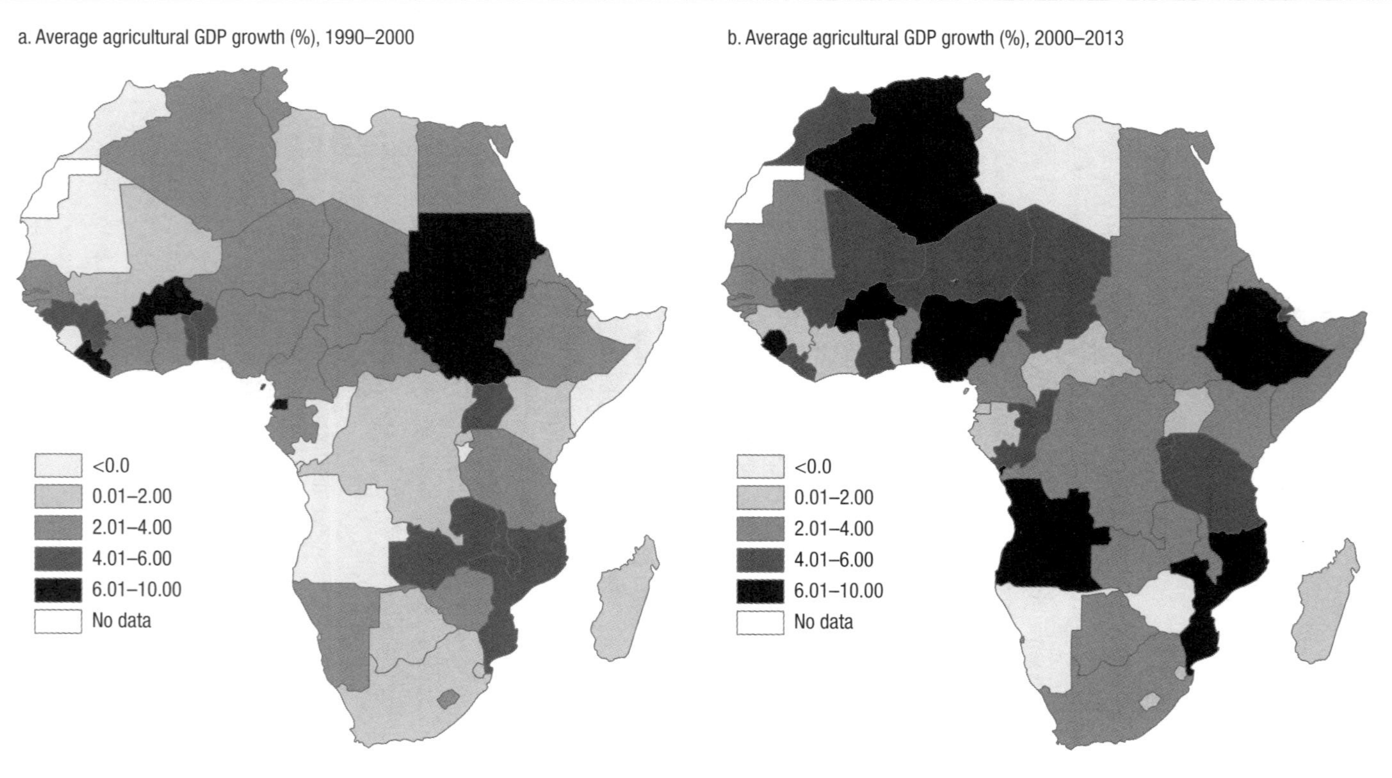

Source: Calculated by authors based on UNSD (2013).

Note: The lighter/darker a country's shading, the lower/higher its growth rate; the contrast between the two maps signals the extent to which growth has not only accelerated, but also spread geographically.

FIGURE 1.2 GDP and AgGDP growth by subregion, 1995–2003 and 2003–2010 averages

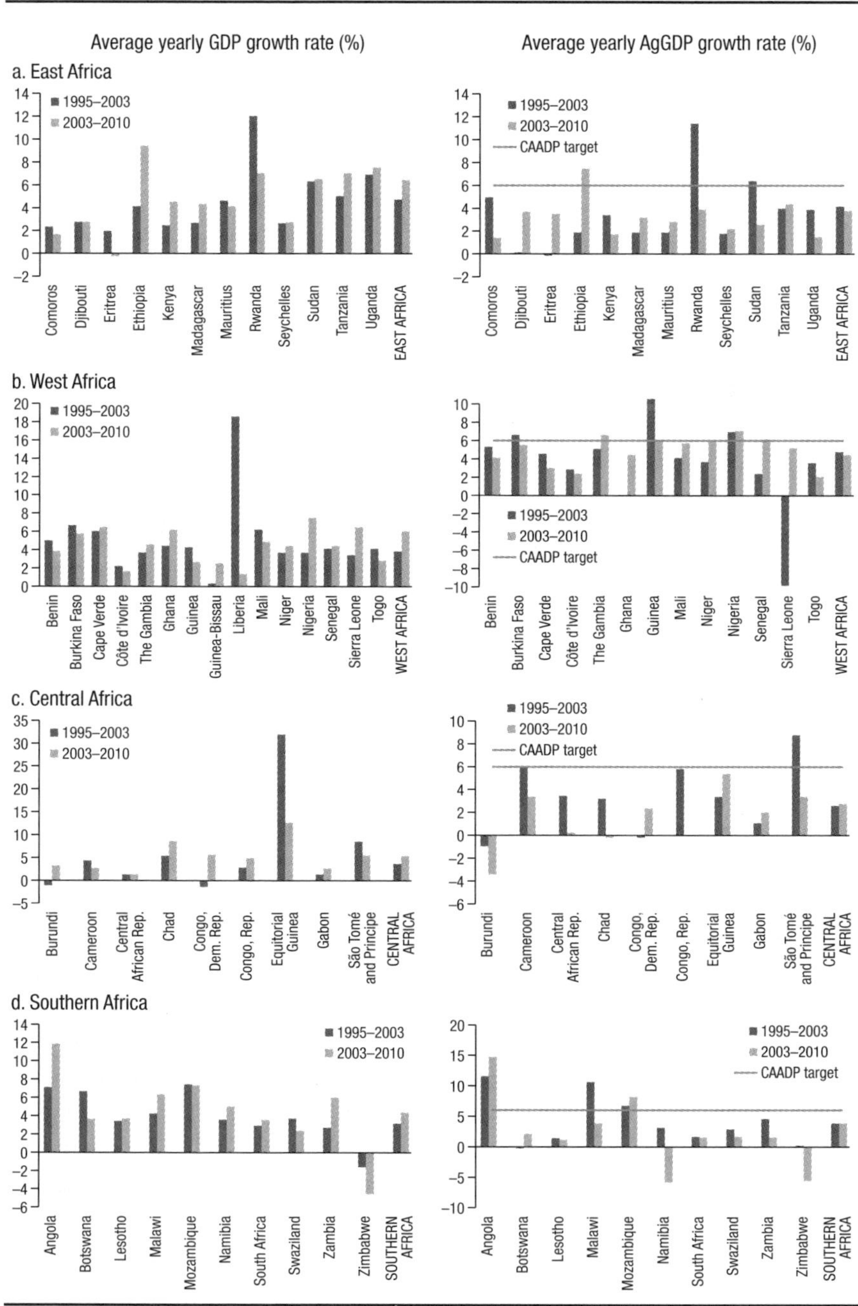

Source: ReSAKSS (2013b).

Notes: AgGDP = agricultural gross domestic product; CAADP = Comprehensive Africa Agriculture Development Programme; GDP = gross domestic product.

FIGURE 1.3 Indexes of per capita gross agricultural production by subregion, 1961–2012

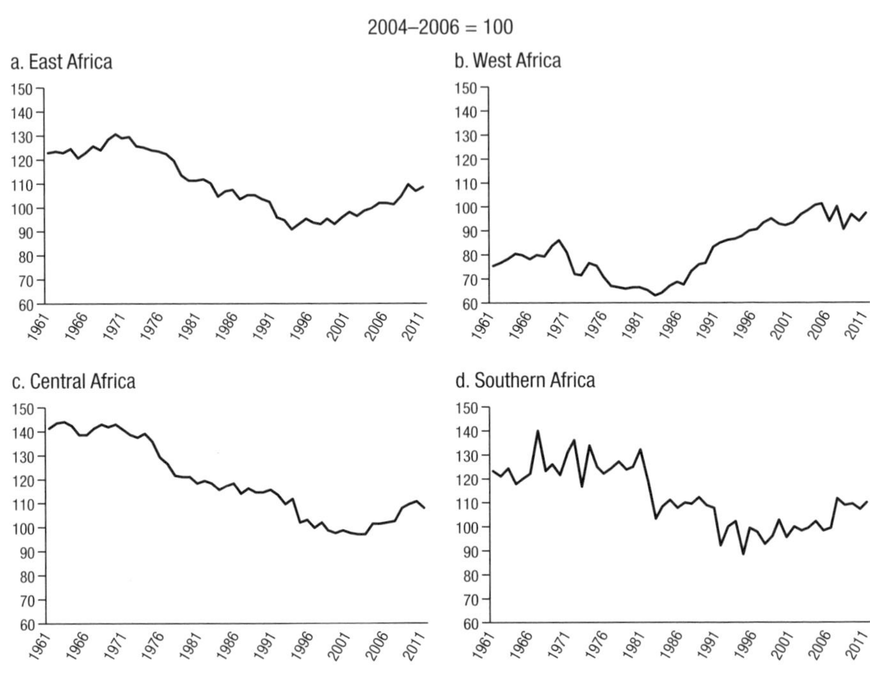

Source: FAO (2014).

The difficulty in raising the pace of agricultural growth above that of population growth has led to a rapid increase in agricultural import expenditures by African countries. The total value of agricultural imports rose tenfold between 2001 and 2011 to nearly US$80 billion.[2] While African countries have quadrupled the value of agricultural exports during the same period to more than $40 billion, the agricultural trade deficit has widened significantly, and is about to reach the same value as overall agricultural exports (Badiane, Makombe, and Bahiigwa 2014). Faster income growth and rapid urbanization have contributed to the rapid increase in food demand, in general, and food imports, in particular. In the face of limited scope to raise agricultural output in other major world regions significantly, failure to further accelerate and sustain growth in the agricultural sector would have two major negative consequences for African countries. On one hand, they stand to miss the opportunity to capture a larger share of the steadily expanding demand on African

2 All currency is in US dollars, unless specifically noted otherwise.

and global agricultural markets and, hence, the opportunity to create wealth in Africa and earn foreign exchange. On the other hand, greater dependency on food imports makes African countries more vulnerable to (temporary) shortages in global food markets and, hence, to greater volatility in food prices.

It is clear from the above that the lost decades leading to and just following the structural adjustment era not only caused very slow progress in reducing poverty for most African countries, but also altered the role of African countries in the global food and agricultural economy. As a result, the recent growth performance, and its strategic implications, must be assessed in the context of the progress that has been, and still needs to be, realized in terms of the role of agriculture in African economies and its implications for reducing poverty. This is dealt with in the next section.

Growth Performance and Poverty Outcomes

Africa's overall economic and agricultural growth performance over the past five decades is mirrored in changes in poverty levels over the same timeframe. Although Africa had poverty rates lower than those of East Asia and Pacific and South Asia in the 1980s, rates began to rise in Africa while they declined in the other two regions (World Bank 2014a). In 1981, 51.5 percent of Africa's population lived below the $1.25 a day international poverty line compared with 77.2 percent in East Asia and Pacific and 61.1 percent in South Asia. During 1981–1993, the proportion of people living below the international poverty line increased by 7.9 percentage points in Africa, whereas it decreased by 26.5 and 9.4 percentage points in East Asia and Pacific and South Asia, respectively. As Africa's economic growth began showing signs of improvement in the mid-1990s, poverty rates also began trending downward. In particular, between 1993 and 2010, Africa's poverty rate fell by 10.9 percentage points to 48.5 percent in 2010. Meanwhile, poverty fell more rapidly in other developing regions—by 38.2 and 20.7 percentage points to 12.5 and 31.0 percent in 2010 in East Asia and Pacific and South Asia, respectively. Although the proportion of poor people has trended downward, Africa is the only developing region of the world where the absolute number of poor people living below $1.25 a day continued to rise, from 330.0 million in 1993 to 413.7 million in 2010. Meanwhile, over the same period, the numbers of poor people fell drastically in East Asia and Pacific, by 71.2 percent to 250.9 million people, and in South Asia, by 19.8 percent to 506.8 million people.

The challenges of increasing growth and reducing poverty are best appreciated by looking at progress toward the Millennium Development Goal

FIGURE 1.4 Poverty and Millennium Development Goal targets by subregion, 1990–2015

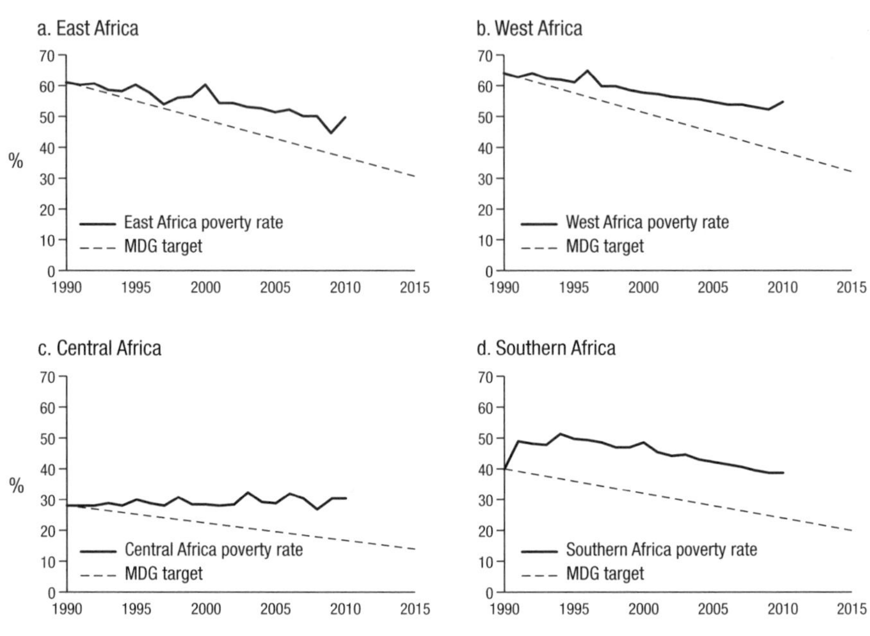

Source: ReSAKSS (2013b).
Note: MDG = Millennium Development Goal.

(MDG) of halving poverty by 2015. While poverty levels have fallen across the board—in particular in the 2000s, with the exception of Central Africa—none of Africa's subregions is on track to meet the poverty goal (Figure 1.4). More important, the remaining gap suggests that considerably more efforts are needed if the goal is to be achieved soon. The slow progress in eliminating poverty is an effect of agricultural growth trends. The stagnation and decline in the agricultural sector during most of the post-independence era has led to a *stunting* of the sector. In other words, poor agricultural performance over decades has led to a rapid decline in the sector's share of GDP in most African economies. As a result, agriculture plays a less significant role in African economies than should have been the case, based on their current levels of development.

To measure the extent of the stunting of the agricultural sector, the relationship between per capita income and the relative size of the sector was estimated and used to compare actual and "expected" shares among African countries (Figure 1.5). AgGDP shares have declined much faster in African countries than would be expected considering the slow income growth. Observed average

FIGURE 1.5 Expected versus actual average AgGDP shares, selected African countries, 2000–2008

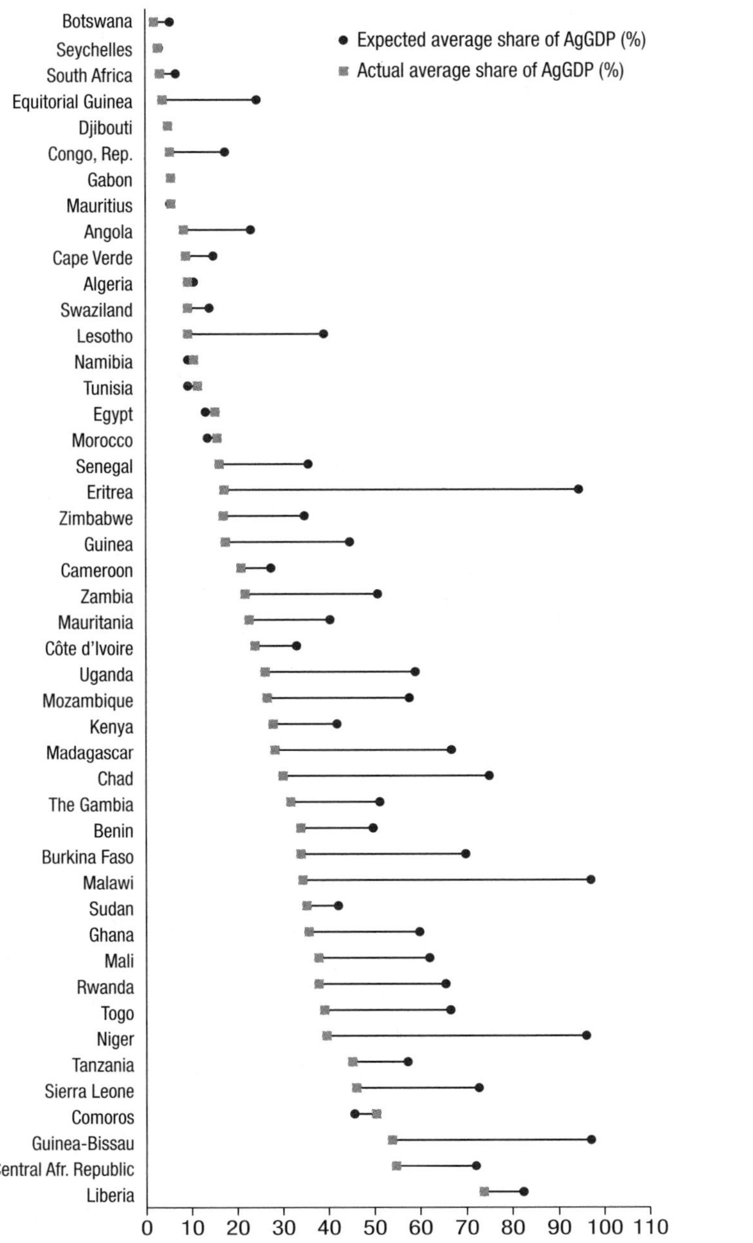

Source: Based on Badiane (2011).

Note: AgGDP = agricultural gross domestic product.

FIGURE 1.6 Agricultural sector underperformance and poverty levels, 1995–2012

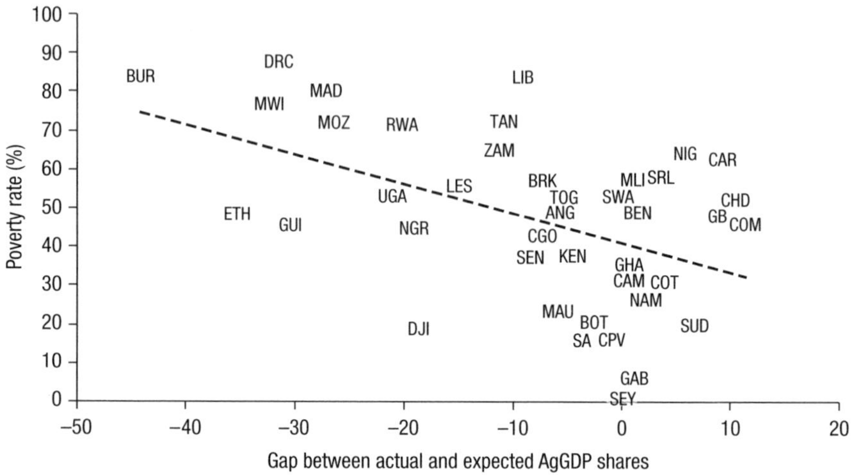

Source: Constructed by authors from World Bank (2014b, 2015).

Notes: The size of the deviation decreases moving away from the origin along the x-axis; as is indicated by the plotline, the level of poverty among countries increases with the size of the performance gap. AgGDP = agricultural gross domestic product; ANG = Angola; BEN = Benin; BOT = Botswana; BRK = Burkina Faso; BUR = Burundi; CAM = Cameroon; CAR = Central African Republic; CGO = Republic of the Congo; CHD = Chad; COM = Comoros; COT = Côte d'Ivoire; CPV = Cape Verde; DJI = Djibouti; DRC = Democratic Republic of the Congo; ETH = Ethiopia; GAB = Gabon; GB = Guinea-Bissau; GHA = Ghana; GUI = Guinea; KEN = Kenya; LES = Lesotho; LIB = Liberia; MAD = Madagascar; MAU = Mauritania; MLI = Mali; MOZ = Mozambique; MWI = Malawi; NAM = Namibia; NGR = Niger; NIG = Nigeria; RWA = Rwanda; SA = South Africa; SEN = Senegal; SEY = Seychelles; SRL = Sierra Leone; SUD = Sudan; SWA = Swaziland; TAN = Tanzania; TOG = Togo; UGA = Uganda; ZAM = Zambia.

shares are around 30 percent, or nearly 20 percentage points, below projected levels. The extent of the stunting of the agricultural sector can also be seen by comparing sectoral shares among African countries with those of other developing regions. Results presented in Badiane (2011) show that agriculture's average share of GDP among African countries is significantly smaller than among South Asian countries of similar income levels. The African average share is barely larger than the average share among countries in East Asia and the Middle East and North Africa, which have per capita incomes three times higher.

The relatively rapid decline of the agricultural sector could not have occurred without having serious effects on regional poverty levels (Figure 1.6). Countries with higher levels of discrepancy between actual and expected AgGDP shares also have higher poverty levels. Achieving greater progress toward poverty reduction requires a reversal of past trends. The real battle is not only to achieve growth, but also to raise productivity levels in the agricultural sector, which is the most effective way of eliminating poverty in the near future (Diao et al. 2012).

TABLE 1.1 Yield gaps for selected commodities

Crop	Nin-Pratt et al. 2011	Hengsdijk and Langeveld 2009	FAO 2013
Maize	36.5	16.5–29.5	10–25
Rice	53.6	14.8	10–40
Millet	29.6	9.2–15.9[a]	10–40
Sorghum	30.5	9.2–15.9[a]	0–25
Cassava	65.4	24.2	25–85[c]
Potatoes	21.5		10–25
Sweet potatoes	56.7		10–25
Beans	47.4	9.3[b]	10–25[d]
Groundnuts	61.5	24.1	10–40
Soybeans	52.7		
Bananas	22.2		25–85[c]
Cotton lint	33.8	60.6	10–25

Source: Compiled by authors from Nin-Pratt et al. (2011), Hengsdijk and Langeveld (2009), and FAO (2013).

Notes: The Nin-Pratt study covered the member countries of the West and Central African Council for Agricultural Research; the Hengsdijk and Langeveld and FAO studies covered various regions. [a]Data refer to tropical cereals. [b]Data refer to dry beans. [c]Data refer to cassava, yams, plantains, and other roots and tubers. [d]Data refer to pulses.

Raising productivity must include closing yield gaps for major commodities, which remain substantial in most of Africa's subregions (Table 1.1). Yield levels in the countries of West and Central Africa are a mere fraction of the realizable potential (Table 1.1, column 2). In some cases, the yield gaps for other subregions are equally large (Table 1.1, columns 3 and 4). As is demonstrated by the widespread existence of large yield gaps, the recent uptick in productivity has not been sufficient to make up for ground lost in the 20–30 years after independence. It appears that countries will need to perform far better than indicated by the recent recovery to make up for the gap in productivity. This stresses the critical importance of technological and institutional innovations in pursuit of continued agricultural growth.

Trends in Agricultural Productivity

Over the past 50 years, trends in overall agricultural productivity exhibit patterns of decline and recovery that are similar to observed sectoral growth trends. The findings from studies of agricultural productivity by several authors all point to faster rates of productivity growth in later periods. Authors may disagree about the exact periods of growth or decline and the exact magnitude of the rate of productivity growth in later periods compared with stagnating or falling productivity levels in earlier periods, but they all agree that the pace of productivity growth has picked up in the past decade or two (Fulginiti, Perrin, and Yu 2004; Ludena et al. 2007; Nin-Pratt and Yu 2008; Alene 2010; Block 2010; Benin et al. 2011). Several of the authors concluded that, although total factor productivity (TFP) has risen in past decades, it has not recovered to the levels recorded in the early 1960s (Box 1.1).

Both wealth creation and competitiveness, and thus long-term growth and poverty reduction, are driven by increases in productivity, which in turn are determined by the pace of technical change. This highlights the critical importance of investments in policies to promote technological and institutional innovations in the agricultural sector. Many of the studies that have looked at changes in agricultural productivity among African countries have also studied the determining factors. While a number of factors are identified by the different authors, many of the findings stress the critical importance of investment in R&D systems.

For example, according to Fuglie and Rada (2013), international and national agricultural research, economic policy reform (increasing the nominal rate of assistance to agriculture), irrigation, reduction in armed conflict, and increased farmer schooling are the main drivers of the change in TFP. Alene (2010) finds that increased R&D expenditures, increased rainfall, and increased trade (a proxy for economic policy reforms) had positive impacts on TFP growth. Alene (2010, 223) finds a 10-year lag between agricultural R&D expenditure growth and agricultural productivity growth: "While a strong R&D expenditure growth of about 2 percent per year in the 1970s led to strong productivity growth after the mid-1980s, stagnation of R&D expenditure in the 1980s and early 1990s led to slower productivity growth in the 2000s." Results reported by Block (2010) suggest that agricultural R&D expenditures had the highest impact on TFP growth: 10-year lagged R&D expenditures could explain 75 percent of TFP growth from 1981 to 2000.

The next-largest effect was that of policy distortions, measured by the black market premium for foreign currency and changes in the relative rate of assistance: policy reforms lessening discrimination against agriculture increased TFP growth. Civil wars decreased TFP growth. In contrast, Fulginiti, Perrin, and Yu (2004) stress the importance of institutional variables. They find that countries with greater political rights and civil liberties performed better, whereas wars and conflicts had negative impacts on TFP. Nin-Pratt and Yu (2008), on the other hand, attribute the recovery of TFP to economic policy reforms that improved incentives to farmers, such as the economic reforms that began in Ghana and Nigeria and later spread to many

BOX 1.1 Total factor productivity

Total factor productivity (TFP) is used to measure changes in the quantity of output that can be produced with a given level of inputs. TFP is expressed as an index, representing changes in productivity relative to a base year.

Unlike partial productivity measures, which estimate the additional output resulting from an increase in one input, such as land or labor, TFP is a measure of total output produced by all inputs. Mathematically, TFP is a ratio of an output index to an input index; each of these indexes measures changes over time in outputs or inputs, meaning that TFP measures changes in the relationship between levels of inputs and outputs. An increase in TFP means that the same level of inputs can be used to produce a greater level of outputs than previously. Increased productivity—this ability to produce more with the same inputs—can result from increasing efficiency or from introducing technical change (or from a combination of both, which is often the case). The scope to increase productivity by increasing efficiency is limited, and researchers agree that technical change through research and development is the only sustainable source of continued productivity growth.

Differences in estimates of TFP result from a number of factors. As the ratio of changes in total output to changes in total inputs, estimating TFP requires methods to aggregate and compare different types of inputs and outputs across time and space. Several methods exist, each deriving somewhat different results and offering different advantages and challenges in terms of the data required, the computational burden, and the assumptions made. The authors cited in this chapter also measure changes over different time periods, use different datasets for input and output variables, and look at different sets of countries, from the whole continent to Africa south of the Sahara, with and without South Africa.

Source: Authors.

other countries, or the devaluation of the CFA (Communauté Financière africaine) franc in 1994, which helped many West African countries increase their exports and improve their agricultural sectors.

Of greater importance from the point of view of poverty reduction are changes in agricultural productivity associated with rising labor productivity as opposed to land productivity. In line with the broad agricultural growth recovery documented above, land and labor productivity have risen considerably in the past couple of decades (Table 1.2). Both land and labor productivity grew at around 3.5 percent per year during 2000–2011, having accelerated considerably from the preceding two decades. Slow or even negative growth in

TABLE 1.2 **Growth trends in area harvested and output per hectare and per agricultural worker, 1980–2000 and 2000–2011**

	Arable Land		Land productivity		Labor productivity	
Subregion	1980–2000	2000–2011	1980–2000	2000–2011	1980–2000	2000–2011
East Africa	0.61	2.25	1.51	0.77	−1.08	0.48
West Africa	1.20	1.78	2.27	5.33	2.08	5.96
Central Africa	0.19	0.54	2.14	3.74	0.13	1.09
Southern Africa	0.95	0.85	0.82	2.88	−2.04	1.85
Total	0.85	1.62	1.71	3.45	−0.06	3.54

Source: Data for arable land are from World Bank (2014b); data for land productivity (agricultural value-added per hectare of arable land) and labor productivity (agricultural value-added per agricultural worker) are from ReSAKSS (no date).

labor productivity from 1980 to 2000 in most subregions gave way to healthier growth rates in the current decade in all subregions, but it is West Africa's rapid labor productivity growth of nearly 6 percent per year that brings the total for SSA to 3.5 percent. West Africa's growth rate, in turn, is strongly influenced by Nigeria, the largest agricultural economy in Africa. Nigeria's high labor productivity growth brought up the subregion's average in both periods, particularly in 2000–2011, when Nigeria's yearly growth rate of 9.25 percent exceeded those of all other West African countries.

Land productivity growth rates of around 1.5 to 2.3 percent (for West Africa) from 1980 to 2000 (with the exception of Southern Africa's rate of less than 1 percent) accelerated in most subregions in the following decade, but dropped by nearly half in East Africa. As with labor productivity, high land productivity growth rates in West Africa reflect Nigeria's influence, particularly in 2000–2011, when Nigeria registered the highest yearly growth rate in the subregion: 7.29 percent. In 2000–2011, East Africa had the lowest land and labor productivity growth rates of all subregions, and West Africa had the highest, with both land and labor productivity growing at more than 5 percent per year. The overall acceleration of land and labor productivity growth is encouraging, but improvement has not been across the board and reflects the need for continued investments to raise productivity.

Area expansion continues to play an important role in output growth in all subregions. During 1980–2000, growth in arable land (or nonpermanent cropland and pasture) generally exceeded growth in labor productivity, with the exception of West Africa. This was reversed in the following decade, with labor productivity growing more than twice as fast as the expansion of arable land in all subregions except East Africa. During this time, East Africa had the fastest arable land growth but the lowest labor and land productivity growth of all subregions. Land productivity has also generally grown faster than arable land, with the exceptions of Southern Africa, during 1980–2000, and East Africa, during 2000–2011. Arable land expansion in Nigeria was slower than the West African subregional average in 1980–2000, and was fairly close to the subregional average in 2000–2011.

The broader trends discussed above further highlight the imperative to promote agricultural S&T and invest in R&D to ensure continued productivity growth. As can be inferred from the magnitude of the rates of growth discussed above, the initial levels of productivity were fairly low with the exception of Southern Africa (Figure 1.7). In 1980 Africa began with the lowest levels of land productivity of all regions of the world, but has raised those levels quite a bit since then. Nevertheless, other world regions also raised their

FIGURE 1.7 Trends in land and labor productivity in Africa versus other world regions, 1980–2010

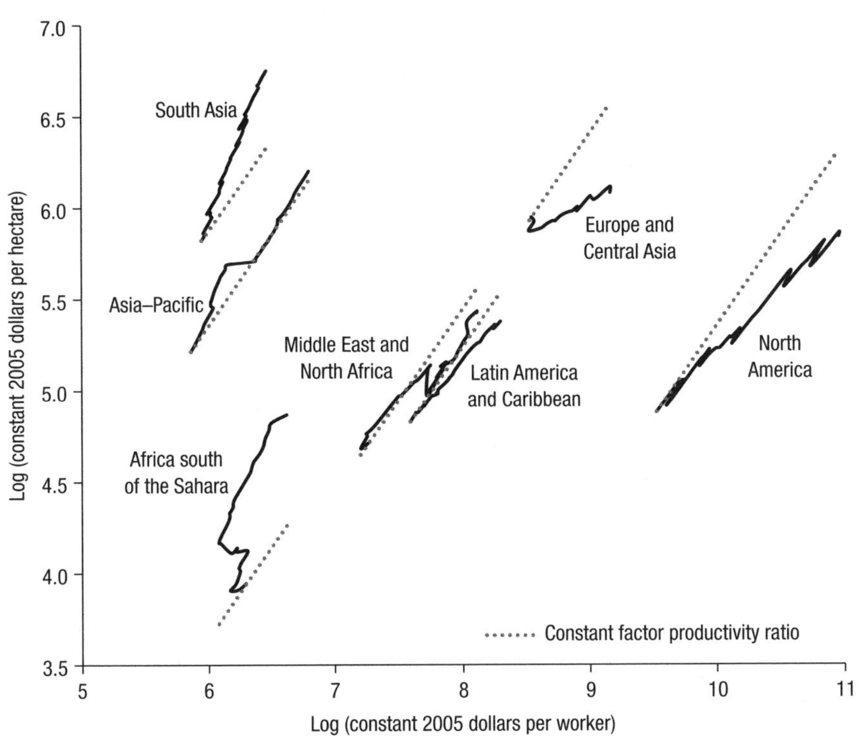

Source: Calculated by authors based on World Bank (2014b).

Note: The position, slope, and length of the lines reflect the magnitude of land versus labor productivity levels and their relative rates of growth over the entire period under consideration.

FIGURE 1.8 Trends in land and labor productivity across Africa's subregions, 1980–2011

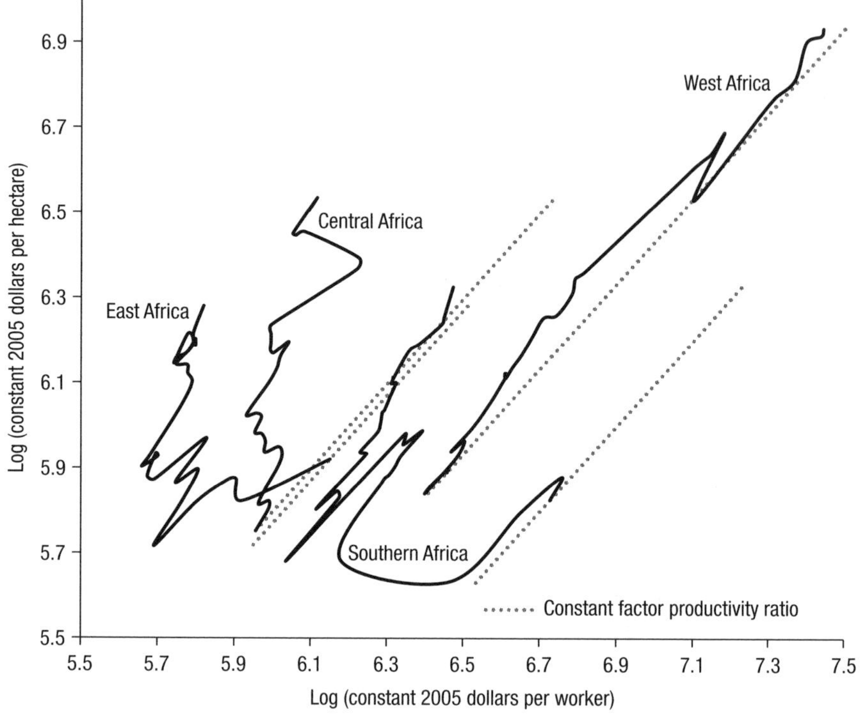

Source: Calculated by authors based on ReSAKSS (no date).

Notes: The position, slope, and length of the lines reflect the magnitude of land versus labor productivity levels and their relative rates of growth over the entire period under consideration. "Southern Africa" includes South Africa.

land productivity over this period, so a considerable land productivity gap remains between Africa and the rest of the world. In terms of raising labor productivity in agriculture, Africa made dismal progress during 1980–2010: labor productivity initially declined during the 1980s, and positive growth only began after 1990.

Overall productivity trends were positive during 1980–2011, despite considerable differences in terms of land and labor productivity levels and rates of growth across subregions (Figure 1.8). Land productivity increased across all subregions, while progress in increasing labor productivity was mixed. Labor productivity increased markedly in West Africa and also rose significantly in Southern Africa after initial decreases, but increased only slightly in Central Africa, and decreased in East Africa. West Africa experienced rapid productivity growth, surpassing the other three subregions and ending the period with the highest levels of both land and labor productivity. Most of this increase was in Nigeria; the rest of the subregion showed levels of land productivity similar to those of Central and East Africa, although labor productivity levels were still higher than in these two subregions. Trends in land and labor productivity growth in Southern Africa are also biased by the inclusion of South Africa, but they reflect the much larger size of average holdings and related technology choices in that subregion. Central Africa, East Africa, West Africa (excluding Nigeria), and Southern Africa (excluding South Africa) still have low levels of labor productivity and remain highly dependent on agriculture. In addition, East and West Africa tend to have a higher concentration of vulnerable areas, in particular the hotspots of the Sahel and Horn of Africa. The challenge of raising TFP in general, and labor productivity in particular, remains a key strategic priority in these subregions.

The analysis in the preceding sections reveals that African countries have achieved a level of growth performance over the past 15 years that is without precedent in the continent's postcolonial era. Improvement has occurred across the board in overall economic and agricultural growth, as well as in total and partial factor productivity in the agricultural sector. Poverty rates have also reversed their rising trends and have been declining since the late 1990s. The analysis also shows that African countries are still facing major challenges that make it necessary not only to sustain the current recovery, but also to accelerate its pace. A closer look reveals that the recent progress has primarily allowed African countries to make up for lost ground during the decades-long period of economic stagnation and decline that preceded the recent recovery. Fifteen years of recovery have moved the per capita food production index back to its level of the early 1960s. The same is observable for TFP levels (Benin et al. 2011). Furthermore, African countries, on average, are not on track to meet the MDG poverty goal by 2015. Although they may reach the goal in the near future, doing so would still leave average poverty levels for the continent at a high 30 percent.

While the recent recovery is both encouraging and proof that progress is possible, much work remains. The poverty reduction agenda through faster, broad-based—and thus agricultural—growth is still "unfinished business." The ultimate contribution of agricultural growth to wealth creation and poverty reduction will depend on the extent to which it is linked to increases in land productivity, and—in particular—labor productivity, especially in the context of rapid population growth. A major component of the future growth and poverty reduction agenda, therefore, needs to deal with the technological, policy, and institutional innovations required to raise the productivity of rural farm and off-farm labor faster than has been the case to date.

BOX 1.2 The Comprehensive Africa Agriculture Development Programme (CAADP)

Launched in 2003, the Comprehensive Africa Agriculture Development Programme (CAADP) is a continentwide framework to facilitate faster agricultural growth and progress toward poverty reduction and food and nutrition security in Africa. It seeks to promote policies and partnerships, raise investments in Africa's agricultural sector, and achieve better development outcomes. CAADP has defined a limited set of clear continentwide goals, including the attainment of a 6 percent yearly agricultural growth rate at the national level. For that purpose, the allocation of at least 10 percent of national budgets to the sector is another target under CAADP. In addition, the program contains the following key values and principles:

1. Leadership and ownership of all aspects of the agenda at all levels by African decisionmakers and their constituencies. CAADP is fundamentally a home-grown agenda to increase the likelihood of alignment with local priorities and concerns.
2. Inclusiveness of all major stakeholder groups to facilitate participation in planning and implementation decisionmaking. Albeit far from perfect, the CAADP effort has invested heavily in creating wide understanding of and support for its goals and action agenda.
3. Partnership and mutual accountability among African governments, their constituencies, and development agencies. A number of dialogue and review platforms have been established at the country, subregional, and regional levels to support this principle.
4. Evidence- and outcome-based planning and implementation to improve the growth and poverty reduction outcomes of agricultural strategies. One of the main innovations of CAADP is the use of locally based empirical economic analysis to support strategic decisionmaking, priority setting, and investment planning in the sector.

This strategy was established on the basis of four pillars to guide investments. The pillars deal with (1) sustainable land and water management, (2) agribusiness development and market access, (3) hunger and social safety nets, and (4) science and technology.

Key milestones in the CAADP implementation process are (1) the organization of a roundtable and signing of a country CAADP compact specifying policy and investment priorities and commitments guided by the analysis discussed above; (2) the design of a comprehensive, multiyear agricultural investment plan by each country; and (3) the organization of a business meeting and an independent technical review to systematically evaluate the technical quality of country investment programs and to discuss funding and implementation modalities. As of November 2015, 46 countries had launched the CAADP process, 42 had held roundtables and signed compacts, and 39 had completed national agricultural investment plans. In addition, four subregional economic communities had signed subregional CAADP compacts. While it is too early to say anything definitive about the impact of CAADP on the agricultural sector in Africa, the broad adoption and implementation of the CAADP agenda is of great significance, offering the opportunity to sustain and deepen the recovery process. If, through CAADP, a large number of countries manage to maintain a 6 percent growth trajectory, living conditions on the continent would change dramatically within a generation.

Sources: Based on Badiane (2011) and IFPRI (2014).

Aware of the interlinked challenges involved in raising agricultural productivity, in 2003 African countries launched CAADP, under which they identified priorities and designed investment plans seeking to invest at least 10 percent of government budgets in agriculture and achieve a sectoral growth rate of 6 percent (Box 1.2). The CAADP framework emphasizes agricultural S&T combined with natural resource management, agribusiness development and market access, social safety nets, and institutional innovation to promote inclusive agricultural policymaking and support an enabling environment for agricultural development. The goals and ambitions declared under CAADP provide a more legitimate benchmark for evaluating the need for future S&T policies and interventions than do goals or targets based on arbitrary projections. CAADP facilitates an assessment of the degree to which African countries are addressing policy and institutional changes, and investing in their agricultural sectors to meet CAADP's goals and targets and to sustain and broaden the current recovery to further reduce poverty. These issues are dealt with in subsequent chapters of this book. The remainder of this chapter reviews and summarizes country-level goals under CAADP, using examples from the Economic Community of West African States (ECOWAS) as background for analysis in subsequent chapters.

Country Goals and Ambitions under CAADP

The most compelling evidence of the need for African countries to invest in technical change and raise TFP in the agricultural sector is the strong linkage between agricultural growth and poverty reduction. While recent growth and productivity trends are encouraging and indicate that countries are on the right trajectory, undoing the cumulative impact of decades-long stagnation

FIGURE 1.9 The projected cumulative effect of an additional 1 percentage point of agricultural growth per year on poverty reduction in the countries of the Economic Community of West African States to 2025

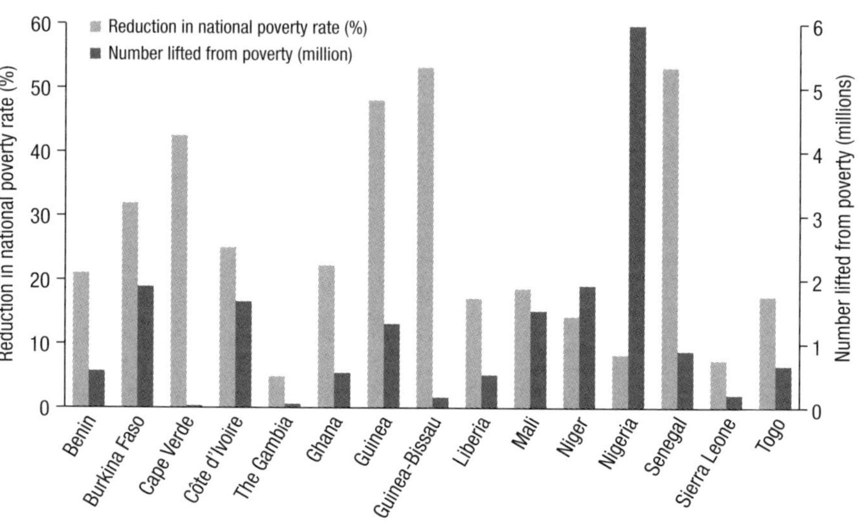

Source: Calculated by authors based on IFPRI simulation results.

and decline stemming from neglect remains an immense challenge. Future progress will depend on the extent to which countries will succeed in sustaining and accelerating the current recovery process. How much is at stake can be seen from the simulated cumulative impact of an additional 1 percentage point of agricultural growth per year on poverty levels by 2025 (Figure 1.9).

The pace of agricultural growth leading to reduced poverty rates—which in some countries can be as high as 50 percent—depends on the initial size of the agricultural sector and its share of employment among poor and vulnerable people. Simulation results indicate that most countries would register a decline in poverty rates of more than 10 percent. The decrease in the absolute number of poor people would be as high as 17.9 million for the subregion as a whole, ranging from 38,000 in Cape Verde to 6.0 million in Nigeria. It is this critical role of agricultural growth in reducing poverty that has motivated CAADP. The program, in other words, is a reflection of the ambition of African countries to achieve higher rates of economic growth and significantly reduce poverty levels across the continent. It is against this goal that future efforts and needs to increase technological and institutional innovations to accelerate technical change should be evaluated. This would carry greater policy relevance than benchmarking countries against arbitrary future outcomes. The question should be whether African countries are making sufficient efforts to meet the required outcomes resulting from their own ambitions, which are couched in the compacts and national agricultural investment plans (NAIPs) prepared by all countries as part of the implementation of CAADP. The compact is a formal document that lays out a country's priorities, as well as its policy and budgetary commitments. The NAIPs present priority investments through which every country aims to accelerate growth and reduce poverty.

Figure 1.10 and Figure 1.11 summarize simulation results of achievable rates of agricultural growth and poverty reduction by 2025 under the NAIP scenario compared with a "business-as-usual" scenario that reflects the continuation of the prevailing trends at the time the compact was signed. As expected, individual country results under the NAIP scenario imply significantly higher rates of agricultural growth such that, in some cases, the magnitude of the difference calls into question whether the NAIP goals can be considered realistic. Most countries have the ambition of raising agricultural growth rates by between 0.5 and 8.5 percentage points (Figure 1.10). In some

FIGURE 1.10 Projected agricultural growth rates under national agricultural investment plan and business-as-usual scenarios in the countries of the Economic Community of West African States to 2025

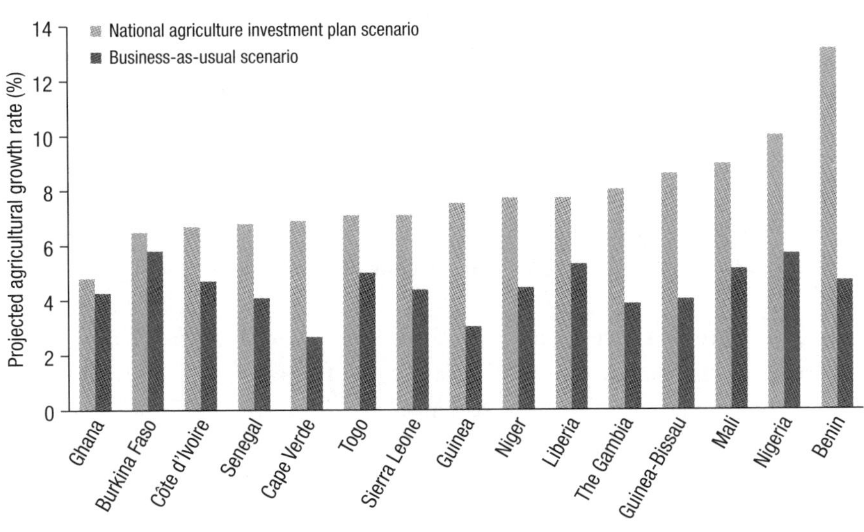

Source: Calculated by authors based on IFPRI simulation results.

FIGURE 1.11 Reduction in national poverty rates under national agricultural investment plan and business-as-usual scenarios in the countries of the Economic Community of West African States to 2025

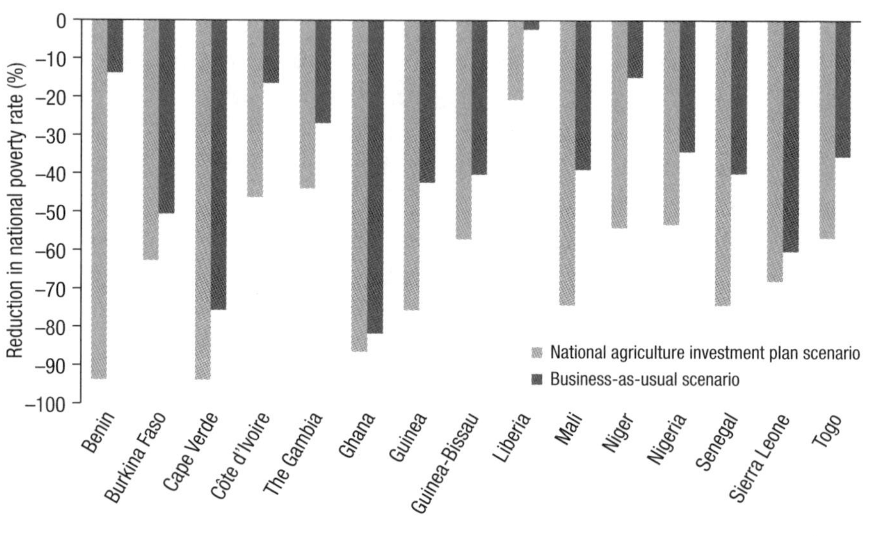

Source: Calculated by authors based on IFPRI simulation results.

cases (such as Benin, Cape Verde, and Guinea), the projected growth rates would mean more than doubling the historical rate of agricultural growth.

Is it realistic to expect that countries can achieve these high rates of growth, given the growth and productivity trends and patterns discussed above? Is the current pace of recovery enough? What additional changes need to occur in terms of technical change and factor productivity growth? What actions do they call for with respect to the different drivers identified? What would be the building blocks of a technological and institutional innovation agenda to support these ambitions? This volume endeavors to provide answers to these questions.

The remainder of the analysis in this chapter deals with the possible growth and investment gaps that are emerging from or have significant implications for the country investment plans (Figure 1.12). For most countries, achieving the CAADP target of 6 percent yearly agricultural growth would be sufficient to at least halve poverty from its 1990 level and achieve the MDG poverty goal, although 10 years later than intended. Several countries would even achieve single-digit poverty rates. About half the countries—including Burkina Faso, The Gambia, Guinea, Mali, Niger,

FIGURE 1.12 **Growth and investment challenges under CAADP and national agricultural investment plans**

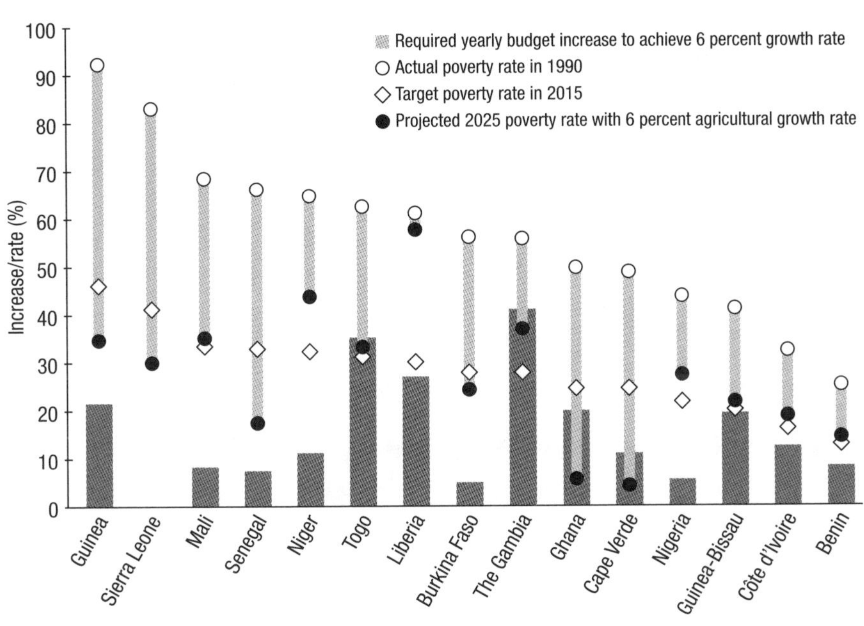

Source: Calculated by authors based on IFPRI simulation results.
Note: CAADP = Comprehensive Africa Agriculture Development Programme.

Nigeria, and Senegal—achieved an average agricultural growth rate of about 6 percent or higher between 2003 and 2010 (ReSAKSS 2013a). For some of these countries, achieving and sustaining a 6 percent rate of agricultural growth would not be enough to halve poverty by 2025 (The Gambia, Liberia, Nigeria, and Niger); the historical growth rates of four of the other countries fall far below that mark (Côte d'Ivoire, Benin, Guinea-Bissau, and Togo). It appears, therefore, that about half of the ECOWAS countries need to realize agricultural growth rates far in excess of the CAADP target or far above their historical growth rates.

Possible considerable gaps in public expenditures are also indicated (Figure 1.12), which countries will have to deal with in their efforts to raise productivity and accelerate growth. Most countries would have to maintain double-digit rates of growth of public expenditure in the agricultural sector. The countries that have the lowest level of actual public expenditure in agriculture—The Gambia, Guinea-Bissau, Liberia, and Togo (ReSAKSS no date)—would need to significantly raise sector expenditures to meet their declared ambitions under CAADP. The "big spenders"—historically, Burkina Faso, Mali, Niger, and Senegal—would need to raise sector expenditures the least, next to Benin and Nigeria.

The gaps identified above in terms of desirable rates of agricultural and expenditure growth primarily refer to ECOWAS member countries for which the data are readily available. However, it would be safe to generalize the results beyond this particular grouping. The gaps imply that countries will face great pressure to increase the effectiveness of the growth process in order to achieve the CAADP goals: they will need to achieve more with limited resources. The best option for doing that is to target investments that would accelerate the pace of technical change the most.

Country Investment Priorities under CAADP

African countries are pursuing a number of priorities to improve poverty outcomes through their CAADP investment plans (Figure 1.13). The distribution of priority areas can be compared with the drivers of productivity growth discussed earlier. While investments in agricultural R&D were identified by several authors as key drivers of productivity growth, only 5 of 15 countries (Nigeria, Senegal, Togo, Guinea, and The Gambia) are planning to make any sizable investments in R&D. It is difficult to imagine how countries intend to promote considerable technical change, when they are underinvesting in a key priority area, such as R&D and systems for developing and supplying modern inputs to farmers.

The other leading priority area that is in line with the productivity drivers is management of water resources. Several authors have shown that investment in irrigation had a positive impact on productivity growth. Four of the countries have planned to allocate around 20 percent or more of investment resources to water management. Agricultural value chains are another area of concentration of investment resources. It is to be anticipated that the other set of productivity drivers—sector policy reforms and incentives—will be addressed here; however, it is not clear how the underinvestment in R&D and input systems is going to affect efforts in the above areas. Social and political stability, another key contributor to lagging productivity growth, is understandably missing from Figure 1.13 because it is addressed outside of the investment planning process. Nevertheless, investment plans should be expected to contribute to greater political and economic stability if they lead to better economic outcomes.

FIGURE 1.13 Share of budgets allocated to priority areas in national agricultural investment plans

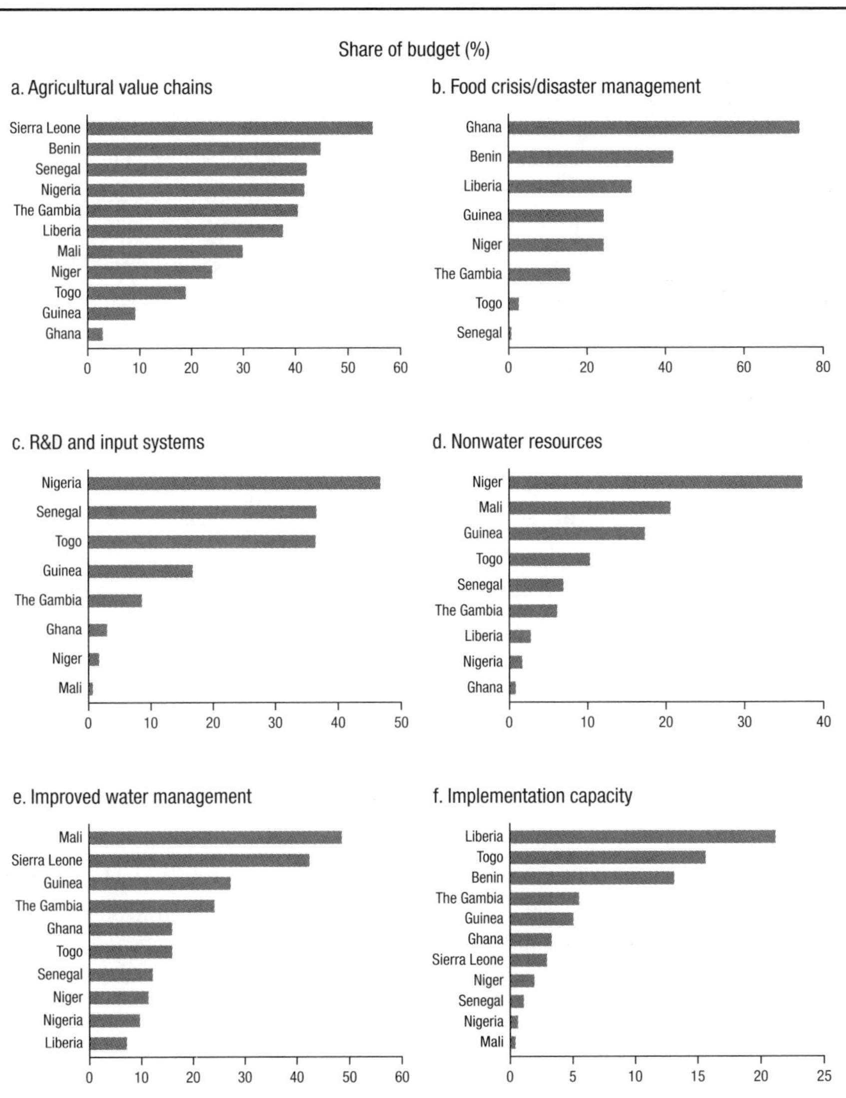

Source: Calculated by authors based on national agricultural investment plans.

Notes: The horizontal bars denote the share of each country's investment budget allocated to each of the six priority areas. R&D = research and development.

Implications for the Science and Technology Agenda

The evidence presented in this chapter indicates how imperative it is that African countries make the necessary efforts to sustain and accelerate the current recovery process. In particular, efforts must be made to foster technological progress more than has been the case in the past. Technological progress is market driven and depends on the development, adoption, and effective use of improved technical inputs and production processes in all segments of value chains. Agricultural market development, however, is constrained by the high costs of transport and, for smallholder agriculture, high transaction costs. Reducing these will require institutional and organizational innovations, such as farmers' organizations, community seed systems, agro-dealer networks, and warehouse receipt systems—in essence, rural capacity that incentivizes the delivery and uptake of new technologies and motivates the adaptation and innovation of these technologies across the extraordinary heterogeneity of African smallholder farming systems. Such an evolving rural innovation system will enable all key categories of actors, including farmers, agribusiness firms, input and service suppliers, research institutes, and other public-sector institutions, to continuously identify technology bottlenecks and generate adequate solutions to overcome them (see Chapter 13, this volume). Improvements in education and training, better access to markets and information (facilitated by the expansion of information and communications technologies, particularly cellular phones), and more fully developed links between farmers and service providers are also needed to facilitate the adoption of productivity-enhancing technologies, while also increasing productivity in their own right.

The new generation of investment plans being prepared under the CAADP agenda could be a good opportunity to boost investment in innovation systems to support growth. The example of ECOWAS member countries suggests, however, that countries may not be making the necessary investments in critical areas, such as R&D and input systems. This has to be corrected if countries want to achieve the growth and poverty reduction ambitions declared through their respective investment plans.

The above evidence calls for a better understanding of the nature and evolution of agricultural R&D systems in Africa in order to identify required actions to increase their contribution to the quest for faster growth and more broadly shared development outcomes. It calls for a closer assessment of the levels and determinants of investments by governments, the global community, and the private sector in agricultural R&D and associated returns in Africa. The question of the adequacy of the stock of human and institutional capital and future needs in this area must be answered in order to map long-term strategies to advance technical progress and raise productivity to reduce poverty. Because it is unlikely that any given African country can meet all of its research and technology needs alone, the inquiry and debate should include the role of, as well as opportunities for, cross-border collaboration and partnership with global centers of expertise, including CGIAR. These and other strategic questions dealing with the future of the agricultural S&T agenda are treated in depth throughout the rest of this book.

References

Alene, A. 2010. "Productivity Growth and the Effects of R&D in African Agriculture." *Agricultural Economics* 41 (3/4): 223–238.

Badiane, O. 2011. *Agriculture and Structural Transformation in Africa.* Stanford Symposium Series on Global Food Policy and Food Security in the 21st Century. Stanford, CA: Stanford University.

Badiane, O., T. Makombe, and G. Bahiigwa, eds. 2014. *Promoting Agricultural Trade to Enhance Resilience in Africa.* ReSAKSS Annual Trends and Outlook Report 2013. Washington, DC: International Food Policy Research Institute.

Benin, S., A. Nin Pratt, S. Wood, and Z. Guo. 2011. *Trends and Spatial Patterns in Agricultural Productivity in Africa, 1961–2010.* ReSAKSS Annual Trends and Outlook Report 2011. Washington, DC: International Food Policy Research Institute.

Block, S. 2010. *The Decline and Rise of Agricultural Productivity in Sub-Saharan Africa since 1961.* NBER Working Paper 16481. Cambridge, MA: National Bureau of Economic Research.

Bojo, J. 1996. "The Costs of Land Degradation in Sub-Saharan Africa." *Ecological Economics* 16: 161–173.

Diao, X., J. Thurlow, S. Benin, and S. Fan, eds. 2012. *Strategies and Priorities for African Agriculture: Economywide Perspectives from Country Studies.* Washington, DC: International Food Policy Research Institute.

FAO (Food and Agriculture Organization of the United Nations). 2013. Global Agro-Ecological Zones database. Accessed November 2013. http://gaez.fao.org/Main.html#.

———. 2014. FAOSTAT database. Accessed May 2014. http://faostat3.fao.org/faostat-gateway/go/to/home/E.

Fuglie, K., and N. Rada. 2013. *Resources, Policies, and Agricultural Productivity in Sub-Saharan Africa.* Economic Research Report 145. Washington, DC: U.S. Department of Agriculture, Economic Research Service.

Fulginiti, L., R. Perrin, and B. Yu. 2004. "Institutions and Agricultural Productivity in Sub-Saharan Africa." *Agricultural Economics* 31 (2/3): 169–180.

Hengsdijk, H., and J. Langeveld. 2009. *Yield Trends and Yield Gap Analysis of Major Crops in the World.* Working Document 170. Wageningen, Netherlands: Wettelijke Onderzoekstaken Natuur & Milieu.

IFPRI (International Food Policy Research Institute). 2014. *2013 Global Food Policy Report.* Washington, DC.

Jayne, T., J. Chamberlin, and M. Muyanga. 2012. *Emerging Land Issues in African Agriculture: Implications for Food Security and Poverty Reduction Strategies.* Stanford Symposium Series on Global Food Policy and Food Security in the 21st Century. Stanford, CA: Stanford University.

Ludena, C., T. Hertel, P. Preckel, K. Foster, and A. Nin. 2007. "Productivity Growth and Convergence in Crop, Ruminant, and Nonruminant Production: Measurement and Forecasts." *Agricultural Economics* 37 (1): 1–17.

Nin-Pratt, A., M. Johnson, E. Magalhaes, L. You, X. Diao, and J. Chamberlin. 2011. *Yield Gaps and Potential Agricultural Growth in West and Central Africa.* IFPRI Research Monograph. Washington, DC: International Food Policy Research Institute.

Nin-Pratt, A., and B. Yu. 2008. *An Updated Look at the Recovery of Agricultural Productivity in Sub-Saharan Africa.* IFPRI Discussion Paper 787. Washington, DC: International Food Policy Research Institute.

ReSAKSS (Regional Strategic Analysis and Knowledge Support System). No date. ReSAKSS database. Accessed February 2014. www.resakss.org/map/.

———. 2013a. "CAADP 6% Growth Target." Accessed February 2014. www.resakss.org/region/western-africa/caadp-targets.

———. 2013b. "Monitoring Progress." Accessed May 2014. www.resakss.org/region/monitoring-progress/africa-wide.

Svendsen, M., M. Ewing, and S. Msangi. 2009. *Measuring Irrigation Performance in Africa.* IFPRI Discussion Paper 894. Washington, DC: International Food Policy Research Institute.

UNSD (United Nations Statistics Division). 2013. "National Accounts Estimates of Main Aggregates: Gross Value Added by Kind of Economic Activity at Constant (2005) Prices—US Dollars." Accessed June 2014. http://data.un.org/Data.aspx?d=SNAAMA&f=grID%3a202%3bcurrID%3aUSD%3bpcFlag%3a0.

World Bank. 2014a. PovcalNet database. Accessed July 2014. http://iresearch.worldbank.org/PovcalNet/index.htm.

———. 2014b. *World Development Indicators.* Accessed July 2014. http://databank.worldbank.org/data/views/variableSelection/selectvariables.aspx?source=world-development-indicators.

———. 2015. *World Development Indicators.* Accessed February 2015. http://databank.worldbank.org/data/views/variableSelection/selectvariables.aspx?source=world-development-indicators.

Xie, H., L. You, B. Wielgosz, and C. Ringler. 2014. "Estimating the Potential for Expanding Smallholder Irrigation in Sub-Saharan Africa." *Agricultural Water Management* 131: 183–193.

Chapter 2

THE EVOLUTION OF AGRICULTURAL RESEARCH IN AFRICA: KEY TRENDS AND INSTITUTIONAL DEVELOPMENTS

Johannes Roseboom and Kathleen Flaherty

The current institutional framework of agricultural research in Africa south of the Sahara (SSA) has been shaped by more than a century of agricultural research activity. In colonial times, the European powers invested substantial resources in agricultural research to unlock the agricultural potential of Africa. Initially, they focused mainly on commercial export crops, such as coffee, tea, cacao, cotton, rubber, oil palm, and sugarcane. Research on local food crops and livestock only started in earnest after World War II (WWII). With the attainment of independence in the late 1950s and early 1960s, most African countries inherited an agricultural research system dominated by foreign scientists. The prospect of replacing them with national scientists presented a major challenge, because the newly independent countries lacked an academically trained cadre and, even more crucially, universities to train them. As a result, the nationalization of agricultural research required a lengthy transition for most African countries. In the early 1980s, some 20 years after independence, roughly 30 percent of the agricultural researchers employed in the region's national agricultural research systems (NARSs) were still expatriates (Pardey, Roseboom, and Anderson 1991), but by the mid-1990s this share had fallen to negligible levels in most countries.

The nationalization process after independence often also resulted in the dismantling of established colonial linkages among agricultural research entities, including various agencies established by Great Britain, France, and Belgium, whose mandates encompassed multiple African colonies. In the first few decades following independence, most African countries were so preoccupied with the nationalization process that they had little political or other incentives to explore regional collaboration. Since the early 1970s, some of this vacuum in cross-country collaboration has been filled by international

The authors would like to thank all the individuals across Africa who contributed to the Agricultural Science and Technology Indicators database, which forms the primary source for the analysis in this chapter. In addition, the authors would like to thank the various reviewers who helped to improve this chapter at different stages.

agricultural research centers operating under the CGIAR umbrella. As part of their work, they manage multicountry research networks and programs. Interest in cross-country collaboration in agricultural research by NARSs only began in earnest in the late 1980s and early 1990s, which eventually led to the establishment of Africa's subregional organizations (SROs) and the Forum for Agricultural Research in Africa (FARA). A major challenge of such NARS-led collaboration is the strong fragmentation of Africa's agricultural research capacity into some 47 NARSs, which differ greatly in size, research ambition, and competency. Moreover, they are embedded in a rich diversity of agroecologies, production systems, cultures, languages, customs, levels of development, political systems, and so on. This makes finding common ground for collaboration in agricultural research and translating that into action all the more challenging.

This chapter provides an overview of the structure of agricultural research in SSA; its development through time; and its struggle with widespread, systemic constraints. It begins with an overview of the region's NARSs, followed by a discussion of supranational agricultural research collaboration, before offering an analysis of the constraints affecting the performance of agricultural research in SSA, including (1) policies related to agriculture and agricultural research and development (R&D); (2) the organization and management of agricultural research; (3) human, financial, and physical resources; and (4) collaboration and coordination.

National Agricultural Research Systems

Overview

NARSs are the principal building blocks of Africa's agricultural research system. Of the 47 low- and middle-income countries that constitute SSA, almost all have a NARS; however, their size and strength vary greatly.[1] As of 2011—the most recent year for which comprehensive data are available through the

1 South Sudan became independent in 2011, raising the number of low- and middle-income countries in SSA to 47. Despite being geographically located within SSA, Equatorial Guinea is excluded because the World Bank classifies it as a high-income country, because of its high oil revenues, and Mayotte and Réunion are excluded because they are French territories, not independent countries. The dataset that underlies most of the analysis in this chapter comprises 45 of the 47 SSA countries; Somalia and South Sudan are excluded because of extreme data constraints. In several African countries—such as Côte d'Ivoire, Democratic Republic of the Congo, Rwanda, Sierra Leone, Somalia, Sudan, and more recently Mali—NARSs have been forced to close down or at least significantly contract their activities for long periods of time because of civil war and military conflict.

FIGURE 2.1 Agricultural R&D spending and staffing trends in Africa south of the Sahara, 1971–2011

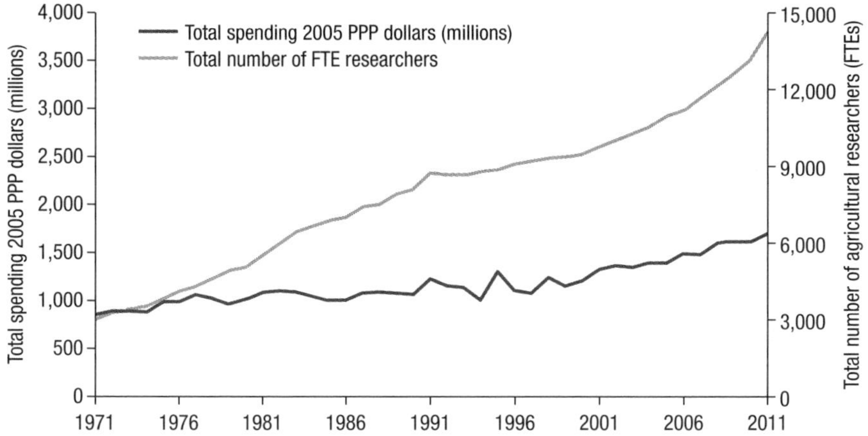

Source: Calculated by authors based on ASTI (2014).
Notes: FTEs = full-time equivalents; PPP = purchasing power parity; R&D = research and development.

Agricultural Science and Technology Indicators (ASTI) initiative, led by the International Food Policy Research Institute—the region's NARSs collectively employed an estimated 14,221 full-time-equivalent (FTE) researchers, representing a fivefold increase over the 1971 total of 2,981 FTEs (Figure 2.1). A further 50,000 staff members supported these researchers (including research technicians, laborers, administrative support staff, and so on), yielding an average support-staff-per-researcher ratio of about 4:1 (calculated by authors from ASTI 2014). This ratio varies widely across individual NARSs, ranging from less than 2:1 to more than 14:1. In response to efforts to improve efficiency, the overall trend is toward declining support-staff ratios. High support-staff ratios often reflect a strong presence of labor-intensive nonresearch activities, such as seed multiplication or agricultural production activities.

NARS spending (in constant 2005 PPP dollars) grew far more slowly than research staff numbers over this period, from $853 million in 1971 to $1,693 million in 2011 (Figure 2.1), indicating a dramatic decline of financial resources per researcher over time, from $286,208 in 1971 to $119,065 in 2011. Most often, declining spending levels per researcher were the result of (1) the transition from employing relatively expensive expatriate researchers to employing less expensive local researchers (a phenomenon that largely occurred in the 1970s and 1980s), (2) minimal adjustment of salaries to

FIGURE 2.2 Number of full-time equivalent agricultural researchers per 100,000 economically active agricultural population in Africa south of the Sahara, 1981–2011

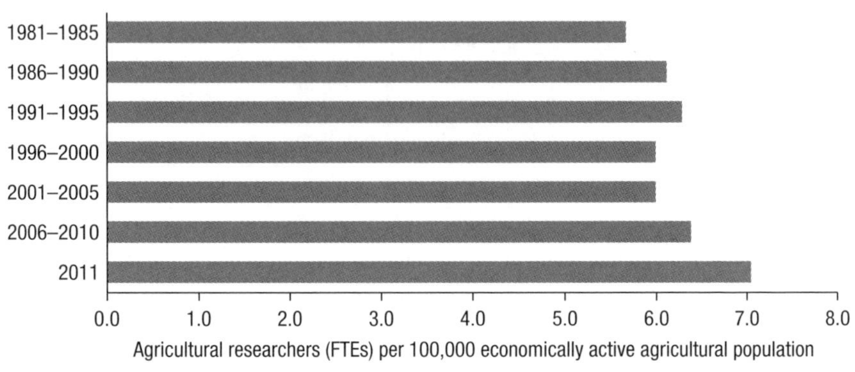

Source: Calculated by authors based on ASTI (2014).
Note: FTEs = full-time equivalents.

compensate for inflation over long time periods, (3) extreme contraction of operating costs and capital investment, and (4) declining support-staff ratios (particularly in more recent years). The combined effect of these factors has left many African NARSs with a poorly paid (and hence poorly motivated) pool of researchers who are required to conduct research with very limited resources—such as out-of-date or dysfunctional scientific equipment and lack of fundamental services.[2] Of course, there are exceptions, both within and across NARSs, but the overall picture is quite grim. The establishment of national agricultural research institutes/organizations (NARIs/NAROs) in the 1970s and 1980s was expected to resolve these problems by removing agricultural research from the government bureaucracy; however, these entities have remained largely dependent on government funding, and are therefore subject to many of the same constraints as before.

In relative terms, the growth in agricultural research staffing only just exceeded growth in agricultural labor. As a result, the number of agricultural researchers per 100,000 economically active agricultural population grew only modestly, from an average of 5.7 FTEs during 1981–1985 to 7.0 FTEs in 2011 (Figure 2.2).

Growth in NARS spending lagged behind growth in agricultural gross domestic product (AgGDP) during most of the 1980s and 1990s. This trend

2 This can include transport, electricity, computer hardware, reliable Internet access, necessary software and databases, and publications.

FIGURE 2.3 Public agricultural R&D spending as a share of AgGDP in Africa south of the Sahara, 1971–2011

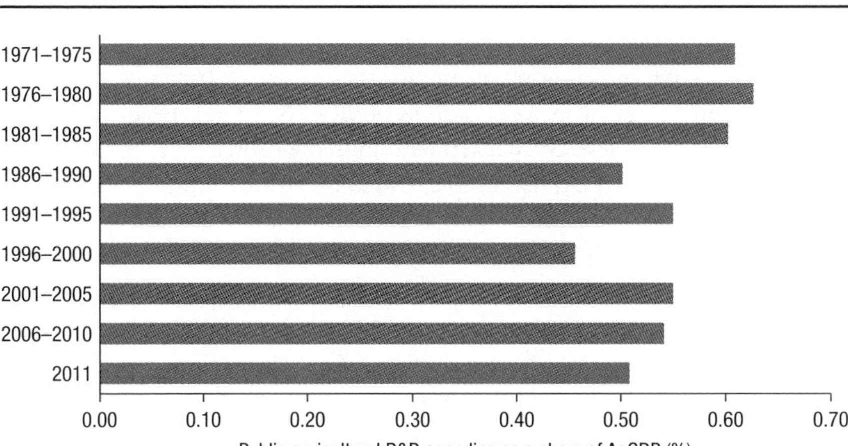

Source: Calculated by authors based on ASTI (2014).
Notes: AgGDP = agricultural gross domestic product; R&D = research and development.

reversed itself during the 2000s, but average spending by NARSs as a share of AgGDP during this decade was still lower than during the late 1970s and early 1980s, and also lower than the recommendation by the New Partnership for Africa's Development (NEPAD) of an investment of at least 1 percent of AgGDP (Figure 2.3).

Fragmentation and Interlinkages

As indicated in the previous section, African NARSs are predominantly small; only nine employed more than 500 FTE researchers in 2011. Nonetheless, African NARSs have progressively grown in size over time in terms of researchers, given that in 1961 the overwhelming majority employed fewer than 100 FTEs (Chapter 8, this volume). Despite this growth in human resource capacity, the fragmentation of agricultural research across the region acts as a considerable constraint on effectiveness. For the most part, small countries have to deal with the same range of agricultural research issues as do large countries, but with a substantially more limited capacity. Consequently, they tend to focus on adapting existing technologies to meet local needs, rather than developing new ones. Only in rare circumstances can small countries afford to undertake such activities as plant breeding. Seeking collaboration with other countries in such situations is imperative. The problem for African countries is that most of their neighbors also have only limited agricultural research capacity, focusing mainly on adaptive research, so opportunities to borrow from each other are also limited.

In addition to fragmentation of agricultural research capacity *across* NARSs, fragmentation also occurs *within* NARSs. Systems typically comprise either an agricultural research department housed within a ministry of agriculture or one or more NARIs, complemented by various entities that conduct a small amount of research, such as universities and nonprofit agencies. NARIs typically operate as semiautonomous bodies that receive most of their funding from the government, but have more autonomy to set internal policy and procedures than do government departments, especially in terms of recruiting and remunerating staff and generating their own revenues. Some of the larger NARSs with multiple institutes have established a council to provide coordination and oversight. Examples include South Africa's Agricultural Research Council, the Agricultural Research Council of Nigeria (ARCN), and Ghana's Council for Scientific and Industrial Research (CSIR).

Interlinkages within NARSs remain a problem in most countries because of a lack of effective coordination mechanisms. This internal fragmentation talso complicates linkages with third parties, such as extension providers, policymakers, and farmers' organizations. Weak linkages between agricultural research and extension have been identified by many as a key bottleneck in agricultural innovation in SSA for decades (Chapter 13, this volume), but they remain largely unresolved.[3]

Institutional Composition

Government research agencies still represent by far the largest component of NARSs in terms of the number of FTE researchers employed (72 percent of the total in 2011; Figure 2.4). The higher-education sector's share of researchers expanded substantially in the past couple of decades from 16 percent in 1991 to 25 percent in 2011. This growth stems from the expansion of existing universities, the establishment of new universities, and the addition of new MSc and PhD programs in agricultural and related sciences (Chapters 8 and 10, this volume). While this growth of the university sector has provided NARSs with more and better qualified researchers (although most of them work only part time on research), it has also contributed to a greater

3 In Rwanda, agricultural research and extension were merged under a single entity in 2009. The only other instance of such a merger occurred in Zimbabwe in 2001, but the two functions were again separated in 2009.

FIGURE 2.4 Institutional categories by country based on share of national agricultural researchers, 2011

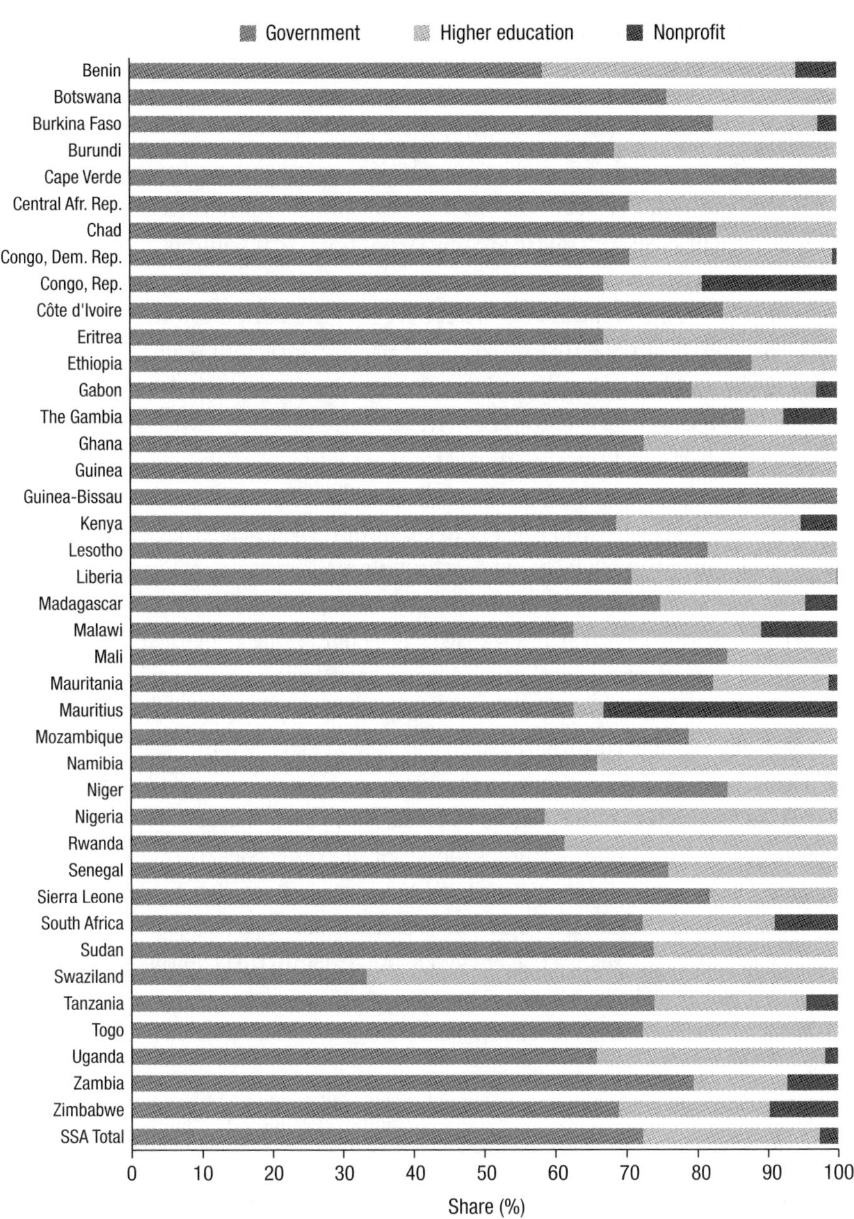

Source: Calculated by authors based on ASTI (2014).

fragmentation of agricultural research capacity across entities, exacerbating challenges related to national coordination. Nigeria's higher-education sector contributes a relatively large share of agricultural researchers to the NARSs, reflecting the adoption of the US land-grant university model (emphasizing university provision of agricultural research and extension) across the Nigerian states in the 1970s and 1980s.[4]

The role of privately funded and implemented agricultural research in SSA is still very limited. Private companies in SSA employ agricultural researchers on a permanent basis, but these are few (Chapter 7, this volume). A complicating factor is that private companies tend to keep information about their research efforts confidential and often have no legal obligation to report on them, with the result that information on private, intramural agricultural research tends to be incomplete. With the exception of South Africa, it is generally believed that this category of private, intramural agricultural research represents only a small fraction of the total agricultural research effort in SSA. More commonly, private companies either outsource their research to public agencies or collectively fund a specialized, nonprofit research agency through a levy or membership fee (Chapter 7, this volume).

The first type of extramural private agricultural research remains largely unnoticed because it is captured statistically under public agricultural research.[5] The second type of extramural private agricultural research is captured under the nonprofit research category. On average, this category represented about 2 percent of SSA's agricultural research capacity in 2011 (Figure 2.4). While many countries have no agencies in this category, 33 percent of FTE researchers in Mauritius fell in this category because of the dominance of the Mauritius Sugar Industry Research Institute, funded for many decades through a cess collected by the sugar industry (until 2012).[6] The track record of these industry-funded, nonprofit research agencies is considerably better than that of government or university research agencies (Kangasniemi 2002). However, this success has not translated into an expansion of this model to other agricultural

4 Swaziland's higher-education sector stands out as contributing an exceptionally large share of agricultural researchers to its NARS, but this is mainly a reflection of the limited capacity of the agricultural research department within the Ministry of Agriculture, not an unusually high contribution by the universities.

5 It is only when detailed data on the funding of public agricultural research agencies are available that this "private" research component becomes visible.

6 In 2012 all sugar industry–related agencies financed by cesses were merged into the newly established Mauritius Cane Industry Authority operating under the Ministry of Agro-Industry and Food Security. The sugar industry of Mauritius has been heavily affected by EU sugar reform and trade policies, which ended the preferential access of Mauritius sugar to the EU market.

industries. In fact, the share of nonprofit agricultural research agencies has slightly declined since 1991, and the number of researchers employed at nonprofit agencies has changed little since 2000.

Supranational Agricultural Research Collaboration

The colonial research infrastructure built by Great Britain, France, Portugal, and Belgium during the first half of the 20th century was typically headquartered in Europe and (1) oversaw agricultural research across colonies not only in Africa, but also in other parts of the world; (2) provided backup support for the work in the colonies; and (3) managed a corps of European agricultural scientists who rotated among the various colonies and headquarters. Funding for agricultural research was mainly derived from the colonies themselves (for example, through levies on agricultural export commodities), although in later years (in particular after WWII) metropolitan investment in African agricultural research became more important. Federal agricultural research agencies spanning multiple African colonies were introduced by France in the 1920s and 1930s (and reinforced after WWII) and by Great Britain in the 1940s and 1950s. Examples of federal agricultural research agencies established during the colonial time by Great Britain are the East African Agriculture and Forestry Research Organization, the Tea Research Institute of East Africa, the West African Cocoa Research Institute, and the West African Rice Research Institute. In the case of France, the chain of tropical commodity institutes headquartered in France operated in the field through research centers and stations with either local or multicountry mandates. Nearly all of these federal agricultural research entities eventually lost their supranational mandate after independence and were disbanded or integrated into the NARS of the host country.

The first wave of such post-independence collaboration came in the form of regional agricultural research networks that emerged during the 1980s and 1990s, often operating under the auspices of CGIAR centers or other external agents, such as the Food and Agriculture Organization of the United Nations (FAO), France's CIRAD, or Canada's International Development Research Centre (IDRC) (Roseboom, Pardey, and Beintema 1998). Characteristic of this period was that the drive for this new supranational collaboration came mainly from outside and not so much from within. This often resulted in a top-down approach, with the external agent in charge and the national partners as relatively passive participants (so-called central source networks in contrast to collaborative networks).

Subregional Organizations

It took a while for the "internal drive" toward supranational collaboration in agricultural research in SSA to emerge, very much supported by development partners. Important in this process has been the establishment of three SROs[7]:

1. **Centre for Coordination of Agricultural Research and Development for Southern Africa (CCARDESA).** CCARDESA's predecessor, the Southern African Centre for Cooperation in Agricultural Research (SACCAR), was established in 1984 by the Southern African Development Community (SADC), which currently comprises Angola, Botswana, Democratic Republic of the Congo, Lesotho, Madagascar, Malawi, Mauritius, Mozambique, Namibia, Seychelles, South Africa, Swaziland, Tanzania, Zambia, and Zimbabwe. Unfortunately, SACCAR was phased out between 1997 and 2001. Some of SACCAR's tasks were continued by SADC's Food, Agriculture and Natural Resources Directorate (SADC–FANR), but at a substantially lower level of activity and within a restrictive administrative setting. A plan to re-establish a semiautonomous SRO for the subregion was developed in 2007/08 (SADC 2008a, 2008b) and approved by the SADC Council in 2010. CCARDESA operates under the SADC secretariat, but is financially and administratively autonomous. It officially commenced operation in August 2011, but is still in its initial stages of development. CCARDESA is intended to address relevant agricultural R&D issues by (1) coordinating the implementation of regional agricultural R&D programs; (2) facilitating collaboration among NARS stakeholders; (3) promoting public–private partnerships in regional agricultural R&D; and (4) improving the generation, dissemination, and adoption of agricultural technologies in the subregion through collective efforts, training, and capacity building.

2. **West and Central African Council for Agricultural Research and Development (CORAF/WECARD).** Established in 1987, CORAF/WECARD initially only covered French-speaking African countries and was dominated by French advisors. In some ways, it tried to maintain the French colonial agricultural research structures. For the first few years the secretariat was based in Paris; it was only transferred to Dakar, Senegal, in 1990. That year it was also decided to include the subregion's English- and Portuguese-speaking countries, which eventually occurred

7 This section is largely based on Roseboom (2011).

in 1995. Currently, 22 countries in West and Central Africa are members of CORAF/WECARD: Benin, Burkina Faso, Cameroon, Cape Verde, Central African Republic, Chad, Republic of the Congo, Côte d'Ivoire, Democratic Republic of the Congo, Gabon, The Gambia, Ghana, Guinea-Bissau, Guinea-Conakry, Liberia, Mali, Mauritania, Niger, Nigeria, Senegal, Sierra Leone, and Togo.

CORAF/WECARD underwent a major restructuring in 2007/08, when it adopted a programmatic approach comprising eight programs: (1) livestock, fisheries, and aquaculture; (2) staple crops; (3) nonstaple crops; (4) natural resource management; (5) biotechnology and biosafety; (6) policy, markets, and trade; (7) capacity strengthening and coordination; and (8) knowledge management. While the last two programs do not conduct research, they support the others in the effective delivery and dissemination of research results. Budget constraints meant that it took several years before all eight programs became operational.

3. **Association for Strengthening Agricultural Research in Eastern and Central Africa (ASARECA).** Established in 1994, ASARECA currently has 11 member countries: Burundi, Democratic Republic of the Congo, Eritrea, Ethiopia, Kenya, Madagascar, Rwanda, Sudan, South Sudan, Tanzania, and Uganda.[8] ASARECA's program structure is almost identical to CORAF/WECARD's, comprising (1) staple crops, (2) high-value nonstaple crops, (3) livestock and fisheries, (4) agrobiodiversity and biotechnology, (5) natural resource management and biodiversity, (6) policy analysis and advocacy, and (7) knowledge management and upscaling. In contrast to CORAF/WECARD, capacity strengthening is no longer separate, but is instead integrated across the programs.

Initially, the SROs only aimed to coordinate the work of the different regional agricultural research networks and programs within their mandated areas. By the late 1990s, however, they started to develop their own subregional agricultural research strategies. This was a first clear sign that the SROs and their members were starting to take charge of the supranational agricultural research agenda in their subregions. From merely coordinating regional agricultural research networks, the SROs have assumed the following five

---

8 Madagascar and Tanzania are members of both CCARDESA and ASARECA, and the Democratic Republic of the Congo is a member of all three SROs. Comoros, Djibouti, São Tomé and Príncipe, Somalia, and Réunion do not belong to any of the SROs; South Sudan joined ASARECA in December 2011.

functions over the past decade: (1) advocacy and policy formulation, (2) capacity strengthening, (3) knowledge management and information exchange, (4) coordination of agricultural research in each subregion, and (5) promotion of supranational agricultural research activities in selected priority areas. The fifth function is usually seen by the SROs as the one with the biggest growth potential.

The SROs have no research capacity of their own; they focus on mobilizing their NARS members to conduct agricultural research that is of regional interest through commissioned or competitive agricultural research grant schemes (CARGs). In the case of competitive bidding, the SROs are experiencing difficulty in keeping the instrument focused on supranational research priorities, as national researchers have a tendency to see these regional competitive funding schemes as an opportunity to fund their own national research priorities. Stronger upfront priority setting (that is, defining the supranational research agenda) and far more specific calls for proposals are needed to keep the instrument focused on supranational research priorities. In addition to CARGs, the SROs are promoting the development of national agricultural research centers of excellence that conduct research of supranational relevance (see the section on regional agricultural productivity programs below).

ASARECA and CORAF/WECARD have adopted important institutional innovations over the past decade, which have also influenced the establishment of CCARDESA: (1) the introduction of CARGs for joint research activities, (2) the adoption of a programmatic approach, and (3) the introduction of a multidonor trust fund (MDTF) managed by the World Bank.

Forum for Agricultural Research in Africa

In addition to the three SROs, FARA was established in 2001 to promote and coordinate supranational collaboration in agricultural research across Africa (including North Africa). FARA took over from the Special Program for African Agricultural Research (SPAAR), which was first conceived by the World Bank and other donors at a CGIAR meeting in Tokyo in 1985. SPAAR, which was primarily a donor instrument, became operational in 1987 and was hosted by the World Bank until 2001. FARA very much differs from SPAAR, however, in terms of ownership: SPAAR was controlled by donors, whereas FARA is controlled by the SROs and NARSs. The mandates of FARA and the SROs strongly overlap, requiring coordination and a clear division of labor based on the "subsidiarity" principle, which dictates that responsibility should be adopted at the lowest level possible. The idea is that FARA should take on responsibilities only that are better dealt with collectively than

TABLE 2.1 Expenditures by the Forum for Agricultural Research in Africa and the subregional organizations, 2006–2011

Organization	2006	2007	2008	2009	2010	2011	2012
	US dollars (millions)						
Forum for Agricultural Research in Africa (FARA)							
Secretariat	2.3	3.3	4.3	3.7	5.5	7.3	na
Programs	3.0	5.5	7.8	12.8	14.7	11.6	na
Association for Strengthening Agricultural Research in Eastern and Central Africa (ASARECA)							
Secretariat	0.2	0.4	1.0	1.4	1.6	2.1	2.5
Programs	7.9	12.6	4.6	7.6	13.5	12.3	10.9
West and Central African Council for Agricultural Research and Development (CORAF/WECARD)							
Secretariat	0.4	0.6	0.7	0.7	1.0	na	1.4
Programs	0.5	0.8	0.9	1.9	7.8	na	4.8

Source: Compiled by authors from ASARECA (various years), CORAF/WECARD (various years), and FARA (various years).

Notes: CCARDESA is not included in the table because it only began operating in 2011. Between the demise of its predecessor, the Southern African Centre for Cooperation in Agricultural Research (SACCAR) in the late 1990s and CCARDESA's 2011 establishment, the Food, Agriculture and Natural Resources of SADC (SADC-FANR) took on some of SACCAR's Directorate former activities. Although no specific data were available for SADC-FANR, it is estimated to have spent about US$3–$5 million per year on agricultural research. The projected yearly budget for CCARDESA is about $11 million by 2016, of which $1.8 million is allocated to the secretariat and the rest for its programs. na = data were not available.

by the SROs individually. In practice, however, this principle is logistically challenging and creates political friction, not least because the SROs are at different levels of development.

What FARA and the SROs have in common is their dependence on donor funding, which is unsustainable in the long run. Mobilizing more funding from within the region should be high on the agendas of both FARA and the SROs in the coming years. Moreover, high dependency on donor funding leaves FARA and the SROs vulnerable to the capriciousness of donor policies. Despite substantial efforts by FARA and the SROs to comply with donor requirements, such as the adoption of a programmatic approach and greater accountability, the amount of donor funding that has come forward in recent years has been substantially below expectations (Table 2.1).

Regional Agricultural Productivity Programs

As previously discussed, the effectiveness of African agricultural research is seriously constrained because of the high level of fragmentation of research capacity across countries. As a result, most African countries have very limited capacity in any scientific or technological domain, which both constrains the attainment of national research goals—because resources are spread too thinly over a broad range of topics—and encourages duplication of effort with countries often pursuing the same, rather limited, research agenda. The concept of supranational collaboration is an attempt to improve the effectiveness of agricultural research by pooling resources and talent so that participating countries can jointly pursue a more ambitious research agenda of common interest. One approach to this, as discussed above, is through the CARGs for supranational agricultural research projects operated by the SROs. The current limitation of these schemes is their complete dependence on donor funding in the absence of mechanisms to collect a regional tax or levy to finance such schemes (which itself would pose administrative and coordination challenges).

An alternative approach being piloted through regional agricultural productivity programs is to convert selected national agricultural research programs into initiatives with a supranational mandate, such that the host country becomes a "center of excellence" (that is, specialization) on the target commodity or topic. The funding for these programs is provided by the government hosting the program, in full knowledge that substantial benefits from the research may—and desirably will—spill over into neighboring countries, but this is purposefully designed to be a reciprocal process. By cooperating, countries aim to share both the costs and the results of the research conducted, thereby leveraging their resources on the one hand, and reducing wasteful duplication on the other. This mechanism of mutual interdependence has the added advantage of circumventing the need for a regional tax or levy, whose collection requires a yet to be realized level of economic and political integration among African nations. Since 2007, this approach has been implemented through regional agricultural productivity programs facilitated by the World Bank through a series of loans to national governments, but depending strongly on the coordination capacity of the SROs to bring the various partners together.

The programs have been established as program loans that can be adapted both horizontally (that is, to include more countries) and vertically (that is, through potential extensions after the completion of the first phase). Key in all three programs is the investment in the development of national centers of excellence to promote supranational collaboration. Particularly in the case of West Africa, a substantial component of the investment actively targets the promotion of research spillovers to the other participating NARSs, by training researchers and conducting joint research projects between the center of excellence and the satellite research agencies (Table 2.2).

TABLE 2.2 Agricultural productivity programs in East, West, and Southern Africa

Project/ phase	Approval date	Participating countries	Centers of excellence, specialization, or leadership	Budget (US$)
A. East Africa Agricultural Productivity Program				
Phase 1	June 2009	Ethiopia, Kenya, Tanzania	Wheat, dairy, and rice	90 million
Phase 1a	November 2009	Uganda	Cassava	30 million
Phase 1b	Contemplated but not implemented	Other East African countries		
B. West Africa Agricultural Productivity Program				
Phase 1a	March 2007	Ghana, Mali, Senegal	Roots and tubers, irrigated rice, and drought-tolerant cereals	49.4 million
Phase 1b	September 2010	Burkina Faso, Côte d'Ivoire, Nigeria	Onions, mangoes, bananas and plantains, and fisheries	122.2 million
Phase 1c	March 2011	Benin, Guinea, Liberia, Niger, Sierra Leone, The Gambia, and Togo	Maize, livestock, and mangrove rice	118 million
Phase 2a	May 2012	Ghana, Senegal (Mali dropped out at the last moment due to war)	Roots and tubers, and irrigated rice	131.8 million
C. Agricultural Productivity Program for Southern Africa				
Phase 1	March 2013	Malawi, Mozambique, Zambia	Maize, rice, and food legumes	94.6 million

Sources: Compiled by authors; Roseboom (2011).

It is still too early to assess the impacts of these programs and of the approach in general. Nevertheless, the assumption that this type of mutual interdependence will work is not without risk. If one of the partners pulls out or fails to deliver, the whole constellation of mutual interdependence may collapse. For the moment, the programs receive ample funding through World Bank loans, but once this stops there is no guarantee that countries will continue their collaboration. Moreover, the model assumes spillover potential that in practice may not materialize because of differences in agroecologies, market development, consumer preferences, and so on.

CGIAR's Contribution

The CGIAR Consortium comprises a group of 15 international agricultural research centers that address global development challenges and are funded by a large group of bilateral and multilateral donors. In 2012, the Consortium's global turnover reached US$756 million, of which more than half was spent

FIGURE 2.5 CGIAR expenditures in Africa south of the Sahara, 2000–2011

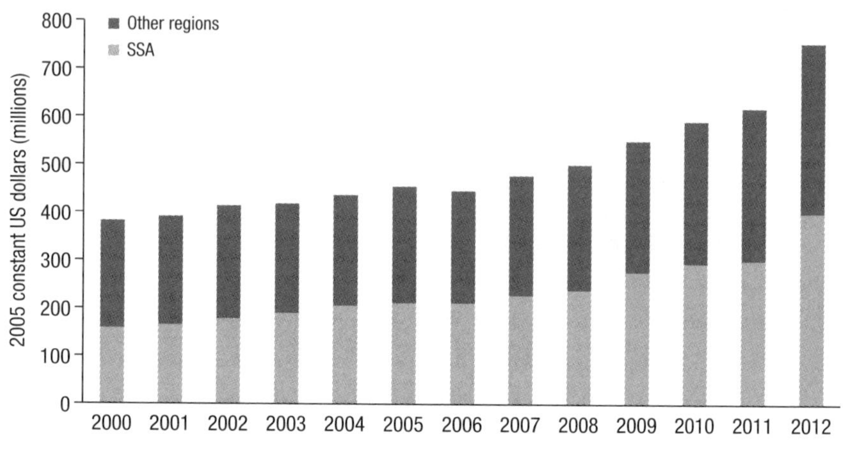

Source: Calculated by authors based on CGIAR (2000–2012, various years).
Note: SSA = Africa south of the Sahara.

on SSA (Figure 2.5). CGIAR investment has shifted toward SSA in the past two decades; SSA's share of total CGIAR spending increased from 40 percent in 1992, to 43 percent in 2002, to about 50 percent in more recent years. SSA receives more support from CGIAR than any other region, although not all CGIAR centers are equally active there. Four centers are headquartered in SSA, and two of them—AfricaRice and the International Institute of Tropical Agriculture (IITA)—focus exclusively on SSA. At the other end of the spectrum are the Center for International Forestry Research and the International Rice Research Institute, which spend only 5 and 7 percent of their resources, respectively, on SSA (Table 2.3). The number of people employed by CGIAR centers focusing on SSA totaled an estimated 763 international staff members and 4,384 local staff members in 2012 (CGIAR 2012).

Compared with African NARSs, CGIAR centers are well funded; however, from the African perspective, they are largely donor-driven, top-down agencies with weak accountability toward local counterparts. Moreover, coordination among the different CGIAR centers and programs is also quite weak, leading to duplication and sometimes outright competition (CGIAR 2005). Much has been done in recent years to improve collaboration among CGIAR centers, as well as between them and their local SSA counterparts through consultations and joint priority setting. For example, CGIAR developed two region-specific medium-term plans in 2005/06 (one for West and

TABLE 2.3 CGIAR center staff focused on Africa south of the Sahara, 2012

CGIAR center	Share SSA (%)	Total number of CGIAR staff (head counts)		Estimated CGIAR staff in SSA (head counts)	
		International	Local	International	Local
Africa Rice Center (AfricaRice)[a]	100	51	224	51	224
Bioversity International	36	62	148	22	53
International Center for Tropical Agriculture	43	88	744	38	320
Center for International Forestry Research	37	73	133	27	49
International Maize and Wheat Improvement Center	51	191	811	97	414
International Potato Center	63	88	744	56	469
International Center for Agricultural Research in the Dry Areas	13	89	324	12	42
International Crops Research Institute for the Semi-Arid Tropics	49	76	1,162	37	568
International Food Policy Research Institute	50	149	328	75	164
International Institute of Tropical Agriculture[a]	100	116	958	116	958
International Livestock Research Institute[a]	61	108	485	66	295
International Rice Research Institute	29	129	1,127	37	326
International Water Management Institute	60	113	202	68	122
World Agroforestry Centre[a]	62	60	393	37	244
WorldFish	47	53	290	25	136
Total		1,446	8,073	763	4,384

Source: Calculated by authors based on CGIAR (2012).

Note: [a] These centers are headquartered in SSA. SSA = Africa south of the Sahara

Central Africa and one for East and Southern Africa), thereby unifying the different activities of CGIAR centers within these regions (Mokwunye 2010). A renewed effort to improve the overall coordination of agricultural research activities in SSA among CGIAR centers, SROs, and NARSs was launched in 2010, known as the Dublin Process. A great deal of potential overlap exists between the CGIAR agenda and the emerging supranational NARS agenda (Chapters 14 and 15, this volume). The best outcome would be for these organizations to collaborate to their mutual benefit, given that CGIAR has the scientific capacity, but NARSs have the political legitimacy.

Collaboration with International Partners

In addition to the multilateral agricultural research efforts by CGIAR's centers, bilateral collaboration with overseas research partners constitutes a substantial contribution to African agricultural research. Initially, such collaboration was predominantly with the former colonial powers as a carryover of the pre-independence research structures (for example, the origins of the French CIRAD and the UK Natural Resources Institute (NRI) can be traced back to colonial times); however, other developed countries, such as the governments of the United States, Canada, Australia, Japan, and various European countries have increasingly offered scientific collaboration programs on a bilateral basis, whereby advanced research agencies and universities provide some of their time and resources to address agricultural research problems of specific relevance to SSA. Often such collaboration is combined with capacity-building activities.

Collaborative research support programs (CRSPs), recently rebranded Feed the Future Innovation Labs, funded fully or partly by the US Agency for International Development (USAID), are a good example. Initiated in 1977, this successful scheme mobilizes the capacities of US land-grant universities (in collaboration with counterpart agencies in developing countries) to address issues of food security, human health, agricultural growth, trade expansion, and sustainable use of natural resources in the developing world. Currently, 10 thematic CRSPs are in operation, each mobilizing multiple US universities and counterpart agencies in developing countries. Of the 19 countries currently targeted by CRSPs, 12 are in SSA. In Europe, many national governments as well as the European Union offer funding opportunities to agricultural research agencies and universities in Europe to conduct collaborative research with overseas partners in developing countries. The Platform for African–European Partnership on Agricultural Research for Development (PAEPARD), hosted by FARA, has the specific mandate to strengthen agricultural research collaboration between Europe and Africa, and make it more demand driven.

More recently, also "South–South" collaboration in agriculture has begun to shape up, including collaboration in agricultural research and technology transfer; the two most prominent of these collaborators are Brazil and China (Chapter 6, this volume). As part of their foreign policies, both countries began to offer support in agricultural research and technology transfer to African countries in 2006.[9] In the case of Brazil, this mainly takes the form of collabo-

9 See Scoones, Cabral, and Tugendhat (2013) for a more in-depth discussion of the contributions of China and Brazil to African agriculture.

rative research and scientific capacity-building projects funded by the Brazilian Cooperation Agency (ABC) and implemented by the Brazilian Agricultural Research Corporation (Embrapa). About 25 of these projects were operating across 15 African countries in 2011, some involving multiple countries (Embrapa 2011). China's contribution is mainly through agricultural technology demonstration centers, 15 of which have been established across SSA in the past few years; another 10 are in the planning phases (FAC 2013). Under this model, the recipient country provides the land, and China provides the infrastructure and operating expenses for the first three years (including the costs of Chinese experts to train local counterparts). After this start-up phase, the centers are expected to be run as commercial, self-supporting demonstration farms.

Key Constraints Affecting Agricultural Research

The following four key constraints have been distilled from the literature as holding back the performance of agricultural research in SSA, each of which is discussed in more detail in the subsequent sections (FARA 2006a, 2013a, 2013b; USAID 2013): (1) inadequate policies; (2) weak organization and management; (3) insufficient financial, human, and physical resources; and (4) poor collaboration and coordination.

Policy Constraints

Structural macroeconomic reform dominated the policy agenda of SSA during the 1980s and 1990s, liberating the economy from excessive, and often inappropriate, government intervention. In retrospect, this emphasis on macroeconomic reform was often at the expense of more specific policies, including agricultural (innovation) policies. Around the turn of the 21st century, the realization emerged that the right macroeconomic context was crucial, but it was not enough to eradicate poverty. The launching of the United Nations Millennium Development Goals (MDGs) was a turning point, introducing a far more qualified approach to economic growth. For example, the poverty reduction strategies that African countries developed together with the World Bank in the early 2000s all addressed the need to improve agricultural policies. However, less than half of them addressed the need to invest in agricultural innovation. Moreover, the emphasis was mostly on extension and advisory services—only 4 of 24 African poverty reduction strategies explicitly mentioned the need to invest in agricultural research (Roseboom, Beintema, and Mitra 2003).

Another important policy initiative that has helped to move agriculture back to the top of Africa's development policy agenda is the Comprehensive

Africa Agriculture Development Programme (CAADP), launched by the African heads of state in 2004 (NEPAD 2003; Chapter 1, this volume). After an initial slow start, CAADP has assumed a leading role in strengthening agricultural policies and investment plans across Africa in recent years. Agricultural innovation is very much part of CAADP's agenda, as is reflected in the fact that agricultural research, technology dissemination, and adoption constitute one of the program's four key thematic pillars. FARA was asked to be Africa's lead agent on this theme, and to this end developed a Framework for African Agricultural Productivity (FAAP) in 2004. This framework argued for a substantial increase in investment in agricultural research, extension, and education, and an institutional reform agenda putting farmers at the center of agricultural innovation (FARA 2006b). Moreover, FAAP proposed the development of subregional agricultural productivity programs and, in particular, lobbied for additional investment in agricultural research addressing subregional priorities. FAAP is being implemented by FARA in close collaboration with the SROs, subregional economic communities,[10] national governments, international organizations, and donors.

Of the 28 national agricultural investment plans that have been developed as part of CAADP to date, only a few have allocated significant resources to agricultural research, extension, and education. As with the MDG poverty strategies, other priorities seem to attract more funding. Policymakers should not only direct more funding to agricultural research, but also help allocate those resources most wisely at various levels (individual research programs and organizations, countries, subregions, regionally, and worldwide). In addition, donors prioritize where they want to place their support for agricultural research. It all results in a complex patchwork of priority setting and decision-making in efforts to steer the allocation of resources (Table 2.4).

The higher the level of aggregation of the strategy or plan, the more abstract the document usually becomes, and the less concrete it is in terms of setting priorities and allocating resources. These documents (which are often impressive in terms of content and length) are useful for consensus building but lack real buy-in and backing through funding; hence, follow-up on their implementation is usually weak. Going down the list of aggregation, the subregional agricultural productivity programs are the first to be backed by substantial funding (that is, World Bank loans to national governments). These

10 These include the Common Market for Eastern and Southern Africa, East Africa Community, Economic Community of Central African States, Economic Community of West African States, and SADC.

TABLE 2.4 Steering Africa's agricultural research agenda

Level	Strategy/planning documents
Global	*International Assessment of Agricultural Knowledge, Science and Technology for Development (McIntyre et al. 2009):* Assessment initiated by the World Bank and FAO, involving broad consultations across the world; the final version of the report was not supported by Australia, Canada, or the United States because of opinions in the report on biotechnology.
Regional	*Science Agenda for Agriculture in Africa (FARA 2013b):* Initiated by the participants of the Dublin Process and facilitated by FARA; an independent panel organized various consultations and formulated the agenda.
	Framework for African Agricultural Productivity (FARA 2006b): Initiated by FARA under the umbrella of CAADP; the first version focused on crops, with livestock added later.
	Realizing the Promise and Potential of African Agriculture (IAC 2004): Initiated by then UN Secretary General Kofi Anan and facilitated by the InterAcademy Council; an international study panel assisted by consultants organized various consultations to formulate this strategy.
Subregional	*SRO strategies/operating plans (ASARECA 2006, 2013; CORAF/WECARD 2007; CCARDESA 2013):* The SROs have adopted a programmatic approach and have prioritized commodities/themes of focus; SROs do not have their own research capacity, but instead mobilize their members to execute the identified research priorities through commissioned or competitive research grants, although the volume of funding available has been limited and rather volatile to date.
	East, West, and Southern Africa agricultural productivity programs (EAAPP/WAAPP/APPSA): Developed by the World Bank in close collaboration with the respective SROs and the participating countries.
National	Very few countries have developed a *national agricultural research strategy* encompassing all the entities constituting its NARS; the NARI/NARO strategy usually acts as a substitute. One exception is Uganda's NARO, which is the apex body for the NARS. Some countries (Ghana, Nigeria, and South Africa) have an (agricultural) research council that oversees a range of agricultural research institutes, but even these councils do not cover the entire NARS. The growing importance of universities requires improved coordination—for example, by promoting joint research planning and projects—between government and higher-education agencies, whose research agendas usually strongly overlap.
Organizational	Organizations need *strategy documents* that articulate both their mission and pathway to achieving that mission within a time horizon of 5–10 years, with revisions if needed. A strategy is often accompanied by a *master/operating plan* detailing what is needed to implement the mission, including infrastructure and equipment, staffing, organizational structure, internal management processes, and governance issues.
Programmatic	*Programmatic planning documents* set out research objectives and priorities (usually in the form of projects) that will be undertaken within a time horizon of 3–5 years, often revised yearly through the addition of another year. A widespread problem across African agricultural research organizations is weak adherence to planning processes. It is not uncommon for research program plans to be nonexistent or out of date, and where plans do exist, there is usually great discrepancy between what is planned and what is actually implemented. Lack of funding usually plays a major factor.

Source: Authors.

Notes: ASARECA = Association for Strengthening Agricultural Research in Eastern and Central Africa; CAADP = Comprehensive Africa Agriculture Development Programme; CCARDESA = Centre for Coordination of Agricultural Research and Development for Southern Africa; CORAF/WECARD = West and Central African Council for Agricultural Research and Development; IAC = InterAcademy Council; FAO = Food and Agriculture Organization of the United Nations; FARA = Forum for Agricultural Research in Africa; NARI = national agricultural research institute; NARO = National Agricultural Research Organization (Uganda); NARS = national agricultural research system; SRO = subregional organization.

subregional programs have prioritized only a selected number of commodities (Table 2.2).

A major weakness of African agricultural research is poor strategizing and planning at the research program level. This usually goes hand in hand with weak monitoring and evaluation (M&E) of research activities. Strengthening the planning capacity at the research program level is crucial for all strategizing and planning higher up in the system to be effective (Chapter 12, this volume).

CGIAR centers do not fit logically within the hierarchy presented in Table 2.4; they have their own research strategies and plans, which add to the complexity of the patchwork. Some centers operate only within Africa, while others operate on a more global scale. In addition, the most recent CGIAR reorganization in 2010 transformed CGIAR from a loose conglomerate of individual centers into a stronger integrated entity (Chapter 15, this volume).

Weak Organization and Management

An assessment of African NARSs by FARA in 2005 (FARA 2006a) revealed that 54 percent of the agricultural research agencies surveyed in SSA lacked a long-term strategic plan. Of those that had a strategic plan, about a quarter had no implementation plan. Moreover, 38 percent of the respondents rated their capacity to undertake research priority setting, program planning, and M&E as inadequate.

Based in part on this assessment, FARA and the SROs initiated the program Strengthening the Capacity for Agricultural Research and Development in Africa (SCARDA) in 2007. In preparation, a more in-depth assessment of the 12 focal institutes that participated in the pilot phase of the program was undertaken in 2007/08. Organization and management topics that were most frequently mentioned as a weakness included research planning and priority setting, M&E, financial management, human resource management, and mobilization of funding. SCARDA offered various training opportunities addressing (some of) these issues (Annor-Frempong, Roseboom, and Ojijo 2012), but the program has been dormant since the completion of the pilot phase in 2010, with the exception of some subregional activities. More investment in this line of capacity building is definitely warranted. An important lesson learned from SCARDA and similar initiatives is that management processes require permanent maintenance. Many African agricultural research organizations have received the necessary assistance with strategic planning, priority setting, financial management, M&E, and so on in the past, but they often fail to maintain and update such processes.

Since the 1970s, the World Bank has played an important and dominant role in the development of agricultural research in SSA. Three distinctive waves of World Bank investment can be identified:

1. During the 1970s, 1980s, and 1990s, the World Bank strongly supported the creation of national agricultural research organizations or institutes, which usually required major investments in infrastructure and human resources, as well as in the design of the new organizational governance structure.
2. From the mid-1990s onward, however, World Bank projects shifted their emphasis to improving the performance of existing organizations. Topics that dominated this new agenda included redefining the role of government to include greater private-sector participation and the promotion of public–private partnerships; decentralization, either geographically or to lower levels of government; stakeholder participation; new funding mechanisms, particularly competitive funding schemes; and systemwide linkages. As documented by Chema, Gilbert, and Roseboom (2003), many of the reforms introduced during this era were strongly influenced by new public management theory promoting the use of private business management concepts in a public-sector context (including competition, client orientation, production targets, and contracts).
3. The current wave of World Bank investment in agricultural research in SSA, which began in 2007, comes in the form of subregional agricultural productivity programs, as previously discussed.

Insufficient Human, Financial, and Physical Resources

The Global Forum on Agricultural Research recommends that developing countries invest at least 1.0–1.5 percent of AgGDP in agricultural research (Lele et al. 2010). Although there is no theoretical underpinning for this recommendation, the lower end of the target (1 percent) has been a widely accepted norm for public investment in agricultural research by developing countries for some time, but most African countries fall short of this target (Figure 2.3). In addition, high volatility of donor funding causes serious boom-and-bust problems (Chapter 4, this volume). Underinvestment in agricultural research is further exacerbated by the fact that the few resources available are spread too thinly over a very wide range of research topics and, hence, are not making much progress in any of them. What matters is not only the absolute volume of funding, but also how it is allocated across the different inputs needed to implement a research program. Common imbalances are understaffing combined with low salaries and poor working conditions, which undermine staff motivation, along with insufficient capital and operating budgets to properly maintain infrastructure and equipment and support the day-to-day conduct of research. Lack of operating budgets also constrains much-needed on-farm, adaptive research trials.

Many African NARSs are currently experiencing a wave of retirement of relatively well-qualified and experienced agricultural researchers (Chapter 8, this volume) who were trained overseas in the 1970s and 1980s and were subsequently recruited into the newly established NARIs. Thereafter, recruitment was often scaled down or frozen for long periods, in part because of the structural adjustment programs of the 1990s. The growth in the number of agricultural researchers employed dropped from an average 6.3 percent per year in the 1970s, to 4.1 percent per year in the 1980s, to 1.3 percent per year in the 1990s. In 2000–2011, growth bounced back to a moderate average level of 3.5 percent per year (calculated by the authors from ASTI 2014). This slow growth has skewed the age distribution of agricultural researchers in many African NARSs, and with the current exodus of retiring researchers, significant expertise is being lost. New recruits have predominantly been educated locally, but a major concern is that the quality of that education is substandard, particularly in terms of knowledge of research methodologies and statistics. Several focal institutes that participated in the SCARDA program indicated a strong need for on-the-job training of newly recruited researchers on these particular topics (Annor-Frempong, Roseboom, and Ojijo 2012).

Relatively little is known about the adequacy of agricultural research infrastructure in SSA. The World Bank has traditionally been an important provider of funding for agricultural research infrastructure, but the agricultural productivity program model only focuses on specific commodities and issues, leaving large components of NARSs behind. With the exception of China, most bilateral donors are reluctant to invest in agricultural research infrastructure.

Weak Coordination and Collaboration

Another often-cited constraint to the functioning of African NARSs is the weakness of institutional linkages within and between national and supranational entities, between agricultural research and agricultural extension, and between research providers and end users. These issues are discussed in turn below.

1. **Weak linkages within and between national and supranational agricultural research entities.** Many African NARSs are weak or outright

dysfunctional at the system level. NARS coordination mechanisms (ranging from ad hoc committees to agricultural research councils) are not uncommon, but they are often constrained by lack of resources, commitment, and goals. Interestingly, cross-border linkages in agricultural research in SSA have improved significantly over the past decade as a result of the activities of FARA, the SROs, and the regional agricultural productivity programs.

2. **Weak linkages between agricultural research and extension providers.** This classic system constraint remains problematic in most African countries. A recent survey of agricultural researchers in Ghana and Nigeria revealed little interaction of researchers with agricultural extension or with other innovation actors (Ragasa, Abdullahi, and Essegbey 2011). Underinvestment in agricultural extension is perhaps even worse than in agricultural research. Moreover, political interference and frequent changes in extension modalities have added to the complexity of creating proper linkages between research and extension.
3. **Weak linkages between agricultural research providers and end users (farmers, traders, and processors).** The concept of the agricultural innovation system has introduced a far more holistic and interactive approach to agricultural innovation and has placed end users at the center (World Bank 2012). An important innovation is the introduction of innovation platforms, which incorporate all the different actors, including agricultural research and extension, to define mutual goals and facilitate collaboration and coordination (Chapter 13, this volume).

Conclusion

Despite substantial growth in agricultural research capacity across Africa over the past 50 years, agricultural research intensity ratios—such as agricultural researchers per 100,000 economically active agricultural population and agricultural research expenditures as a percentage of AgGDP—have been relatively weak and stagnant on average. In other words, African countries have continued to underexploit the potential of (scientific) knowledge to boost agricultural productivity. Despite CAADP's efforts to promote stronger investment in agriculture (including agricultural research), on average, evidence does not indicate significant improvement in the relative intensity of agricultural research in the region. What matters is not only the volume of inputs invested in agricultural research, but also how these inputs are being used. Weak organization and management continue to be widespread among African agricultural research organizations, and more capacity building in this area is warranted. Many organizations also expect that the adoption of more integrative research approaches (such as innovation platform and value-chain approaches) have the potential to improve the impact of agricultural research. This, however, requires substantial capacity building of existing and future researchers in these new research methods and approaches.

At a higher level, a topic that has attracted significant interest and support over the past decade is stronger collaboration among African NARSs in the form of joint research programs and regional centers of excellence. Such collaboration is expected to eventually lead to higher research impact. While optimizing the use of agricultural research resources is to be lauded, investments in agricultural research undoubtedly need to be drastically increased. Taking into account where an additional dollar has the biggest impact, priority should be given to investment in CGIAR, NARSs in countries with large agricultural sectors, and cross-country collaborative research. This does not mean that local adaptive research should be neglected (it is needed to exploit the benefits of more upstream research), but only that the potential returns to such research are generally lower.

References

Annor-Frempong, I., J. Roseboom, and N. Ojijo. 2012. *A Pilot Study on Institutional and Organizational Changes in Selected National Agricultural Research and Education Agencies in Sub-Saharan Africa.* Accra, Ghana: Forum for Agricultural Research in Africa.

ASARECA (Association for Strengthening Agricultural Research in Eastern and Central Africa). 2006. *ASARECA's Strategic Plan 2007–2016: Agricultural Research-for-Development in Eastern and Central Africa.* Entebbe, Uganda.

———. 2013. *Operational Plan 2014–2018.* Entebbe, Uganda.

———. Various years. *ASARECA Annual Report.* Entebbe, Uganda.

ASTI (Agricultural Science and Technology Indicators). 2014. ASTI database. Washington, DC: International Food Policy Research Institute. www.asti.cgiar.org.

CCARDESA (Center for Coordination of Agricultural Research and Development in Southern Africa). 2013. *Medium Term Operational Plan 2014–2018.* Gabarone, Botswana.

CGIAR. 2005. *Report of the CGIAR Sub-Saharan Africa Task Forces: The Tervuren Consensus.* Washington, DC: CGIAR Secretariat.

———. 2011. *A Strategy and Results Framework for the CGIAR.* Montpellier, France: CGIAR Consortium Office and the Fund Office.

———. 2012. *CGIAR Financial Report 2011.* Montpellier, France: CGIAR Consortium Office and the Fund Office.

———. 2000–2012. Financial reports. Accessed April 2014. www.cgiar.org/resources/cgiarfinancial-reports/.

Chema, S., E. Gilbert, and J. Roseboom. 2003. *A Review of Key Issues and Recent Experiences in Reforming Agricultural Research in Africa.* Research Report 24. The Hague: International Service for National Agricultural Research.

CORAF/WECARD (West and Central African Council for Agricultural Research and Development). 2007. *CORAF/WECARD Strategic Plan 2007–2016.* Dakar, Senegal.

———. Various years. *CORAF/WECARD Annual Report.* Dakar, Senegal.

Embrapa (Brazilian Agricultural Research Corporation). 2011. *Africa: A Continent Full of Opportunities for Agricultural Research.* Brasilia.

FARA (Forum for Agricultural Research in Africa). 2006a. *Agricultural Research Delivery in Africa: An Assessment of the Requirements for Efficient, Effective, and Productive Agricultural Research Systems in Africa. Main Report and Strategic Recommendations.* Accra, Ghana.

———. 2006b. *Framework for African Agricultural Productivity.* Accra, Ghana.

———. 2013a. *A Discussion Paper on the Development of a Science Agenda for Agriculture in Africa: A Long Term Strategic Framework.* Accra, Ghana.

———. 2013b. *Science Agenda for Agriculture in Africa (S3A). A Report of an Expert Panel.* Accra, Ghana.

———. Various years. *FARA Annual Progress Report.* Accra, Ghana.

Future Agricultures Consortium. 2013. *Can China and Brazil Help Africa to Feed Itself?* FAC-CAADP Policy Brief 12. Accessed April 2014. www.future-agricultures.org/research/cbaa/7899-can-china-and-brazil-help-africa-feed-itself#.U7kyZahM5FQ.

IAC (InterAcademy Council). 2004. *Realizing the Promise and Potential of African Agriculture.* IAC Report. Accessed June 2014. www.interacademycouncil.net/File.aspx?id=26635.

Kangasniemi, J. 2002. "Financing Agricultural Research by Producers' Organizations in Africa." Chapter 5 in *Agricultural Research Policy in an Era of Privatization*, edited by D. Byerlee and R. Echeverría. Wallingford, UK: CABI Publishing.

Lele, U., J. Pretty, E. Terry, and E. Trigo. 2010. *Transforming Agricultural Research for Development.* Report for the Global Conference on Agricultural Research (GCARD) 2010. Rome: Global Forum on Agricultural Research.

McIntyre, B., H. Herren, J. Wakhungu, and R. Watson, eds. 2009. *International Assessment of Agricultural Knowledge, Science and Technology for Development.* Washington, DC: Island Press.

Mokwunye, U. 2010. *Regional Review of Africa's Agricultural Research and Development.* Report for the Global Conference on Agricultural Research (GCARD). Montpellier, France.

NEPAD (New Partnership for Africa's Development). 2003. *Comprehensive Africa Agriculture Development Programme (CAADP).* Midrand, South Africa.

Pardey, P., J. Roseboom, and J. Anderson. 1991. "Topical Perspectives on NARSs." Chapter 8 in *Agricultural Research Policy: International Quantitative Perspectives*, edited by P. Pardey, J Roseboom, and J. Anderson. Cambridge, UK: Cambridge University Press.

Ragasa, C., A. Abdullahi, and G. Essegbey. 2011. *Measuring R&D Performance from an Innovation Systems Perspective: An Illustration from the Nigeria and Ghana Agricultural Research Systems.* Conference Working Paper 14 from the ASTI–IFPRI/FARA Conference "Agricultural R&D: Investing in Africa's Future: Analyzing Trends, Challenges, and Opportunities," Accra, Ghana, December 5–7. Accessed April 2013. www.asti.cgiar.org/pdf/conference/Theme3/Ragasa.pdf.

Roseboom, J. 2011. *Supranational Collaboration in Agricultural Research in Sub-Saharan Africa.* Conference Working Paper 5 from the ASTI–IFPRI/FARA Conference "Agricultural R&D: Investing in Africa's Future: Analyzing Trends, Challenges, and Opportunities," Accra, Ghana, December 5–7. Accessed April 2013. www.asti.cgiar.org/pdf/conference/Theme4/Roseboom.pdf.

Roseboom, J., N. Beintema, and S. Mitra. 2003. *Building Impact-Oriented R&D Institutions.* Background paper commissioned by the InterAcademy Council (IAC) Study Panel on Science and Technology Strategies for Improving Agricultural Productivity and Food Security in Africa, Amsterdam.

Roseboom, J., P. Pardey, and N. Beintema. 1998. *The Changing Organizational Basis of African Agricultural Research.* EPTD Discussion Paper 37. Washington, DC: International Food Policy Research Institute. Accessed April 2013. www.ifpri.org/sites/default/files/publications/eptdp37.pdf.

SADC (Southern African Development Community). 2008a. *Proposal for the Establishment of a Sub-Regional Organisation on Agricultural Research for the Southern African Development Community (SADC) Region, Proposed to Be Known as the Centre for Agricultural Research and Development for Southern Africa (CARDESA).* Gabarone, Botswana: SADC Secretariat.

———. 2008b. *SADC Multi-country Agricultural Productivity Programme (SADC MAPP).* Programme Document. Gabarone, Botswana: SADC Secretariat.

Scoones, I., L. Cabral, and H. Tugendhat. 2013. "New Development Encounters: China and Brazil in African Agriculture." *IDS Bulletin* 44 (4): 1–19. Brighton, UK: Institute of Development Studies.

USAID (United States Agency for International Development). 2013. *Towards USAID Re-engaging in Supporting National Agricultural Research Systems in the Developing World.* Washington, DC.

World Bank. 2012. *Agricultural Innovation Systems: An Investment Sourcebook.* Washington, DC.

Chapter 3

ECONOMIES OF SIZE IN NATIONAL AGRICULTURAL RESEARCH SYSTEMS

Keith Fuglie and Nicholas Rada

Many national agricultural research systems (NARSs) in Africa south of the Sahara (SSA) continue to face significant challenges in securing the necessary human and financial resources to deliver improved technologies to farmers. Despite comprising more than 10,000 scientists and an aggregate investment of more than 1.5 billion purchasing power parity (PPP) dollars, national agricultural research in the region is fragmented among more than 30 separate NARSs, ranging in size from fewer than 50 to more than 2,000 full-time-equivalent (FTE) scientists and spending less than 2 million to more than 400 million PPP dollars per year on average (ASTI 2014). Spending on agricultural research as a share of each country's agricultural gross domestic product (AgGDP) also varies widely, with ratios ranging from less than 0.2 percent to more than 4.0 percent. The majority of countries have ratios of less than 0.5 percent, but some of the smallest countries have the highest ratios, raising the concern that small-country NARSs may not be economically viable (Stads and Beintema 2012). On the other hand, the presence of CGIAR research centers and regional research networks may act as a counterweight to this small-country problem. By linking with international and regional research systems and focusing on local adaptive research, small countries may be able to earn reasonable returns to research investments.

This chapter expands on Fuglie and Rada (2013) by examining size efficiencies in the economic returns to national agricultural research in SSA. The approach is to empirically estimate how past investments in NARSs affected agricultural productivity in SSA counties, and to examine whether there are systematic differences in the rates of return to agricultural research and development (R&D) among small, midsize, and large countries. In this context, national agricultural research spending is treated as an investment in "knowledge capital." However, because of the length of time required for research

The views expressed in this chapter are those of the authors and not necessarily those of the Economic Research Service or the United States Department of Agriculture.

investments to generate the technologies farmers ultimately adopt, these long-lasting impacts accrue with a time lag.

Drawing on data from a panel of SSA countries, an econometric model was used to estimate the average percentage changes in national agricultural output resulting from a 1 percent increase in the stock of NARS knowledge capital, which in economic terms is known as the *elasticity* of national agricultural research. These estimates, combined with information on research intensity ratios (spending on agricultural research as a share of AgGDP) and the time lag between R&D spending and its economic impact, provide the basis for determining the economic returns to agricultural research. The model was also used to test whether national and international agricultural research efforts in SSA are complementary. If they are, linkages with international research organizations, such as the CGIAR Consortium, can raise returns to national investment in NARSs.

The next section describes the size of NARSs in SSA, based on Agricultural Science and Technology Indicators (ASTI) data, and long-term trends in agricultural output and total factor productivity (TFP) growth in the region, based on Fuglie (2011). TFP provides a comprehensive measure of how efficiently a country's agricultural land, labor, materials, and capital produce agricultural outputs (see Chapter 1, this volume). Growth in TFP is a gauge of technical change and provides an important performance indicator of a country's R&D system. The subsequent section briefly describes the model used to explain the different patterns of TFP growth across SSA countries as a function of R&D knowledge capital and the broader "enabling environment" for technology dissemination. Finally, results are presented and discussed in terms of their implications for agricultural science and technology (S&T) policy in the region.

National Agriculture R&D Investment and Technical Change

National capacities in research have been shown to be strongly correlated with long-term growth in a country's agricultural productivity (Evenson and Fuglie 2010). As a result, agricultural productivity in most of the world has accelerated in recent decades, but SSA remains a laggard (Fuglie 2008). Even though the first agricultural research institutes in Africa date to the colonial era, SSA countries as a whole have been relative latecomers in establishing national systems of agricultural research (Eicher 1990). By the 1980s, most SSA countries had functioning NARSs, although with large variation in their scientific resources. As the data in Table 3.1 show, many of these

TABLE 3.1 Agricultural research investment of the countries of Africa south of the Sahara

Country by size of agricultural sector	Gross agricultural output (2005 US dollars, millions)	Agricultural R&D spending as a share of AgGDP (%)
Large countries		
Nigeria	22,460	0.24
Sudan	6,399	0.31
Ethiopia	6,078	0.32
Tanzania	4,485	0.34
Côte d'Ivoire	4,422	1.05
Kenya	4,351	1.50
Uganda	4,304	1.05
Congo, Democratic Republic	3,991	na
Ghana	3,458	0.57
Midsize countries		
Cameroon	2,680	0.78
Madagascar	2,615	0.47
Mali	1,814	0.96
Zimbabwe	1,580	0.51
Niger	1,558	0.59
Mozambique	1,429	0.69
Malawi	1,413	1.38
Burkina Faso	1,390	0.74
Guinea	1,280	0.40
Benin	1,243	0.56
Rwanda	1,167	0.32
Small countries		
Senegal	977	1.65
Burundi	968	0.60
Zambia	819	0.90
Central African Republic	647	0.39
Togo	563	0.74
Sierra Leone	444	0.26
Namibia	387	2.97
Mauritania	358	0.53
Swaziland	259	2.13
Congo Republic	250	1.36
Mauritius	242	2.79
Gabon	217	0.50
Botswana	216	3.50
Eritrea	202	1.19
Guinea-Bissau	178	na
Lesotho	122	5.43
The Gambia	98	0.73

Sources: Data are yearly averages for the period 1981–2008. Gross agricultural output is from FAO (various years). Data on agricultural R&D spending as a share of AgGDP are from ASTI (2014), supplemented for some countries with data from Pardey, Roseboom, and Anderson (1991). Data were not available for Angola, Chad, Liberia, Somalia, and the small island countries.

Notes: AgGDP = agricultural gross domestic product; FAO = Food and Agriculture Organization of the United Nations; R&D = research and development; na = data were not available.

systems are quite small. Although there is actually very little empirical evidence on the appropriate size and organization of NARSs, Ruttan (1982, 173) considered "an agricultural research and training capacity of 250 to 500 postgraduate-level agricultural scientists and technicians . . . an essential component of any serious effort to enhance agricultural productivity." If smaller than this, the system would risk spreading its scientific and technical resources too thinly across multiple commodities and specialized fields. Only a few SSA countries—mostly the large ones—have come close to Ruttan's suggested capacity level.

The "small-country problem" becomes evident when the capacity of a NARS is compared with the size of the agricultural sector the NARS is attempting to serve. While a large country like Nigeria can easily justify a sizable investment in agricultural research, the case becomes less clear for countries like Senegal and Zambia. If Nigeria's 1,200 agricultural scientists produce enough technologies to raise the output of its US$22 billion[1] agricultural sector by just a few percentage points, the returns to its investment in research will be large. On the other hand, the 180 scientists doing agricultural research in Senegal and Zambia serve sectors that produce less than $1 billion in gross output. Scientific resources in such countries are spread thinly across many commodities and subject areas. Moreover, while the scientific effort to develop a new crop variety or other technology may be similar regardless of country size, the potential economic impact of successful new technologies will only be proportional to the size of the country's agricultural sector.

The problem becomes even more prominent in small countries like Botswana and Lesotho, where support for just a few dozen agricultural scientists already accounts for more than 3 percent of AgGDP. An empirical assessment of the small-country problem thus requires answers to two questions. First, does the relationship between a country's investment in agricultural knowledge capital and the growth rate in its agricultural productivity change with country size? And second, is the value created by that productivity growth large enough to justify the research investment?

Estimates of SSA's agricultural TFP growth over the past 50 years are presented in Table 3.2. Unlike partial productivity measures, such as labor or land productivity (output per worker or output per hectare), TFP accounts for all conventional inputs used to produce economic outputs

1 All currency is in US dollars, unless specifically noted otherwise.

(land, labor, capital, and materials). While increases in labor or land productivity may be attributed to increased use of other inputs, increases in TFP reflect efficiency improvements resulting from adoption of improved technologies and farm practices, which in turn are influenced by enabling environment factors, such as infrastructure, political stability, and sound economic policies. For SSA as a whole, agricultural TFP growth was virtually nonexistent in the 1960s and 1970s, with some countries showing negative TFP growth. But regional average productivity appeared to pick up in the mid-1980s, and since then regional TFP has been growing by about 1 percent per year.

Apart from South Africa, Kenya stands out as having sustained steady agricultural TFP growth of about 1 percent per year during the 1961–2010 period. Several countries, including Angola, Cameroon, Ghana, Malawi, Sudan, and possibly Nigeria, seemingly entered a sustained productivity growth path in the 1980s and 1990s. Each increased its TFP by at least 75 percent between 1981 and 2010, although some of this was recovery from declines in the 1960s and 1970s. Based on data from the Food and Agriculture Organization of the United Nations (FAO), Nigeria's agricultural TFP nearly doubled between 1981 and 2010, but Fuglie's (2011) revisions for agricultural labor force and output growth reduces this to 36 percent.[2]

Other patterns of TFP growth are evident (Table 3.2). A few countries appeared to have been on a sustained TFP growth path, but then productivity stagnated or declined. Côte d'Ivoire and Zimbabwe experienced positive TFP growth for several decades, but Zimbabwe's productivity deteriorated sharply from around 1997, and Côte d'Ivoire's stagnated after 2000. In both countries, the reversal in TFP growth correlated with periods of macroeconomic mismanagement or civil unrest. Another set of countries, notably Angola and Mozambique after 1991, showed strong TFP growth (or recovery) after a prolonged period of decline during protracted civil wars. Finally, a number of SSA countries have shown no significant change in agricultural TFP over the past 50 years. Ethiopia, Chad, Togo, countries in Central Africa other than Cameroon, and scattered other countries fall into this "no productivity growth" category.

2 Fuglie (2011) suggests that FAO may significantly underestimate the amount of agricultural labor employed in Nigeria (FAO assumed zero growth since the late 1960s). The revised estimates for this chapter assume that Nigeria's use of labor is similar to the rest of SSA (about 2 percent per year).

TABLE 3.2 Indexes of agricultural output and total factor productivity for subregions and countries of Africa south of the Sahara

Subregion/country	Average output, 2006–2010 (2005 dollars, billions)	Gross agricultural output index (1961 = 100)					Agricultural total factor productivity (TFP) index (1961 = 100)					Average TFP growth, 1961–2010 % per year
		1971	1981	1991	2001	2010	1971	1981	1991	2001	2010	
Central Africa	10.06	127	153	201	208	278	91	85	92	85	102	–0.12
Cameroon	4.60	150	179	222	301	486	102	94	108	118	175	1.17
Central African Republic	0.90	133	166	203	286	340	90	82	93	108	111	0.21
Congo Republic	0.38	124	141	161	216	311	85	91	87	96	99	–0.03
Congo, Democratic Republic	3.84	119	143	194	159	164	87	87	92	90	87	–0.29
Gabon	0.27	118	162	205	239	273	97	78	80	90	88	–0.26
Eastern Africa	22.38	148	170	234	278	399	107	110	114	120	130	0.45
Burundi	1.06	122	132	168	154	173	85	85	87	81	72	–0.69
Kenya	6.73	133	189	277	327	472	98	118	125	134	161	0.99
Rwanda	1.82	150	222	252	286	439	102	120	105	115	122	0.42
Tanzania	6.97	136	186	234	297	429	98	105	114	125	133	0.60
Uganda	5.80	174	152	201	268	308	126	128	128	128	115	0.30
Horn of Africa	22.09	128	164	177	260	387	97	102	99	108	136	0.48
Ethiopia, former	8.89	119	137	147	203	304	86	97	89	91	107	0.15
Somalia	1.64	140	178	176	200	225	108	107	111	124	132	0.58
Sudan	11.52	132	176	201	356	525	92	89	94	115	156	0.93
Sahel	11.81	126	149	197	279	420	89	93	99	109	118	0.32
Burkina Faso	2.24	129	157	287	442	574	86	80	96	132	110	0.19
Chad	1.44	106	115	151	216	249	84	91	92	93	93	–0.15
The Gambia	0.13	126	108	102	141	179	88	63	50	58	51	–1.40
Mali	3.14	128	173	230	320	511	81	97	112	117	142	0.73
Mauritania	0.46	109	122	141	169	198	86	99	97	99	107	0.15
Niger	3.00	125	164	190	319	520	77	69	78	87	111	0.22
Senegal	1.35	100	109	127	149	208	77	78	81	77	92	–0.17

Southern Africa	15.53	137	146	166	218	304	108	97	100	119	145	0.47
Angola	2.90	133	91	103	178	376	78	52	55	81	128	0.52
Botswana	0.25	144	150	171	159	198	125	107	134	109	125	0.47
Lesotho	0.13	114	126	129	144	143	88	95	97	83	95	–0.10
Madagascar	3.40	131	149	175	182	244	101	97	101	104	120	0.38
Malawi	2.74	151	209	236	415	670	105	108	103	165	190	1.34
Mauritius	0.25	118	120	131	138	138	110	112	111	108	107	0.13
Mozambique	2.08	135	118	106	181	216	101	74	74	96	94	–0.13
Namibia	0.43	148	129	137	137	147	127	113	108	95	92	–0.16
Swaziland	0.28	156	230	277	265	296	146	190	205	230	249	1.90
Zambia	1.31	145	173	246	290	449	106	115	124	141	174	1.16
Zimbabwe	1.52	157	182	209	259	210	112	115	120	130	118	0.35
Western Africa	17.85	137	164	234	332	437	99	92	108	124	129	0.68
Benin	2.00	123	151	251	445	552	88	94	111	138	151	0.86
Côte d'Ivoire	5.76	163	255	355	491	525	102	105	109	136	130	0.54
Ghana	6.03	131	109	184	307	432	87	61	93	111	120	0.38
Guinea	1.92	118	135	180	254	325	100	105	115	110	112	0.23
Guinea-Bissau	0.28	77	106	147	208	282	78	69	93	92	111	0.21
Liberia	0.40	153	186	143	221	229	95	88	84	100	83	–0.38
Sierra Leone	0.68	131	148	169	159	280	100	90	89	95	108	0.15
Togo	0.79	125	139	194	273	329	94	85	77	87	97	–0.06
Nigeria	34.62	133	127	242	373	441	90	75	99	136	149	0.84
Nigeria (revised)	28.84	131	130	223	328	359	87	72	81	101	98	–0.04
All Africa south of the Sahara	134.35	135	152	210	282	387	102	101	113	128	140	0.73
All Africa south of the Sahara (revised)	128.57	135	153	207	275	367	101	100	109	122	131	0.56
South Africa	12.36	137	183	190	206	264	103	109	141	187	256	1.90

Source: Fuglie (2011), updated with FAO data (various years).

Notes: Total factor productivity (TFP) growth is the difference between aggregate output and input growth. Output is the Food and Agricultural Organization of the United Nations (FAO) index of gross agricultural output. Input data are also from FAO and include crop area harvested, number of agricultural workers, size of livestock herds, number of tractors in use, and quantity of fertilizer nutrients applied. The "revised" estimates series for Nigeria calculated for this chapter use different measures of output and agricultural labor. For output, production estimates of grains, oilseed, and cash crops are from the United States Department of Agriculture (USDA 2013); estimates of roots and tubers and legumes since 1994 are from Nigeria's national statistics reported by FAO's CountryStat (FAO 2012); and estimates for other commodities use data from FAO (various years). The revision for agricultural labor assumes a 2 percent yearly growth rate from the FAO estimates for 1961–1966.

Evaluating Agriculture's Total Factor Productivity Growth

The econometric analysis considered how (1) agricultural research, both by national governments and the CGIAR Consortium, and (2) the enabling environment for dissemination of new technology, affected agricultural TFP growth in SSA countries. While investments in research provide an obvious mechanism for TFP growth through technical change, the enabling environment (which includes measures of economic policy, farmer education and health, transportation infrastructure, and governance) helps to establish the necessary conditions for technology adoption and economic growth. This environment allows farmers to access new technologies and markets, increases their returns to savings and investments, and provides them with incentives to reallocate resources to the most profitable enterprises.

Investment in Agricultural Research "Knowledge Capital"

The effects of research on productivity require time to accrue, but they can be long-lasting (Alston, Norton, and Pardey 1998). For the purpose of the analysis, research investments were treated as the creation of knowledge capital, and estimates of the capital stock of agricultural research were constructed as the weighted sum of past yearly spending, based on the distributed lag structure estimated by Alene and Coulibaly (2009). In this model, each country's national agricultural research stock is a weighted sum of the current and past 16 years of research spending. To account for the time required for new technologies to be developed and disseminated, the weights on yearly expenditures reflect an inverted U-shape; they start small and gradually rise to a peak in year 8, and then gradually decline as technologies become obsolete. For investments by NARSs, ASTI provides data on yearly spending in public agricultural research systems from 1981 until 2008 for 35 countries of SSA, including South Africa. Data from Pardey, Roseboom, and Anderson (1991) facilitated an extension of the series back to 1961. With a 16-year time lag and yearly R&D investment data available from 1961, the research stock could be estimated from 1977. As a result, the multivariate analytical model of determinants of agricultural TFP growth focuses on the 29-year period from 1977 to 2005.

By far, the biggest agricultural research effort in the region comes from the NARSs (an estimated 14,264 FTE researchers in 2011, including South Africa), which are mostly publicly funded (see Chapters 2 and 4, this volume). The private component of the NARSs (that is, privately funded and executed R&D) is still very small but is on the rise in most African countries (the only exception being South Africa). Small local markets, weak protection of intellectual property rights, little government support in the form of R&D subsidies and tax deduction facilities, and lack of qualified staff discourage private R&D investment (see Chapter 7, this volume). The other important player in African agricultural research is the CGIAR Consortium, with an estimated 648 "internationally recruited" FTE researchers focusing on African agriculture in 2011. The CGIAR budget allocated to Africa has been between 40 and 50 percent of CGIAR's total budget for the past two decades (see Chapter 2, this volume).[3] Other contributions to African agricultural research come from developed countries, such as France through its agricultural research for development center, CIRAD, and the United States through its collaborative research support programs linking US land-grant universities with counterparts in the developing world.[4] Many other countries have similar, although smaller, programs. More recently, Brazil and China have started to provide assistance to African agricultural research (see Chapter 2, this volume).

It is possible to derive a measure of the knowledge stock of CGIAR research in SSA using data on yearly expenditures from CGIAR Financial Reports (CGIAR various years). However, because the expenditure data are not country specific, modeling CGIAR's contribution to SSA's agricultural productivity in this way would only allow an examination of its impact on the region as a whole. To determine the impacts of CGIAR research by country, the model uses the share of national crop area affected by CGIAR technologies, as estimated by Fuglie and Rada (2013). "Area affected" by CGIAR technology includes crop area under improved crop varieties, crop area affected by biological control, and agricultural land area under natural resource management technologies developed by CGIAR centers. This area affected was divided by the total harvested area for all crops to give the share of this agricultural area in a country affected by CGIAR technologies (see Fuglie and Rada 2013 for sources of information on CGIAR technology adoption in SSA).

The area affected by CGIAR research is a measure of technology dissemination, rather than research input. As such, it is likely to be affected by other variables, including investment in NARSs and the enabling environment. To

3 By the late 1990s, about 20 percent of the region's crop area was sown with improved varieties developed by CGIAR centers (Evenson and Gollin 2003). In addition to improved crop varieties, CGIAR had a major impact by developing and disseminating biological control agents for cassava pests (Zeddies et al. 2001).

4 Recently, the United States Agency for International Development's collaborative research support program has evolved into "Feed the Future Innovation Labs for Collaborative Research."

address this problem, a two-stage estimation procedure was used. In the first stage of the regression, CGIAR technology adoption was modeled as a function of the other model variables, as well as CGIAR research stock and the share of crop area planted to cassava. In the second stage, the predicted values of technology adoption from this first regression were used to create a new variable to be included in a model of TFP growth. The coefficient of this new variable gives an unbiased estimate of CGIAR's impact on agricultural productivity in SSA. Since in this model NARS research capital stock affects both the adoption of CGIAR technology and TFP, the model captures interactions between these institutions. If the effect of NARS research capital on the area affected by CGIAR technology is negative, it implies that CGIAR research may be "crowding out" NARSs (in other words, NARSs may be leaving food-crop research to CGIAR and focusing their resources on other commodities). On the other hand, if NARS research capital is positively associated with CGIAR technology adoption, it implies NARS and CGIAR research are complementary, and that returns to investment in NARSs will be higher as a result of CGIAR's presence.

A particular focus in this chapter is whether the returns to agricultural research in small countries are similar to those in large countries. To address this issue, SSA countries were classified into three groups (small, midsize, and large) based on the size of their agricultural sectors, with about one-third of the sample countries falling into each group. The regression model allows the estimate of the research elasticity—as previously described, the percentage change in agricultural output resulting from a 1 percent increase in national agricultural research capital stock—to be different for each group. If the research elasticity for small countries is smaller (or not significant from zero) than the research elasticity for larger countries, then returns to research in small countries could be quite low or even negative. This specification of the model tests whether small-country NARSs produce sufficient economic benefits to justify their investments.

The Enabling Environment for Agricultural Innovation

While research develops and adapts new technologies for local agricultural systems, conditions for rapid diffusion of improved technologies depend on the broader enabling environment for agricultural innovation. The agricultural innovation system includes attention to producer incentives, marketing infrastructure, farmer education and health, property rights, and conditions enabling collective action and information sharing (World Bank 2011a; OECD 2013). Because countries that prioritize agricultural R&D may also be more likely to invest in the broader innovation system, an econometric model of TFP growth needs to include them to avoid overattribution of productivity growth to R&D investment alone.

The model includes five variables representing the enabling environment for agricultural innovation:

- The World Bank's nominal rate of assistance (NRA) to agriculture captures the effects of *agricultural and economic policies* on producer incentives. The NRA—reported yearly for 18 SSA countries, including South Africa—provides a comprehensive measure of the price distortions caused by government policies. Included are commodity price interventions, input subsidies, import/export taxes, exchange rate over- and undervaluations, and direct taxes on agricultural producers. The NRA gives the net effect of these policies on prices paid and received by farmers as a percentage of what prices would be in a market free of these policy interventions (Anderson and Masters 2009).
- *Farmer education* is measured by the average years of schooling of the labor force (Barro and Lee 2010). While this measure is not specific to agricultural workers, it should capture major differences in farmers' educational levels over time and across countries, given that most of the region's labor force is employed in agriculture.
- Another dimension of human capital is *farmer health*. While many health problems, such as malaria and malnutrition, are pervasive in the region, the spread of HIV/AIDS may be the most significant *change* in the overall health status of the general population over the past several decades. HIV/AIDS is expected to reduce agricultural productivity primarily through its effects on labor supply. Therefore, the health status of the labor force was modeled as the proportion of the population estimated to be infected with HIV/AIDS.
- *Transportation infrastructure* is measured by road density (kilometers of roads per square kilometer of land area), using data from the International Road Federation (2006).
- Finally, *governance* is represented in the model by the number of years a country underwent significant armed conflict, as defined by the Uppsala Conflict Data Program (Gleditsch et al. 2002).

The estimates of R&D knowledge capital stocks, CGIAR technology dissemination, and TFP growth are available yearly for 32 SSA countries for the

1977–2005 period. Adding the enabling environment variables reduces the country and time coverage resulting from missing observations on these factors for some countries. To maintain econometric degrees of freedom, the estimation strategy was to examine the enabling environment variables piece-wise in separate specifications to test their impact on the R&D elasticity estimates. The model is described in further detail in Appendix 3A.

Results

Table 3.3 presents regression estimates showing how research and the enabling environment influenced agricultural TFP growth in SSA. In all of the model specifications, higher investments in NARSs and larger shares of a country's crop area affected by CGIAR technology were significant drivers of productivity improvement. The estimate of the R&D elasticity for national agricultural research is quite stable across the different model specifications. The estimated value of the elasticity of NARS research stock in these regressions ranges from 0.026 to 0.034, which says that a 1 percent increase of the stock of knowledge capital raises agricultural output (net of any change in agricultural inputs) by around 0.03 percent. NARS research stock increased productivity not only by providing its own technologies to farmers, but also by raising adoption of technologies developed by the CGIAR centers. The model indicates that enlarging the crop area affected by CGIAR technology raises average productivity on this cropland by between 46 percent and 84 percent. Including the impact of NARSs on CGIAR technology adoption raises the R&D elasticity of NARSs from around 0.03 to 0.04 (see Appendix 3A for details on this calculation).

The effect of country size on the estimated value of the R&D elasticity is negligible, according to the estimates in Table 3.3. While the coefficient on the R&D stock variable indicates the regional average elasticity of research for all countries, the coefficients on the interaction terms between the R&D stocks and the dummy variables reflecting country size indicate whether countries of a certain size have an R&D elasticity significantly different from the regional average. None of these interaction coefficients is statistically significant in any of the model specifications. In other words, the same elasticity for national agricultural research applies to SSA countries, regardless of the size of their agricultural sectors.

Despite the statistically significant relationship between national agricultural research and productivity growth, the estimated regional average elasticity value of 0.04 is relatively small compared with findings from other studies

TABLE 3.3 Determinants of agricultural total factor productivity, allowing the research elasticity to vary among small, midsize, and large countries

	Model specification			
Variable	(1)	(2)	(3)	(4)
Share of crop area affected by CGIAR technology (%)	0.461***	0.839***	0.812***	0.514***
	(6.6690)	(6.63)	(8.08)	(6.21)
NARS knowledge capital stock (log value)	0.0264***	0.0333***	0.0337***	0.0292***
	(4.75)	(3.06)	(4.73)	(4.88)
NARS knowledge capital stock (log value) × Midsize-country dummy variable	6.85E-04	–0.0066	–0.0033	–0.0018
	(0.14)	(–1.18)	(–0.57)	(–0.31)
NARS knowledge capital stock (log value) × Small-country dummy variable	0.00068	0.00722	0.00620	0.00092
	(0.12)	(1.10)	(1.13)	(0.16)
Share of population infected by HIV/AIDS (%)	–0.174*	–0.851***	–0.513***	–0.248**
	(–1.79)	(–4.76)	(–4.65)	(–2.23)
Number of years of armed conflict since 1977	–0.00745***	–0.00864***	–0.00703***	–0.00863***
	(–6.47)	(–7.08)	(–5.24)	(–6.09)
Nominal rate of assistance to agriculture (%)		0.346***		
		(6.33)		
Road density (km/km^2, log values)			–0.0303***	
			(–5.45)	
Average years of schooling of labor force				0.0053
				(1.13)
Constant	4.568***	4.594***	4.461***	4.537***
	(278.89)	(116.20)	(167.40)	(226.70)
Number of observations	899	467	611	783
R^2	0.1048	0.294	0.193	0.136
Adjusted R^2	0.0988	0.283	0.184	0.128

Source: Authors.

Notes: T-statistics are shown in parentheses; significance tests are indicated by *** $p < 0.01$, ** $p < 0.05$, and * $p < 0.1$.

for NARSs in other parts of the world. Craig, Pardey, and Roseboom (1997), using a panel of 67 developing countries, including 24 SSA countries, found an average NARS elasticity of 0.09. In a study of 12 Asian countries, Fan and Pardey (1998) estimated an average NARS elasticity of 0.17; Rada and Valdes (2012) estimated an elasticity of 0.20 for Brazil's primary agricultural research agency, the Brazilian Agricultural Research Corporation (Embrapa); and recently Rada and Schimmelpfennig (2015) estimated an elasticity of 0.15 for India's public agricultural research system. All of these estimates are significantly larger than what the present study finds for SSA, suggesting that there is substantial room for improvement in the performance of African NARSs. Extensions of this analysis could examine whether NARS research has had a higher or lower payoff for certain commodities (crops versus livestock, for example) and whether other country characteristics (including those of NARS) are having a measurable influence on returns to research.

The extent to which investment in NARS in SSA has paid off requires a comparison of the cost of investment with the benefits from productivity growth. Economic returns to agricultural research depend not only on the research elasticity, but also on the research intensity ratio and the time lag between research expenditures and TFP growth. Note that for a given value of the research elasticity, a country with a lower research intensity ratio will earn a higher return to research. Since smaller countries in SSA tend to have higher intensity ratios, they are likely to have lower rates of return to research than larger countries, even if their R&D elasticity is the same. To see how this works, consider a simple, intuitive example.

Suppose a country has a research intensity ratio of 0.01 (in other words, it spends the equivalent of 1 percent of its AgGDP on research). Then, assuming a research elasticity of 0.04, a 100 percent increase in that country's stock of knowledge capital would eventually raise AgGDP by 4 percent (that is, multiplying the R&D elasticity by 100 percent). Since doubling its present spending on R&D would cost another 1 percent of AgGDP, the country would earn roughly $4 in higher GDP per $1 in new R&D spending. On the other hand, a country with a research intensity ratio of 0.03 (that is, doubling its R&D spending from current levels would cost the equivalent of 3 percent of AgGDP) would earn $4 in agricultural GDP growth per $3 in new research spending. Instead of a 4:1 benefit–cost ratio, this country attains only a 1.25:1 benefit–cost ratio. Of course, the calculation of the rate of return to research must also consider the time lag between research spending and output growth, so the actual returns would be somewhat lower than in this example (for a description of these procedures, see Appendix 3A). This example illustrates how returns to research are sensitive not only to the value of the research elasticity, but also to the research intensity ratio.

Benefit–cost analysis of national agricultural research for different-size countries in SSA is shown in Table 3.4, including the benefit–cost ratio, the internal rates of return, and a "modified" internal rate of return suggested by Lin (1976). While the internal rate of return (IRR) is commonly reported in studies on the impacts of agricultural research, it may overstate the return on investment relative to other investment opportunities because it assumes the

TABLE 3.4 Returns to agricultural research in Africa south of the Sahara, 1977–2005

National and international agricultural research systems	Internal rate of return without CGIAR (% per year)	Internal rate of return with CGIAR (% per year)	Modified internal rate of return with CGIAR (% per year)	Benefit–cost ratio (dollar per dollar)
Large countries	34.0	40.8	19.6	4.4
Midsize countries	23.6	28.9	16.3	2.6
Small countries	12.9	17.0	12.1	1.6
All-country average	23.8	29.3	16.1	3.1
CGIAR	—	57.7	23.2	6.2

Source: Authors.

Notes: The modified internal rate of return (Lin 1976) assumes the initial cost of capital and the return on reinvestment of earnings is 10 percent. The benefit–cost ratio discounts future benefits at a yearly rate of 10 percent.

benefits from research can be continuously reinvested at this same high rate of return. The modified internal rate of return assumes the benefits can be reinvested at a rate more consistent with broader investment opportunities in an economy (Alston et al. 2011; Chapter 11, this volume).

Large countries earned an internal rate of return of 34.0 percent, on average, and a modified internal rate of return of 19.6 percent (assuming a 10 percent return on the reinvestment of benefits). Small countries earned a mean internal rate of return of only 12.9 percent, the modified rate of return being only slightly lower. Assuming a 10 percent real discount rate, agricultural research yielded a benefit–cost ratio of 4.4 for large countries, but only 1.6 for small countries. For midsize countries, the mean internal rate of return was 23.6 percent, and the modified internal rate of return was 16.3 percent, giving a benefit–cost ratio of 2.6. Notably, even though the estimated returns to research are much higher in larger countries, the returns estimated for most small countries are nevertheless sufficiently high to justify investment in these NARSs. In fact, with a research elasticity of 0.04, and taking into account the time lag between research and its impact on growth, there appears to be economic justification for all SSA countries to increase their R&D spending on agriculture to at least 1.5 percent of AgGDP.

Another finding from the model is that international agricultural research in SSA complements NARSs. Countries with larger NARSs achieved more rapid dissemination of technologies from CGIAR research centers (see Table 3A.1 for these results). Having CGIAR technologies to draw from raised the return to national agricultural R&D spending. For SSA countries, on average, the returns to agricultural research without CGIAR would have been about 23.8 percent, compared with 29.3 percent with CGIAR.

Note that these are social, not private, rates of return. They include the benefits of research to farmers (in the form of higher profits) and consumers (through more food at lower prices). Private returns to innovators would be considerably lower (perhaps even negative), given the difficulty a private innovator would have in appropriating the benefits of research. To appropriate the benefits of research an innovator would need to be able to charge a price premium for the technology (to recoup the costs of research) and protect the intellectual property embedded in the technology from being freely used by others.

Finally, the analysis found that the enabling environment had important effects on agricultural productivity. The spread of HIV/AIDS and the presence of significant armed conflict were significant constraints to raising productivity in the region. Economic reforms stimulated more rapid productivity growth. Most SSA countries had negative NRAs during the 1977–2005 period, meaning their economic policies taxed agriculture to keep food prices low for consumers. During the 1980s and 1990s, several countries enacted structural reforms that improved the prices farmers received, reflected by an increase in the NRA. Improved economic incentives for agricultural producers led to greater private investment and, subsequently, to higher TFP. Although schooling did not affect TFP directly in the model, increased education was found to increase the rate of adoption of new CGIAR technologies (Table 3A.1). The effect of better transportation infrastructure on TFP was mixed: more roads increased dissemination of CGIAR technology adoption (raising TFP in these areas) but caused more, and perhaps less fertile, cropland to be brought into production (lowering average TFP). Further discussion of these findings can be found in Fuglie and Rada (2013).

Conclusions and Implications

Despite the significant challenges facing agricultural research systems in SSA, the slow accumulation of new knowledge and improved agricultural technologies and farming practices from research appears to have made a significant contribution to improving agricultural productivity. While there are important data challenges in measuring agricultural productivity trends, most studies agree that agricultural TFP in SSA was stagnant or declining in the 1960s and 1970s, but turned positive in the mid-1980s. While not all countries achieved productivity improvement, the ones with a rising stock of knowledge capital from agricultural research were more likely to be in the growth club.

On average, investments in NARSs earned on the order of $3 in benefits for every dollar spent on R&D. Large countries have earned higher returns to R&D than small countries. Nonetheless, even in small countries, returns were still high enough to justify the investment. For SSA countries, tying into regional and international agricultural research networks and maintaining a policy environment that is receptive to technologies developed elsewhere seems to be critical. In fact, the CGIAR Consortium has played an important role in raising agricultural productivity growth in SSA. Results suggest that spending by CGIAR in the region has generated a modified internal rate of return of around 23 percent per year, or about $6 in benefits for every dollar spent on research. Moreover, results indicate that national and international agricultural research efforts in SSA are complementary: countries that have made a greater national investment in agricultural research are better able to deliver new technologies to farmers emanating from CGIAR centers.

Despite their achievements, NARSs in SSA remain relatively weak and underfunded, with a relatively low research-to-productivity elasticity of only 0.04 percent. Higher and more stable funding, stronger scientific and human resource capacities, and better enabling environments for dissemination of new agricultural technologies will further raise the efficiency and performance of agricultural R&D investments in Africa.

Looking forward, there is reason for cautious optimism about prospects for productivity growth in agriculture in SSA. During the past decade, both CGIAR and national governments have increased spending on agricultural research, and greater availability of antiretroviral therapy and other measures have reduced the scourge of AIDS. If momentum on policy reform and conflict reduction can be sustained, that too will continue to be a source of renewed growth for African agriculture. However, accelerating the rate of agricultural growth in SSA to something approaching the 6 percent target set under the Comprehensive Africa Agriculture Development Programme will require a much greater rate of technological change in the region's agriculture than it has experienced in the past. That goal will require significant reform and strengthening of national systems for agricultural innovation.

Appendix 3A. Estimating Returns to Agricultural Research

The first step in estimating returns to agricultural research is to empirically estimate how spending on agricultural research affects agricultural productivity. For this step, a simultaneous equations model of how research investments and the enabling environment affected TFP growth was estimated. The model was specified as follows:

$$CGIAR\_Area_{ct} = \alpha_1 Ln(CGIAR\ R\&D\ Stock_t) + \alpha_2 Ln(NARS\ R\&D\ Stock_{ct}) + \gamma X_{ct} + \eta_{1t}, \text{ and} \quad (2a)$$

$$Ln(TFP_{ct}) = \beta_2 Ln(NARS\ R\&D\ Stock_{ct}) + \sum_{s=2}^{S}[\delta_s D_s Ln(NARS\ R\&D\ Stock_{ct})] + \beta_1 CGIAR\_Area_{ct} + \gamma X_{ct} + \eta_{2t}, \quad (2b)$$

where the subscripts c and t are for country and year, respectively, and η_1 and η_2 are random error terms. The technology variables are *TFP*, *CGIAR_Area* (share of cropland affected by CGIAR–related technologies), and the R&D or knowledge stocks from past spending by NARSs and CGIAR in SSA (*NARS R&D Stock and CGIAR R&D Stock,* respectively). D_s is an indicator variable for country size, where D_1 takes on a value of 1 for large countries and 0 otherwise, $D_2 = 1$ for midsize countries and 0 otherwise, and $D_3 = 1$ for small countries and 0 otherwise. Other explanatory variables (the enabling environment) and the constant term are contained in the X vector.

The error terms η_1 and η_2 include measurement error and omitted variables. Since omitted variables that affect technology dissemination are also likely to affect TFP, *CGIAR_Area* and η_2 are likely to be correlated. An implication of this correlation is that multiple regression of equation (2b) will produce biased estimates of the parameters of this regression. To avoid this potential bias, predicted values from the estimation of equation (2a) are used for the *CGIAR_Area* variable in the estimation of equation (2b). This two-stage procedure provides unbiased estimates of the parameters in the equations.

The estimates of equation (2a) are presented in Table 3A.1. Estimates of equation (2b) are presented in Table 3.3.

The values of the estimated parameters α_1, α_2, β_1, β_2, and δ_s are used to derive elasticities of research, or the percentage change in productivity (or output, holding inputs fixed) given a 1 percent change in the size of the national or CGIAR research stock. Taking the derivation of the system of equations (2a) and (2b) with respect to *Ln(NARS R&D Stock)* gives $(\alpha_2\beta_1 + \beta_2 + \sum_{s=2}^{S}\delta_s D_s)$. This is the total elasticity of national agricultural research. It measures the percentage change in *TFP* resulting from a 1 percent change in *NARS R&D Stock*. The first term $(\alpha_2\beta_1)$ measures the impact of national agricultural research investment in helping to adapt and disseminate CGIAR technologies within the country; the second term (β_2) captures the direct effect of national research on productivity independent of CGIAR; and the third term $(\sum_{s=2}^{S}\delta_s D_s)$ adjusts the elasticity according to the size of the country's agricultural sector. The elasticity of CGIAR research is given by $\alpha_1\beta_1$, which is the

TABLE 3A.1 Factors influencing dissemination of CGIAR-related technologies in Africa south of the Sahara

	Model specification			
Variable	**(1)**	**(2)**	**(3)**	**(4)**
CGIAR knowledge capital stock (log value)	0.0688***	0.0710***	0.0479***	0.0592***
	(16.15)	(11.08)	(9.675)	(11.85)
NARS knowledge capital stock (log value)	0.0164***	0.00372	0.0180***	0.0122***
	(6.93)	(0.855)	(6.524)	(4.414)
Share of cropland in cassava (%)	0.645***	0.636***	0.899***	0.619***
	(26.11)	(11.92)	(18.98)	(23.41)
Share of population infected by HIV/AIDS (%)	0.222***	0.595***	0.268***	0.0988*
	(4.965)	(8.694)	(5.762)	(1.889)
Number of years of armed conflict since 1977	–0.00151***	–0.00333***	–0.000552	0.000275
	(–2.855)	(–5.709)	(–0.951)	(0.412)
Nominal rate of assistance to agriculture (%)		0.0871***		
		(3.642)		
Road density (km/km2, log values)			0.00595**	
			(2.523)	
Average years of schooling of labor force				0.0106***
				(6.128)
Constant	–0.298***	–0.249***	–0.205***	–0.283***
	(–16.31)	(–8.629)	(–9.051)	(–13.60)
Observations	928	496	640	783
R^2	0.56	0.538	0.555	0.567
Adjusted-R^2	0.557	0.533	0.551	0.564

Source: Fuglie and Rada (2013).

Notes: km = kilometers; km^2 = square kilometers. T-statistics are shown in parentheses; significance tests are indicated by *** $p < 0.01$, ** $p < 0.05$, and * $p < 0.1$.

marginal effect of CGIAR research on technology diffusion multiplied by the impact of diffusion on TFP.

Let the NARS research-to-TFP elasticity be given by ε, and let the stock of R&D knowledge capital (either NARS or CGIAR) be given by S. Note that a change in TFP is equivalent to a change in gross output Y when everything else (that is, inputs) is held constant. So, the research-to-TFP elasticity can be defined equivalently as the research-to-output elasticity:

$$\varepsilon \equiv \frac{\partial lnY}{\partial lnS} = \left(\frac{\partial Y}{\partial S}\right)\left(\frac{\bar{S}}{\bar{Y}}\right),$$

where the bars over S and Y imply average values for these variables. Rearranging these terms to isolate the impact of a change in the research stock on output gives the marginal product of the research stock:

$$\frac{\partial Y}{\partial S}\left(\frac{\bar{Y}}{\bar{S}}\right)\varepsilon.$$

The effect of a one-time increase in research expenditure R on subsequent output is considered in order to derive the internal rate of return. Recall from the assumption on the time-lag structure of research that research spending in year t affects the research stock (and thus output) for 17 years. The effect is not constant over time, however, but is given by the "weights" $\lambda_i (i = 0...16)$, where $\sum\lambda_i = 1$. Note that research stock at t is given by $S_t = \sum_{i=0}^{16} R_{t-i}$. Thus, the impact on output from a change in research expenditure in year t is as follows:

$$\frac{\partial Y}{\partial R_t} = \left(\frac{\partial Y}{\partial S}\right)\left(\frac{\partial S}{\partial R_t}\right) = \left(\frac{\bar{Y}}{\bar{S}}\right)\varepsilon \sum_{i=0}^{16} \lambda_i.$$

This gives a stream of increments to output over the period from t to $t + 16$ from a one-time increase in research spending R at time t. The ratio $(\bar{Y}/\bar{S})$ is constant and indicates the size of the agricultural sector relative to the size of the research system.

The internal rate of return (*irr*) to research is given by the discount rate that equates the present value of costs (each dollar of expenditure on research in time t) to benefits (the increments to output brought about by this research over the current year and subsequent 16 years):

$$1 = \left(\frac{\bar{Y}}{\bar{S}}\right)\varepsilon \sum_{i=1}^{17} \frac{\lambda_i}{(1 - irr)^i}.$$

Assuming that the elasticity of research-to-output ε is constant across all SSA countries, the returns to research will be correlated with the $(\bar{Y}/\bar{S})$ ratio. In other words, if two countries have similarly sized research systems, the country with the larger agricultural sector will receive higher returns from an increase in its research investment. Similarly, for two countries with equally sized agricultural sectors, the country with the (initially) larger research system will receive smaller returns from increased research investment. While this conclusion is consistent with the notion of diminishing returns to research (at least in the short run), it is possible that ε could vary across countries.

The internal rate of return is a useful complement to other investment-ranking tools, such as the benefit–cost ratio and net present value, because unlike those measures, it does not require assumptions about the cost of capital or the discount rate for future costs and benefits. It solves for the interest rate that equates the present value of costs to the present value of benefits over the life of the project. It provides a valid criterion for judging whether to invest in a project when the decision rule is to invest if the project returns at least above some "hurdle" rate, such as the cost of investment capital. But if the choice is to rank projects according to the ones likely to provide the highest return, Lin (1976) suggests that the internal rate of return overstates the relative returns to an investment because it implicitly assumes that benefits can be reinvested at this same rate of return. To select from a set of alternative projects the one likely to generate the highest social welfare, the benefits should be considered as part of the investment and reinvested over the life of the project. The modified internal rate of return (*MIRR*) assumes a rate of return in which future benefits are reinvested. For an initial investment in research of $1 in year 0, the *MIRR* solves the following equation:

$$(1 + MIRR) = \left(\frac{\bar{Y}}{\bar{S}}\right) \varepsilon \textstyle\sum_{i=0}^{N} \lambda_i (1 + r)^{N-i},$$

where $N = 17$ is the time span of the project, and r is the appropriate reinvestment rate for the benefit stream during this time span. The right-hand side of the equation measures the future value of all net incomes from the initial investment in research when benefits from the research are reinvested at rate r. Taking the *Nth* root of this value (converting returns from future value to present value) and subtracting 1 yields the *MIRR* (Lin 1976):

$$MIRR = \sqrt[N]{\left(\frac{\bar{Y}}{\bar{S}}\right) \varepsilon \textstyle\sum_{i=0}^{N} \lambda_i (1 + r)^{N-i} N} - 1.$$

A drawback of the *MIRR* is that, unlike the *IRR*, it is not free of assumptions about interest rates.

References

Alene, A., and O. Coulibaly. 2009. "The Impact of Agricultural Research on Productivity and Poverty in Sub-Saharan Africa." *Food Policy* 24: 198–209.

Alston, J., M. Andersen, J. James, and P. Pardey. 2011. "The Economic Returns to U.S. Public Agricultural Research." *American Journal of Agricultural Economics* 93 (5): 1257–1277.

Alston, J., G. Norton, and P. Pardey. 1998. *Science Under Scarcity: Principles and Practices for Agricultural Research Evaluation and Priority Setting*. Ithaca, NY: Cornell University Press.

Anderson, K., and W. Masters, eds. 2009. *Distortions to Agricultural Incentives in Africa*. Washington, DC: World Bank.

ASTI (Agricultural Science and Technology Indicators). 2014. ASTI database. Washington, DC: International Food Policy Research Institute. www.asti.cgiar.org.

Barro, R., and J. Lee. 2010. *A New Data Set of Educational Attainment in the World, 1950–2010.* Working Paper 15902. Cambridge, MA: National Bureau of Economic Research.

CGIAR. Various years. *CGIAR Financial Report*. Washington, DC.

Craig, B., P. Pardey, and J. Roseboom. 1997. "International Productivity Patterns: Accounting for Input Quality, Infrastructure, and Research." *American Journal of Agricultural Economics* 79: 1064–1076.

Eicher, C. 1990. "Building African Scientific Capacity for Agricultural Development." *Agricultural Economics* 4: 117–143.

Evenson, R., and K. Fuglie. 2010. "Technology Capital: The Price of Admission to the Growth Club." *Journal of Productivity Analysis* 33: 173–190.

Evenson, R., and D. Gollin, eds. 2003. *Crop Variety Improvement and Its Effects on Productivity.* Wallingford, UK: CABI Publishing.

Fan, S., and P. Pardey. 1998. "Government Spending on Asian Agriculture: Trends and Production Consequences." In *Agricultural Public Finance Policy in Asia*. Tokyo: Asian Productivity Organization.

FAO (Food and Agriculture Organization of the United Nations). Various years. FAOSTAT database. Accessed April 2016. http://faostat3.fao.org/home/E.

———. 2012. CountrySTAT Nigeria. Accessed June 1, 2013. http://countrystat.org/home.aspx?c=NGA.

Fuglie, K. 2008. "Is a Slowdown in Agricultural Productivity Growth Contributing to the Rise in Commodity Prices?" *Agricultural Economics* 39 (supplement): 431–441.

———. 2011. "Agricultural Productivity in Sub-Saharan Africa." In *The Food and Financial Crises in Sub-Saharan Africa*, edited by D. Lee and M. Ndulo. Wallingford, UK: CABI Publishing.

Fuglie, K., and N. Rada. 2013. *Resources, Policy and Agricultural Productivity in Sub-Saharan Africa*. Economic Research Report 145. Washington, DC: Economic Research Service, U.S. Department of Agriculture.

Gleditsch, N., P. Wallensteen, M. Eriksson, M. Sollenberg, and H. Strand. 2002. "Armed Conflict 1946–2001: A New Dataset." *Journal of Peace Research* 39: 697–710. (Version 4-2008 of the dataset.)

International Road Federation. 2006. *World Road Statistics*. Geneva. Accessed July 2011. www.irfnet.org.

Lin, S. 1976. "The Modified Internal Rate of Return and Investment Criterion." *The Engineering Economist* 21 (4): 237–247.

OECD (Organisation for Economic Co-operation and Development). 2013. *Agricultural Innovation Systems: A Framework for Analyzing the Role of the Government.* Paris: OECD Publishing.

Pardey, P., J. Roseboom, and J. Anderson, eds. 1991. *Agricultural Research Policy: International Quantitative Perspectives.* Cambridge, UK: Cambridge University Press.

Rada, N., and D. Schimmelpfennig. 2015. *Propellers of Agricultural Productivity in India.* Economic Research Report 203. Washington, DC: Economic Research Service, U.S. Department of Agriculture.

Rada, N., and C. Valdes. 2012. *Policy, Technology, and Efficiency of Brazilian Agriculture.* Economic Research Report 137. Washington, DC: Economic Research Service, U.S. Department of Agriculture.

Ruttan, V. 1982. *Agricultural Research Policy.* Minneapolis: University of Minnesota Press.

Stads, G., and N. Beintema. 2012. "Agricultural R&D in Africa: Investment, Human Capacity, and Policy Constraints." In *Improving Agricultural Knowledge and Innovation Systems: OECD Conference Proceedings.* Paris: Organisation for Economic Co-operation and Development Publishing.

USDA (United States Department of Agriculture). 2013. PSD Online: Production, Supply and Distribution Database. Washington DC: Foreign Agricultural Service. Accessed July 15, 2013. www.fas.usda.gov/psdonline/psdHome.aspx.

World Bank. 2011a. *Agricultural Innovation Systems: An Investment Sourcebook.* Washington, DC.

———. 2011b. *World Development Indicators.*

Zeddies, J., R. Schaab, P. Neuenschwander, and H. Herren. 2001. "Economics of Biological Control of Cassava Mealybug in Africa." *Agricultural Economics* 24: 209–219.

PART 2

Financial Investments

Chapter 4

INVESTMENT IN AGRICULTURAL RESEARCH AND DEVELOPMENT: AN ACCOUNT OF TWO-SPEED GROWTH, UNDERINVESTMENT, AND VOLATILITY

Gert-Jan Stads

Much evidence shows that investments in agricultural research and development (R&D) have tremendously enhanced agricultural productivity around the world over the past five decades, and in turn have led to higher incomes, lower poverty levels, greater food security, and better nutrition (Evenson and Gollin 2003; World Bank 2007; IAASTD 2008). Africa south of the Sahara (SSA) has benefited less from agricultural R&D than other parts of the world, however, because both investment in the development of new technologies and the potential for technology spillovers from elsewhere are low (Chapter 1, this volume; Chapter 2, this volume; Johnson and Evenson 2000). If SSA is to take advantage of the benefits of agricultural R&D, it will need to increase its investments (Chapters 1 and 2, this volume).

African governments have a critical responsibility when it comes to providing sufficient and sustained agricultural R&D funding and for creating a more enabling environment within which agricultural innovation can prosper. Given the substantial time lag between investing in research and reaping its rewards—which usually takes decades, not just years—agricultural research requires a long-term commitment of sufficient and sustained funding. In reality, these long research cycles rarely coincide with short-term election cycles, shifting political agendas, and changes in government budget allocations—all of which have major implications for agricultural research (Alston, Pardey, and Piggott 2006). Decisionmakers have limited incentive to support long-term investment in agricultural research because extracting political credit for doing so is difficult (Chapter 5, this volume).

The author thanks numerous ASTI country collaborators who collected agency-level data across SSA. Without their commitment, ASTI's detailed agricultural R&D expenditure data would not have been available. The author also thanks Kathleen Flaherty, Léa Vicky Magne Domgho, and Michael Rahija for their research assistance, and Nienke Beintema and Mary Jane Banks for their comments on a draft version of this chapter.

This chapter takes stock of recent agricultural R&D investment in SSA using comprehensive datasets collected under the Agricultural Science and Technology Indicators (ASTI) initiative, facilitated by the International Food Policy Research Institute. The chapter presents an analysis of public agricultural research funding and investment trends in SSA, highlighting important cross-country differences. The discussion also focuses on the main drivers of volatility in agricultural R&D funding over time and suggests a number of policy measures that could mitigate both underinvestment and volatility.

Long-Term Trends in Agricultural R&D Spending

In 2011, public agricultural R&D spending in SSA as a whole totaled 1.7 billion 2005 constant purchasing power parity (PPP) dollars. Absolute spending levels varied considerably across countries. About half of the region's agricultural R&D investments were made in just three countries: Nigeria (US$394 million), South Africa ($237 million), and Kenya ($188 million). On the other hand, 19 of the 40 countries for which data were available each spent less than $10 million on agricultural R&D (Figure 4.1).

Agricultural R&D Spending Growth

Following a period of slow growth in the 1980s and 1990s, public agricultural R&D spending in SSA rapidly accelerated from 2000, increasing by 40 percent in real terms by 2011, from $1,208 million to $1,689 million 2005 PPP dollars. A breakdown by country reveals that more than half of this $481 million increase was attributable to just two countries: Nigeria and Uganda. Ghana, Kenya, and Tanzania also recorded relatively high increases in total spending, each accounting for between 5 and 8 percent of regionwide growth during this period.

Although changes in absolute agricultural R&D spending levels among the region's larger countries overshadow those of many of the smaller countries, a closer look at relative shifts in investment levels over time reveals important cross-country differences and challenges. During 2000–2011, 13 of the 27 SSA countries for which a full set of time-series data was available experienced growth in public agricultural R&D spending in excess of 1 percent per year (Figure 4.2). Seven countries experienced near-zero growth rates (of between −0.9 and 0.2 percent), and an additional seven countries experienced considerable negative yearly growth, ranging from −1.2 to −13.6 percent a year. The large number of countries experiencing negative or stagnant yearly growth clearly highlights the challenge of "two-speed growth"

FIGURE 4.1 Absolute levels of agricultural R&D spending, 2011

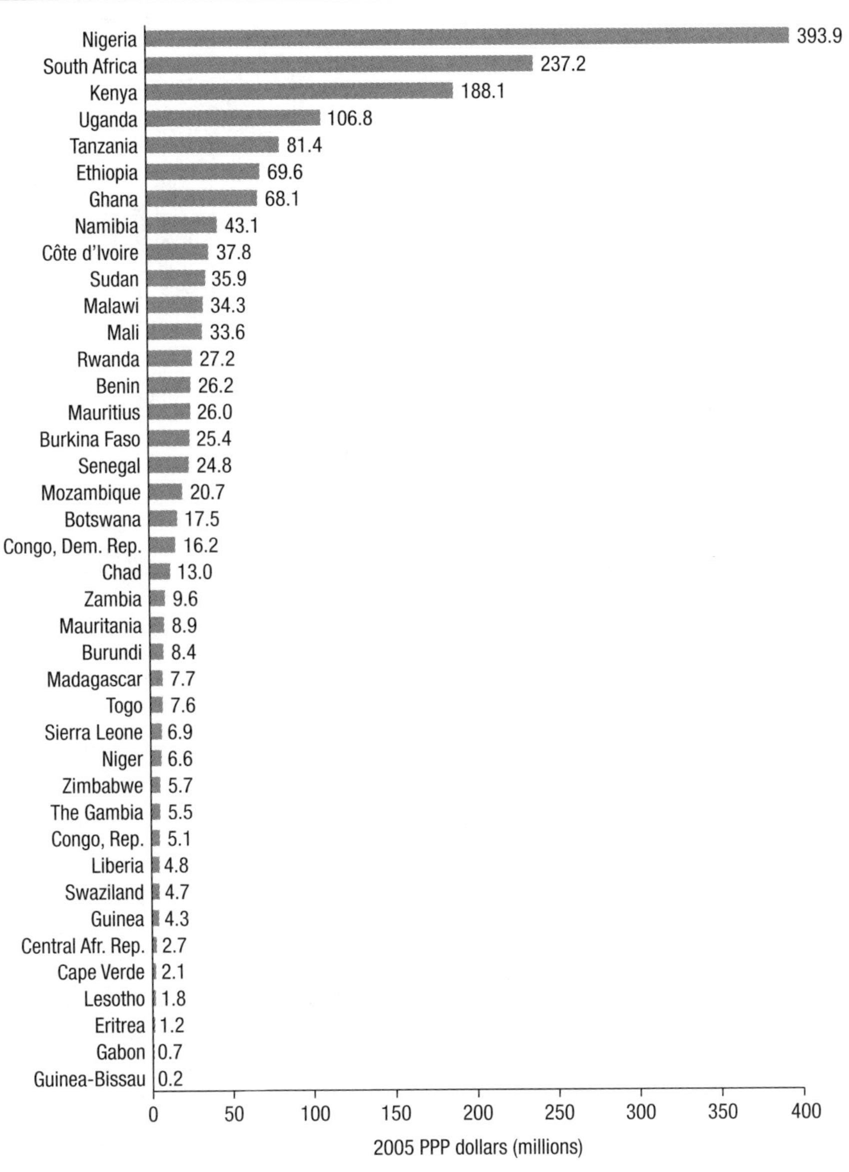

Source: Calculated by author based on ASTI (2014).

Notes: Angola, Cameroon, Comoros, Djibouti, Equatorial Guinea, São Tomé and Príncipe, Somalia, and South Sudan are excluded because data were not available. PPP = purchasing power parity.

FIGURE 4.2 Yearly growth rates in agricultural R&D spending, 2000–2011

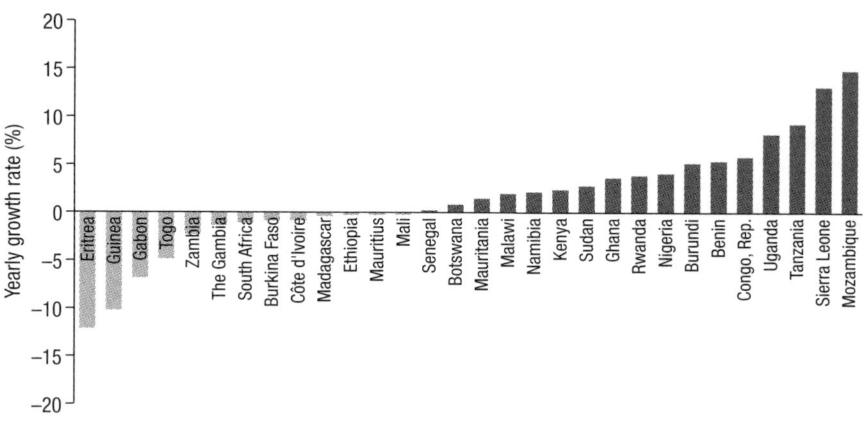

Source: Calculated by author based on ASTI (2014).

Notes: Cape Verde, Central African Republic, Chad, the Democratic Republic of the Congo, Guinea-Bissau, Lesotho, Liberia, Mozambique, Rwanda, Sierra Leone, Swaziland, and Zimbabwe are excluded because complete time-series data for these countries were not available.

in agricultural R&D in SSA: overall spending in the region has grown substantially since the turn of the millennium, but it has been extremely uneven and has evaded many countries. The extremely low (and often declining) long-term investment levels and human resource capacity of some of the region's smallest, often francophone, countries call into question the effectiveness of their national agricultural R&D outputs, and whether they would not be better served by focusing on taking advantage of technological spillovers from their larger neighbors.

Nonetheless, early indications signal the reversal of negative or stagnant spending trends in an increasing number of smaller countries in recent years. The 2007–2008 global food crisis and a number of influential initiatives, including the Comprehensive Africa Agriculture Development Programme (CAADP) and the 2008 L'Aquila Food Security Initiative, have elevated agriculture and agricultural research on political and donor agendas and may in large part explain this shift. In addition, the World Bank launched regional agricultural productivity programs in West, East, and Southern Africa in 2007, 2009, and 2013, respectively. These programs have injected significant funding into agricultural R&D and other agricultural productivity-enhancing measures in SSA (some $636 million in current prices to date (Chapter 2, this volume).

Agricultural R&D Intensity Ratios

In addition to assessing absolute levels of agricultural R&D investment, relative public agricultural R&D investments can be compared by measuring a country's total public agricultural R&D spending as a share of its agricultural gross domestic product (AgGDP); this measure is known as an agricultural R&D intensity ratio. Despite tremendous growth in agricultural R&D spending in recent years, SSA's agricultural R&D intensity ratio has steadily declined, from 0.59 percent in 2006 to 0.51 percent in 2011 (Chapter 4, this volume). This indicates that—notwithstanding the injection of significant funds through regional initiatives, such as the East Africa Agricultural Productivity Program and West Africa Agricultural Productivity Program—regional agricultural R&D spending has not kept pace with growth in agricultural output. In fact, 28 of the 38 SSA countries for which data were available still fall short of the minimum investment target of 1 percent of AgGDP set by the African Union and United Nations (Figure 4.3; Chapter 1, this volume). If SSA is to reach an agricultural research intensity ratio of 1 percent of AgGDP and 5 percent yearly growth in its agricultural output, agricultural R&D spending would need to increase by 10 percent per year over a period of 15 years.

Although R&D intensity ratios provide useful insights into relative investment levels across countries and over time, they do not take into account the policy and institutional environment within which agricultural research occurs, the broader size and structure of a country's agricultural sector and economy, or qualitative differences in research performance across countries; hence, they should be interpreted with care. Fuglie and Rada (Chapter 3, this volume) make the point that, because small countries are unable to take advantage of economies of scale, their benefits from investing in agricultural research are less than those of large countries (all else being equal). Similarly, countries with greater agroecological diversity require higher research investments than countries with limited agroecological diversity. Furthermore, a higher agricultural research intensity ratio can actually reflect reduced agricultural output rather than higher investment. More detailed analysis is therefore needed to ensure a clear understanding of the implications of intensity ratios. Despite these limitations, agricultural R&D intensity ratios reveal that agricultural R&D spending in most SSA countries remains below the minimum target of 1 percent of AgGDP. For most small and medium-sized countries, even this 1 percent investment target is inadequate to support some form of technological autonomy; hence, their research will be limited mainly to adapting technologies developed elsewhere.

FIGURE 4.3 Agricultural R&D intensity ratios, 2011

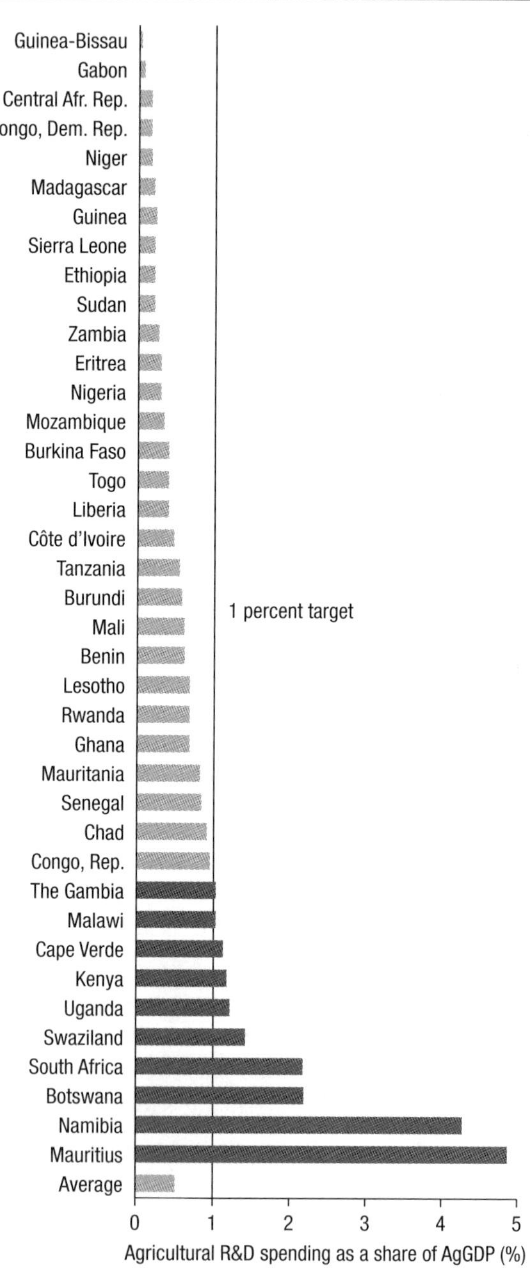

Source: Calculated by author based on ASTI (2014).

Notes: The vertical line represents the 1 percent investment target recommended by the African Union and the United Nations. AgGDP = agricultural gross domestic product; R&D = research and development.

Agricultural R&D Spending Allocations

As established, most of the 40 percent aggregated growth in agricultural R&D spending for SSA during 2000–2011 was driven by just five countries (Ghana, Kenya, Nigeria, Tanzania, and Uganda), but a breakdown of spending by cost category reveals further differences. The rapid increase in Ghanaian agricultural R&D spending, for instance, was almost entirely driven by a major increase in salary levels at the Council for Scientific and Industrial Research, rather than by expanded research activities or greater investment in equipment or infrastructure. Growth in agricultural R&D spending in Nigeria and Tanzania, on the other hand, largely stemmed from increased government commitment to financing R&D programs, equipment, and infrastructure. In Uganda, increased government funding was allocated across salary- and program-related expenditures, as well as capital investments in R&D infrastructure.

The allocation of research budgets across the major categories of salaries, operating and program costs, and capital investments has an important impact on the effectiveness and efficiency of agricultural R&D. No formula can determine the optimal allocation: it depends on numerous factors, including country size, agroecological diversity, the research mandates, and the composition of staffing. That said, when salary-related expenses consume more than three-quarters of a research agency's total budget, a clear imbalance exists, such that too few resources remain to support the costs of operating viable research programs.

In 2011, based on a sample of 168 government and nonprofit agencies in 38 SSA countries for which detailed cost category data were available, 51 percent of available resources was spent on staff salaries, 35 percent was allocated to operating and program costs, and 14 percent was invested in capital improvements (Figure 4.4). These regionwide averages mask a significant degree of cross-country variation. The national agricultural research institutes in small countries like Lesotho (96 percent), Guinea-Bissau (88 percent), and Swaziland (82 percent) spent extremely high shares of their total budgets on salary-related expenses, leaving minuscule resources for the day-to-day running of research programs or the rehabilitation of infrastructure and equipment. This is not exclusively a small-country challenge, however. The main agricultural R&D agencies in larger countries like Ghana, the Democratic Republic of the Congo, and Sudan all spent more than three-quarters of their budgets on salaries. In many countries across SSA, the national government only provides funding for staff salaries, forcing R&D agencies to seek the additional funding needed to conduct R&D programs elsewhere, which can pose real challenges.

FIGURE 4.4 Allocation of agricultural R&D resources across countries, 2011

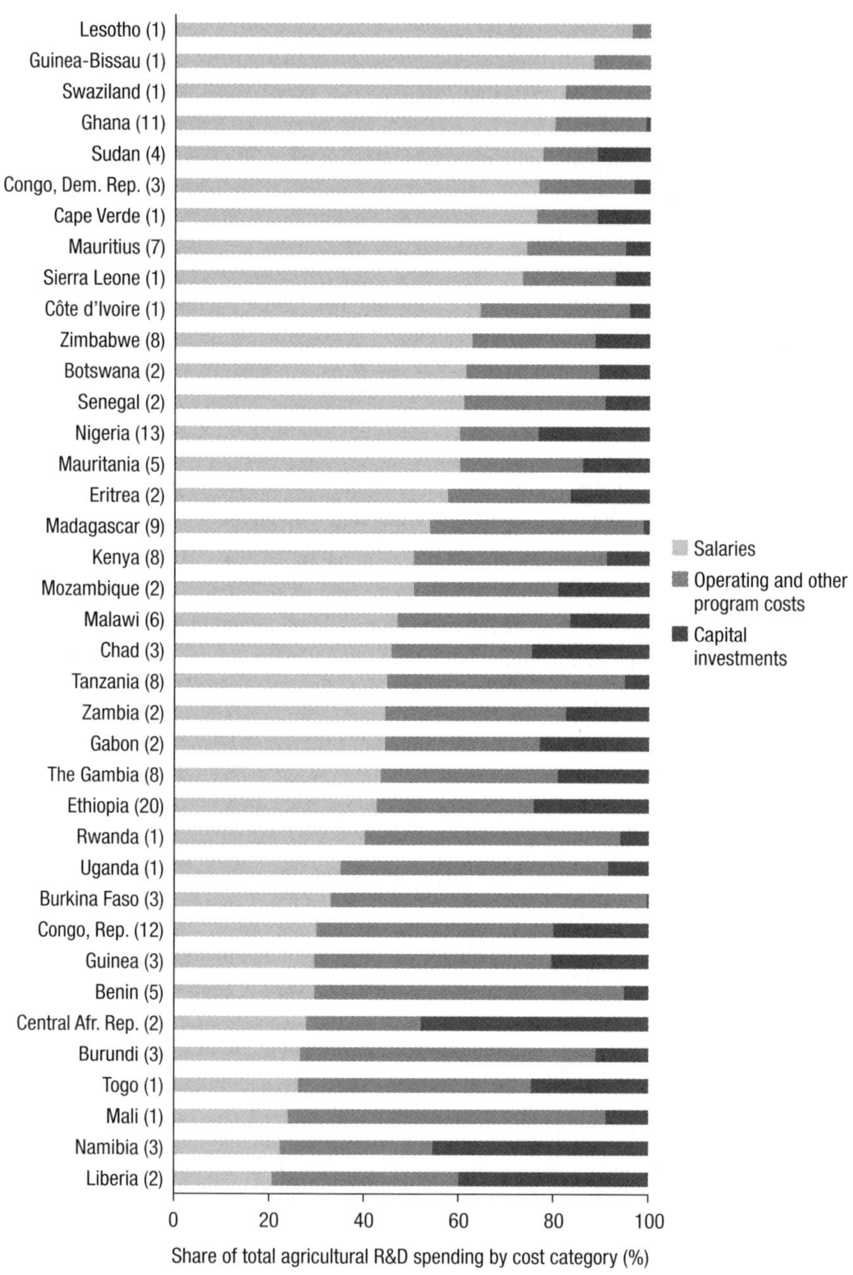

Source: Calculated by author based on ASTI (2014).

Notes: The figures in parentheses indicate the number of government and nonprofit agencies included for each country. Higher-education agencies are not included because data by cost category were not available.

Funding Sources of Agricultural Research and Development

A complete analysis of yearly agricultural R&D investment levels across countries also requires an examination of how agricultural R&D is funded; unsurprisingly, a significant degree of variation exists (Figure 4.5). In some countries, the national government funds the bulk of agricultural R&D activities undertaken by national agricultural research institutes (NARIs), while other countries are extremely dependent on outside funding from donors, development banks, and subregional organizations. In certain countries, R&D agencies manage to generate substantial amounts of funding internally by selling goods and services, while in other countries, the proceeds of such sales are channeled back to the national treasury, discouraging agencies from pursuing this revenue stream. Several countries have established funding systems that mobilize private-sector resources, through either a tax levy or subscription dues.

Overall, in 2011, roughly 60 percent of the funding to NARIs across SSA (excluding Nigeria, South Africa, and a number of smaller countries) was provided by national governments, with donors and development banks representing close to 30 percent. The majority of government funding is usually allocated to salaries, while government funding for operating costs and capital expenditures is generally far lower (Figure 4.6). This clearly raises questions as to the control national governments have over the research agendas of their NARIs.

National Government Funding

Overall, direct institutional funding from a central or local government budget remains the most important source of funding for public agricultural R&D in SSA. In 2011, more than 90 percent of agricultural R&D in countries such as Botswana, Gabon, Namibia, Sudan, and Zimbabwe was funded in this way. Government funding can reach an agricultural R&D agency through a variety of channels. In some countries, staff salaries are directly paid by the Ministry of Finance, while operating and capital costs are paid by the Ministry of Agriculture or another ministry overseeing agricultural research. Some countries may have a Ministry of Science and Technology that allocates research funding through a science fund, either competitively or through direct budget allocations.

A problem that has hindered the performance of agricultural R&D in a number of SSA countries is large discrepancies between approved budgets and actual disbursements of government funding. Gabon, for example, bases

FIGURE 4.5 Relative shares of funding sources at the main national agricultural research institutes and agencies, 2011

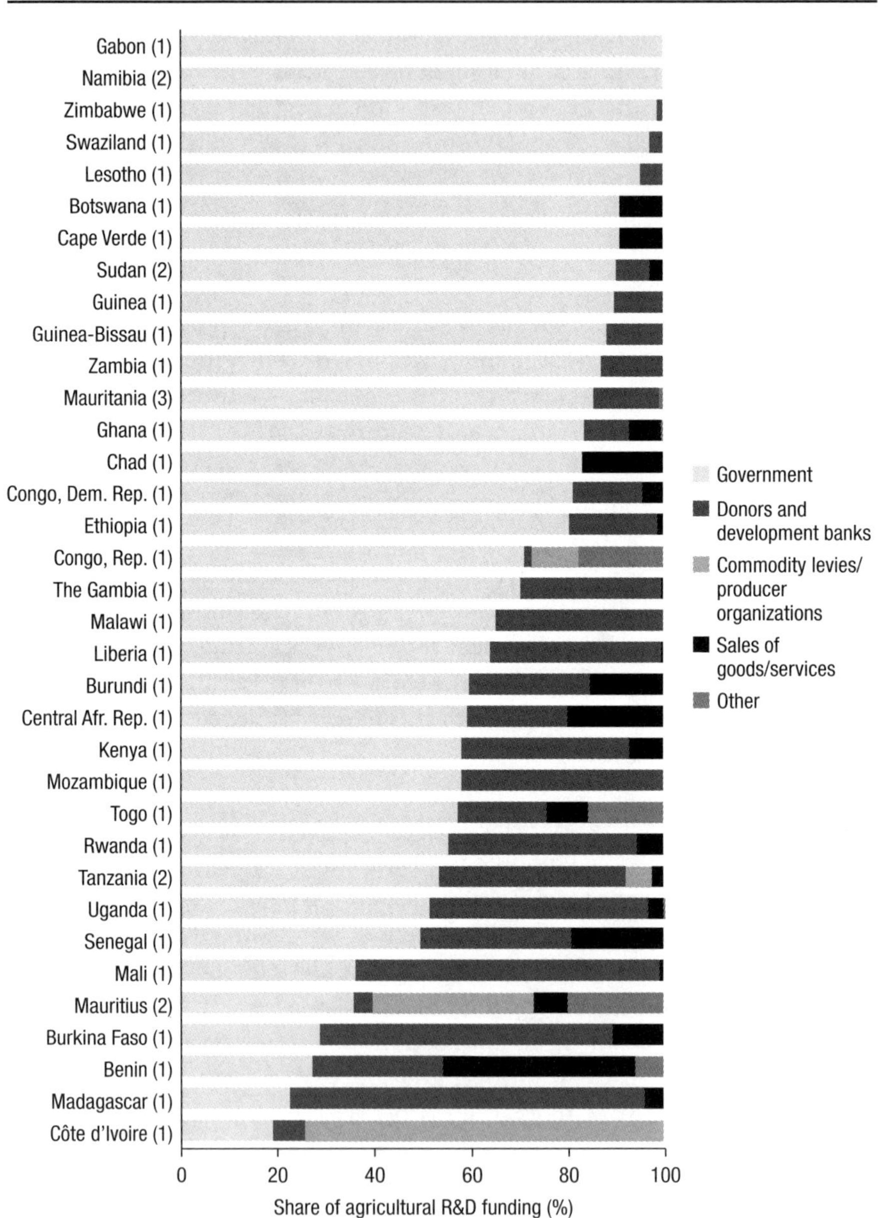

Source: Calculated by author based on ASTI (2014).

Notes: Eritrea, Niger, Nigeria, Sierra Leone, and South Africa are excluded because of a lack of complete data. Figures in parentheses indicate the number of institutes and agencies included for each country.

FIGURE 4.6 Comparison of spending allocations and funding sources at selected national agricultural research institutes, 2011

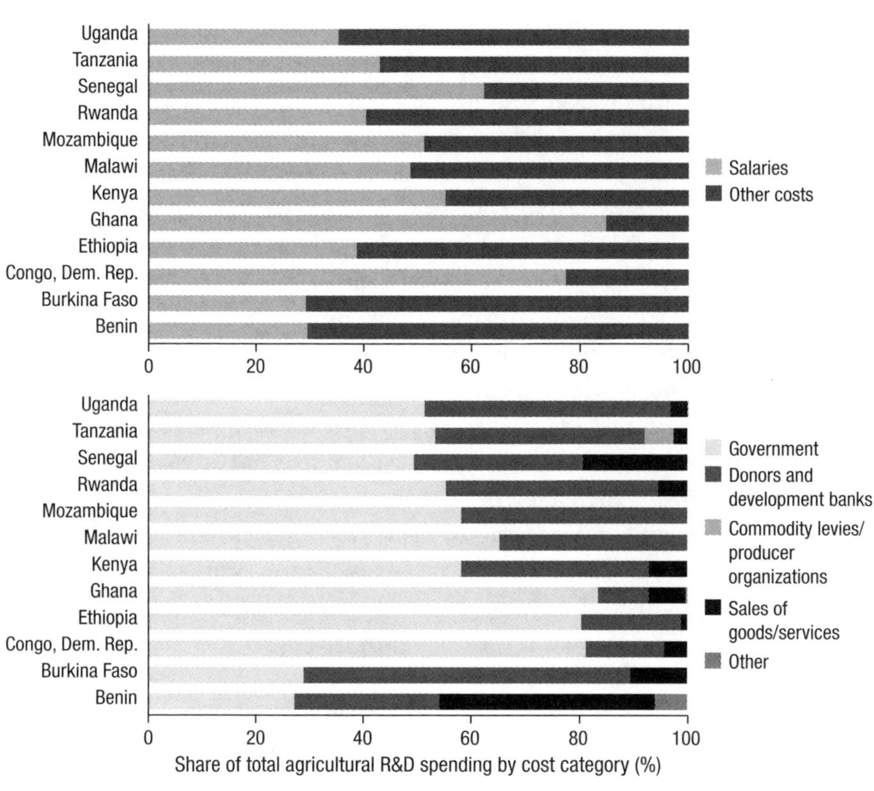

Source: Calculated by author based on ASTI (2014).

its R&D budget estimates on anticipated oil revenues. Fluctuations in the oil price and in the country's production levels can therefore have a major impact on R&D funding. Other countries, including Nigeria, Tanzania, and Uganda, also experienced major discrepancies in budgeted and disbursed funding in recent years. It goes without saying that these discrepancies, including delayed disbursements, can have severely negative consequences for long-term research planning and outputs.

Donors and Development Banks

What distinguishes SSA from other developing regions is its high dependency on donor funding for agricultural research. The principal agricultural R&D institutes in Madagascar, Mali, and Burkina Faso, for instance, derived significant shares of their total funding from donors and development banks in 2011 (73, 63, and 60 percent, respectively). Although overall shares were lower in other SSA countries, donor funding still represents an important source of funding for agricultural R&D agencies in many countries. Moreover, this source of funding seems to have been on the rise in recent years after earlier contractions. Donor funding comprises direct financial support to agricultural R&D agencies by bilateral donors (mostly developed countries); private foundations (such as the Rockefeller Foundation and the Bill & Melinda Gates Foundation); multilateral bodies (such as the European Union, CGIAR centers, or United Nations agencies); regional and subregional organizations (most of which are themselves primarily funded by bilateral, multilateral, and private donors)[1]; and loans and grants from the World Bank and the African Development Bank. In addition, donors can provide in-kind support in the form of technical assistance and equipment and infrastructure, which does not appear in the financial reporting of the recipient country. CIRAD, for example, is a key participant in agricultural research in many francophone countries, but its in-kind contributions are hard to quantify financially. Similarly, China has built some 15 agricultural technology demonstration centers across Africa over the past few years, and another 10 are in the pipeline (Chapter 2, this volume).

Traditionally, the World Bank has been a major contributor to the institutional development of agricultural research in Africa in the form of country projects predominantly financed through loans supplemented by grants. Projects have variously focused purely on agricultural R&D (the more common approach in the 1980s and 1990s) or on agriculture more generally, with an agricultural R&D component (the more common approach in the 2000s). Some projects aimed to reshape a country's entire national agricultural research system, whereas others focused on specific crops, agencies, or general research management and coordination. As of the mid-2000s, the World Bank shifted its country-level approach to agricultural R&D in Africa to a regional approach in the form of regional agricultural productivity programs (Chapters 2 and 14, this volume).

---

1 These organizations are the Forum for Agricultural Research in Africa (FARA), the Association for Strengthening Agricultural Research in Eastern and Central Africa (ASARECA), the Centre for the Coordination of Agricultural Research and Development for Southern Africa (CCARDESA), and the West and Central African Council for Agricultural Research and Development (CORAF/WECARD).

Income Generated through Sales and Services

Research agencies can increase their funding by commercializing their outputs. Some agricultural R&D agencies across SSA manage to derive a significant share of their total funding from the services they render to third parties, such as laboratory analyses or tests done on phytosanitary products, the sale of crop and animal products, and rental of farm equipment. Although the growing importance of the agricultural input and processing sectors, the rise of regional free-trade blocks, and the strengthening of intellectual property legislation have enhanced incentives for the private sector to engage in agricultural R&D, the relative share of business enterprises conducting agricultural R&D in-house remains relatively limited in most SSA countries. Many business enterprises outsource their research needs to public-sector agencies. In some cases, public R&D institutes have entered into long-term alliances with private companies to conduct agricultural R&D on their behalf. In Benin, Burkina Faso, Chad, Senegal, and Togo, for example, cotton companies fund cotton research carried out by national agricultural research institutes on a contractual basis. Similar arrangements between private-sector enterprises and the public sector exist in many other SSA countries (Chapter 7, this volume).

Commodity Levies

Research can also be funded through levies on agricultural production or exports. While these mechanisms empower farmers in setting the research agenda and hence benefiting from the results of the research, there are certain challenges, as outlined by Chapter 7 (this volume) and Echeverría and Beintema (2009). Commodity levies play an important role in financing agricultural R&D in certain African countries. For instance, the Mauritius Sugar Industry Research Institute, the Cocoa Research Institute of Ghana, and Kenya's Coffee Research Foundation are almost entirely funded by a tax on the proceeds of sugar, cocoa, and coffee production, respectively.

In Côte d'Ivoire, the National Agricultural Research Center is structured as a public–private entity, with 40 percent of its funding contributed by the government and 60 percent derived from the private sector. To this end, the Interprofessional Fund for Agricultural Research and Advisory Services (FIRCA) was established in 2002. FIRCA relies on financial contributions not only from the government, but also from cocoa, coffee, oil palm, and rubber producers, who pay subscription dues through their respective producer organizations (Chapter 7, this volume).

Competitive Funding Mechanisms

Competitive funds have gained ground in Africa since the turn of the millennium. Various competitive funds were established around that time as components of World Bank projects in a number of African countries, including Kenya, Mali, Senegal, and Tanzania. These funds typically finance R&D through grants allocated to projects on the basis of their scientific merit and their congruence with broadly defined agricultural R&D priorities. A main concern of these competitive funds is their long-term sustainability, given that most of them are externally funded; once the initial loan or grant has been allocated, the fund becomes defunct. Irrespective of the popularity of competitive funding for research and innovation in other parts of the world, few African countries have adopted competitive funding as an instrument to allocate part of their own national budgets to research and innovation activities (the notable exception being South Africa). Despite exposure to these mechanisms through World Bank and other donor initiatives, the institutional complexity of the competitive funding instrument—such as contractual arrangements—has acted as a disincentive for most African countries, but this may change as research systems grow and strengthen over time.

Various regional and subregional funds have been established in recent years. In 2004, ASARECA launched a competitive grant scheme with multidonor funding to encourage collaborative research initiatives by government, higher education, nonprofit, and private-sector agencies targeting overall agricultural development in East and Central Africa. The fund also aims to promote a more demand-driven and pluralistic approach to increasing agricultural production and empowering end users. CORAF/WECARD implemented a similar fund to expand and diversify scientific and financial partnerships, increase the focus on demand-driven and regional priorities, and improve the quality of research. An array of international competitive funding schemes also exists, but many of them do not exclusively target agriculture, nor are they limited to R&D.

Volatility in Year-to-Year Agricultural Research and Development Spending

Research involves unavoidable time lags from the point at which investments are made until tangible benefits are attained (Alston, Pardey, and Piggott 2006); in the interim, long-term stable funding is required. Agricultural

R&D investments in African countries fluctuate widely over time for an array of reasons that differ greatly across countries (Figure 4.7).[2]

A wide body of literature details the impact of macroeconomic volatility on economic growth and performance in developing countries. This literature focuses primarily on volatility across countries, thereby setting the issue within an international context. Substantial empirical evidence also demonstrates that increased macroeconomic volatility has a negative impact on economic growth, or at a minimum is closely associated with slower growth (Hnatkovska and Loayza 2004; Agion et al. 2005; Fatás and Mihov 2006; Perry 2009). This is unsurprising, given the broad consensus that high macroeconomic volatility likely slows investment (based on expectations of risks and rewards), as well as biasing investments toward short-term returns (Servén 1997). High macroeconomic volatility is also associated with lower investment in human capital, for similar reasons (Krebs, Krishna, and Maloney 2005).

Substantial literature also focuses on the volatility of aid flows to developing countries, which is even higher than the volatility of government revenues, household consumption, or gross domestic product. Aid volatility also reinforces macroeconomic instability and slows economic growth (Bulíř and Hamann 2003; Fielding and Mavrotas 2008; Desai and Kharas 2010). Desai and Kharas (2010) note that some degree of aid volatility is caused by events in recipient countries (for example, regime changes, natural disasters, and civil wars), but that volatility in aid flows is primarily due to donor behavior, including bad planning and shifting priorities.

A number of studies have analyzed fluctuations in R&D expenditures in developed countries, where the predominance of research is conducted by the private sector. These studies find strong evidence that economic growth is positively correlated with R&D expenditure levels, and that long-term basic research will be curtailed in times of economic downturn before applied research (Guellec and Ioannides 1997; Wälde and Woitek 2004). Recent empirical results from Johnstone, Haščič, and Kalamova (2011) provide support for the hypothesis that increased volatility of public R&D spending in environmental technologies has a negative impact on innovation. Their measure includes both direct government expenditures for R&D undertaken by government agencies and public universities, as well as government provision of grants and tax credits

2 This section draws largely on Stads and Beintema (2015), which analyzed agricultural R&D spending and funding volatility in SSA for the 2001–2008 period; new data from 2008 to 2011 enabled the analysis to be updated.

FIGURE 4.7 Fluctuations in yearly public agricultural R&D spending for selected African countries, 1981–2011

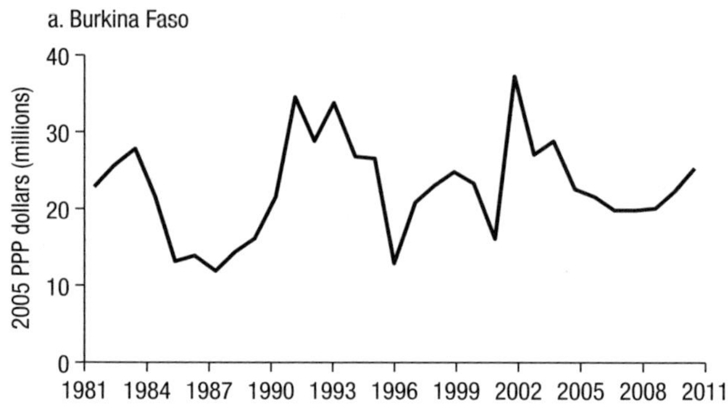

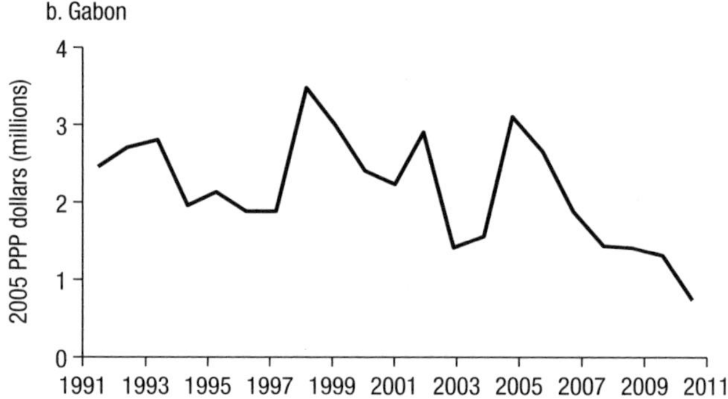

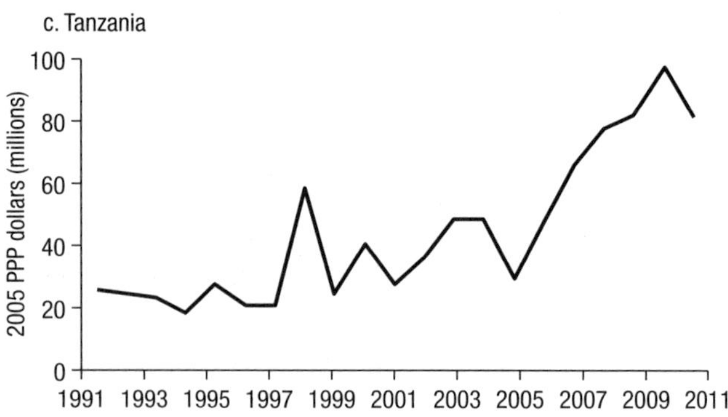

Source: Calculated by author based on ASTI (2014).

Note: Given availability, data for Burkina Faso span 1981–2011. PPP = purchasing power parity.

for R&D undertaken by the private sector. Cullen et al. (2014) also provide evidence of the negative economic effects of R&D volatility. Yet, they also highlight the "entrenchment" argument, where steady R&D investment could reflect problems of moral hazard and a lack of control of the quality of research projects and their impacts. Although Cullen et al. (2014) refer to the private sector, the issues raised are relevant to the public sector as well.

No literature was found on R&D funding volatility in developing countries; however, empirical findings from the literature on macroeconomic and aid volatility suggest that extreme volatility in agricultural R&D funding is similarly harmful to the institutional stability and long-term outputs of agricultural R&D. This is supported by substantial anecdotal evidence. In numerous African examples, agricultural research institutes were plunged into financial hardship upon the completion of multimillion-dollar projects, forcing them to cut research programs and lay off staff. Large fluctuations in yearly investment levels thus hinder the advancement of technical change and the release of new varieties and technologies in the long run, in turn negatively affecting agricultural productivity growth and poverty reduction.

The Volatility Coefficient

In order to measure the degree of volatility in yearly agricultural R&D spending levels across SSA countries, a commonly used method of calculating price volatility in finance and output volatility in macroeconomics was applied to ASTI's agricultural R&D spending data. The resulting "volatility coefficients" quantify volatility in agricultural R&D spending by applying the standard deviation formula to average one-year logarithmic growth of agricultural R&D spending over a certain period (Guellec and Ioannides 1997; Durlauf, Johnson, and Temple 2005).[3] Volatility coefficients were calculated for 30 SSA countries, based on complete time-series data on agricultural R&D expenditures for the 2001–2011 period. Countries with few or no changes in yearly spending levels, or those with steady (positive or negative) growth, have low volatility coefficients. In contrast, countries with erratic fluctuations in yearly spending levels have high volatility coefficients. A value of 0 indicates "no volatility," countries with values between 0 and 0.1 were classified as having "low

3 Growth in agricultural R&D spending (g_s)can be expressed as $g_s = \ln\left(\frac{s_t}{s_{t-1}}\right) s = 1, \ldots, N$, where s is agricultural R&D spending (in constant prices), and t represents the year. Subsequently, the volatility coefficient (V) of agricultural R&D expenditures can be calculated by taking the standard deviation of growth in yearly agricultural R&D spending, that is:

$$V = \sqrt{\frac{1}{N}\sum_{s=1}^{N}(g_s - \mu)^2}, \text{ where } \mu = \frac{1}{N}\sum_{s=1}^{N} g_s.$$

FIGURE 4.8 Comparison of agricultural R&D investment volatility coefficients, 2001–2011

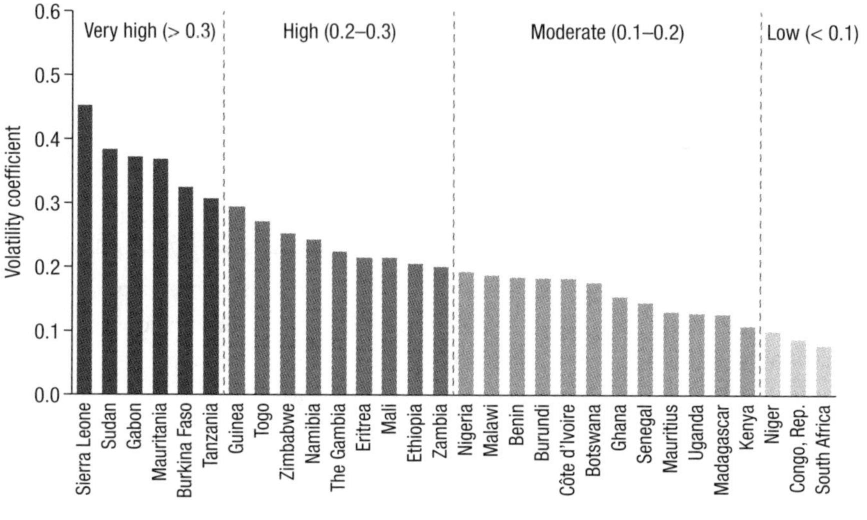

Source: Calculated by author based on ASTI (2014).

Note: The figure excludes Cape Verde, Central African Republic, Chad, the Democratic Republic of the Congo, Guinea-Bissau, Lesotho, Liberia, Mozambique, Rwanda, Swaziland, and Zimbabwe because time-series data did not date back to 2001.

volatility," countries with values between 0.1 and 0.2 were considered to have "moderate volatility," and countries with values above 0.2 fell into the "high volatility" category (Figure 4.8). Note also that these values are weighted averages. The mean volatility coefficient for the 31 countries over the 2001–2011 period was 0.22—twice as high as the mean volatility coefficient for 12 low- and middle-income countries in the Asia–Pacific region (0.11) and 8 Latin American countries (0.11) over the 2000–2008 period (Beintema et al. 2012). Moreover, agricultural R&D spending in SSA was also markedly more volatile than agricultural output in SSA (0.10) during 2001–2011.[4]

Understandably, a large degree of variation was recorded across countries. Those with the highest degree of fluctuation in yearly agricultural R&D spending were Sierra Leone (0.45), Sudan (0.38), Gabon (0.37), Mauritania (0.37), Burkina Faso (0.32), and Tanzania (0.31). In contrast, yearly agricultural R&D spending in such countries as Rwanda, South Africa, and the Republic of the Congo was found to be more stable, with volatility coefficients of just 0.04, 0.08, and 0.09, respectively. It is important to note that

4 Agricultural GDP data were taken from World Bank (2013).

volatility in spending at the agency level is typically higher than at the country level because aggregate fluctuations tend to hide idiosyncratic spending shocks. Similarly, the volatility coefficient for agricultural R&D investments for the 31 sample countries combined—that is, the standard deviation of yearly growth in total SSA agricultural R&D investment during 2001–2011—is just 0.04. This indicates that spending in SSA as a whole is less volatile than spending in the individual countries, which is not surprising.

A closer look at a subsample of 82 agricultural R&D agencies from 25 SSA countries for which complete time-series data by cost category were available for the entire 2001–2011 period shows that volatility in agricultural R&D spending is mainly caused by fluctuations in nonsalary-related expenses, which is also not surprising.[5] Salary expenditures (0.23) were less than twice as volatile as operating and program costs (0.53) and more than seven times less volatile than capital investments (1.66). Although these averages mask some important cross-agency differences, the results were relatively consistent across countries and institutes. Of the sample agencies, 88 percent had a higher volatility coefficient for nonsalary- as opposed to salary-related spending.

Volatility by Funding Source

As described above, agricultural R&D institutes in different SSA countries derive their funding from a variety of sources. Shifts in yearly allocations from one or more funding sources can therefore have a large positive or negative impact on overall agricultural R&D spending levels. Governments, for example, often reduce previously approved yearly budgets in response to shifting priorities or unanticipated contractions in revenues. Donor and development bank funding can also be a major cause of volatile agricultural R&D spending over time. This type of funding is typically short term and ad hoc, and in many instances the completion of large donor-funded projects precipitates abrupt reductions in national agricultural R&D spending. Rising or falling world market prices for cash crops can also have a significant impact on funding levels, especially those derived through a direct tax on the production or export of a certain crop. The Cocoa Research Institute of Ghana, for example, benefited greatly from both an increase in cocoa prices and an increase in the country's cocoa production beginning in 2003–2004, whereas overall funding to the Mauritius Sugar Industry

5 The sample includes 82 large agricultural research agencies in Benin, Botswana, Burkina Faso, Burundi, Côte d'Ivoire, Eritrea, Ethiopia, Gabon, The Gambia, Ghana, Guinea, Kenya, Madagascar, Mali, Mauritania, Mauritius, Namibia, Nigeria, the Republic of the Congo, Senegal, Sudan, Tanzania, Togo, Uganda, and Zambia. Combined, these agencies accounted for 31 percent of total agricultural R&D spending in SSA in 2011.

Research Institute has decreased in recent years because of a loss of preferential access to the EU sugar market and the subsequent decline in national production levels. Global coffee prices have been more volatile over time, which is clearly reflected in the yearly funding Kenya's Coffee Research Foundation receives through commodity levies.

The volatility coefficient is also a useful tool for comparing the relative stability of different funding sources over time and across countries. It is important to note, however, that not all volatility is bad on face value. A sudden injection of government or donor funding to rehabilitate R&D infrastructure is, of course, highly positive. Detailed 2001–2011 time-series data on agricultural R&D funding sources were available for 71 large, public-sector agricultural R&D agencies from 26 SSA countries. A breakdown of volatility by funding source reveals that overall funding from donors and development banks is extremely volatile (1.31)—in fact three times more so than government funding (0.42), which itself is far from stable. Funding from producer organizations and commodity boards, internally generated resources through the sale of goods and services, and other funding sources (0.94) also showed relatively large yearly fluctuations. Interestingly, the mean institute-level volatility (0.38) is lower than the volatility of each of the individual funding sources, indicating that in many cases shocks in one funding source are to some extent absorbed by reverse shocks in other funding sources.[6]

Abundant empirical evidence suggests that volatility in donor funding is costly, particularly in less developed countries with weak institutions, and that measures to reduce volatility would significantly enhance the value of donor aid (Kharas 2008). The fact that donor and development bank funding for agricultural R&D shows a much higher degree of volatility than other funding sources is worrying, given that many national agricultural R&D institutes in SSA, particularly those in low-income countries, derive a substantial share of their total funding from donors and development banks. Although most national governments in SSA publicly recognize the need for rapid agricultural development to reduce poverty, they are struggling to allocate sufficient resources to agricultural R&D. In many countries, the bulk of government appropriations is allocated to salaries, leaving the costs of operating research programs and investing in necessary infrastructure largely dependent on volatile funding from donors, competitive grants, or the private sector. This is

6 Note that, although the data allowed for the calculation of a volatility coefficient for funding derived from commodity levies and producer organizations, this coefficient was irrelevant at the SSA level because only a handful of countries generate agricultural R&D funding this way, so the mean would be skewed.

not to say that competitive salary rates are not crucial to maintaining a critical mass of qualified researchers, but providing these scientists with the necessary resources and infrastructure to enable them to effectively do their work is equally important, and a balance must be struck between the two. This requires long-term, sustainable investment in nonsalary-related expenditures.

On average, agencies that are highly dependent on funding from donors and development banks are more vulnerable to funding shocks than are institutes funded mostly by their governments. Uncertain inflows of funding from donors and development banks have a considerably negative impact on the long-term implementation of R&D programs, and often on much-needed rehabilitation of R&D infrastructure. More than 90 percent of government funding received by Burkina Faso's National Environment and Agricultural Research Institute, for example, is spent on salaries, leaving the funding of actual research programs and rehabilitation of research equipment and infrastructure almost entirely in the hands of donors and development banks. Peaks in capital investments largely coincided with peaks in funding from two consecutive World Bank–funded projects. Upon the completion of these projects, operating and capital budgets were drastically reduced, seriously disrupting day-to-day operations. The situation in Burkina Faso is not unique. R&D programs in a large number of agricultural R&D agencies across SSA are to a large extent driven by donor agendas that do not necessarily correspond to long-term national development agendas. Oftentimes, the gains achieved through major donor-funded projects are quickly eroded in the absence of viable mechanisms to sustain them.

Promoting Sustainable Funding for Agricultural Research and Development Long Term

Despite the fact that agricultural R&D spending in SSA increased by 40 percent in real terms during 2000–2011, overall investment levels in most countries still fall well below the minimum target of 1 percent of AgGDP recommended by the African Union. Higher levels of funding are needed to establish and maintain viable agricultural research programs that achieve tangible results. In fact, growth in agricultural R&D spending will need to exceed growth in agricultural output (the minimum CAADP target is 6 percent) for SSA to move closer to this 1 percent target. On average, agricultural R&D investment commands significant returns, but these returns take time—commonly decades. Consequently, the inherent lag from the inception of research to the adoption of a new technology or the introduction of a new variety calls for sustained and stable R&D funding.

The time-series data presented in this chapter reveal that agricultural R&D funding in many SSA countries has been far from stable over time, and that R&D spending for the region as a whole shows higher volatility compared with spending in other developing regions of the world. Agricultural R&D agencies in SSA, particularly those in the region's low-income countries, are more dependent on funding from donors and development banks than their counterparts in other developing regions, and this type of funding has shown considerably greater volatility over the past decade compared with government funding. In a large number of SSA countries, donors fund the bulk of nonsalary-related expenditures (that is, program and operating costs and capital investments), and there is extensive anecdotal evidence of agencies reverting to financial crisis upon the completion of large donor-funded projects, forcing them to scale down their activities. Too much of the critical decisionmaking about research priorities appears to be devolved to donors, with the result that the research agendas of many agricultural research agencies across SSA—particularly in smaller, low-income countries—can be skewed toward short-term goals that are not necessarily aligned with national and (sub)regional priorities, or are overly focused on commodities with relatively limited economic importance. A new framework is therefore needed whereby governments put forward strategic priorities and donors contribute budgetary support to those programs.

Halting excessive volatility in yearly agricultural R&D investment levels requires a long-term commitment from national governments, donors and development banks, and the private sector. Stable and sustainable levels of government funding are key, not just to secure researcher salaries, but also to enable necessary nonsalary-related expenses. Rather than relying too much on donors and development banks to fund critical research areas, governments need to more clearly identify their own long-term national priorities and design relevant, focused, and coherent agricultural R&D programs accordingly. Donor and development bank funding needs to be closely aligned with these national priorities, and consistency and complementarities among donor programs need to be ensured. Finally, mitigating the effects of any single donor's abrupt change in aid disbursement is crucial, highlighting the need for greater funding diversification, for example, through the sale of goods and services, or by attracting complementary investment from the private sector. The private sector is currently the least developed source of sustainable financing for agricultural R&D in SSA, and its funding potential remains largely untapped in most countries. Cultivating private funding involves providing a more enabling policy environment by national governments in terms of tax incentives, protection of intellectual property rights, and regulatory reforms to encourage the spill-in of international technology.

Collective action by farmers and related agribusinesses (through formal producer organizations) also has the potential to generate additional resources for agricultural research in a number of countries in the region (Chapter 7, this volume). An added benefit of this funding mechanism is that decisionmaking on the use of the resulting funds would generally rest with producers and other stakeholders in the relevant value chain. All of these funding diversification measures are necessary to tackle underinvestment and put an end to the roller-coaster that has characterized agricultural R&D funding in SSA to date.

References

Agion, P., G. Angeletos, A. Banerjee, and K. Manova. 2005. *Volatility and Growth: Credit Constraints and Productivity-Enhancing Investment.* NBER Working Paper 11349. Cambridge, MA: National Bureau of Economic Research.

Alston, J., P. Pardey, and R. Piggott. 2006. "Synthesis of Themes and Policy Issues." In *Agricultural R&D in the Developing World: Too Little, Too Late?* Edited by J. Alston, P. Pardey, and R. Piggott. Washington, DC: International Food Policy Research Institute.

ASTI (Agricultural Science and Technology Indicators). 2014. ASTI database. Accessed May 2014. http://asti.cgiar.org/data/.

Beintema, N., G. Stads, K. Fuglie, and P. Heisey. 2012. *ASTI Global Assessment of Agricultural R&D Spending: Developing Countries Accelerate Investment.* Washington, DC: International Food Policy Research Institute; Rome: Global Forum on Agricultural Research.

Bulíř, A., and A. Hamann. 2003. "Aid Volatility: An Empirical Assessment." *IMF Staff Papers* 50 (1). Washington, DC: International Monetary Fund.

Cullen, G., D. Gasbarro, W. Ruan, and E. Xiang. 2014. *R&D Expenditure Volatility and Stock Return: Earnings Management, Adjustment Costs or Overinvestment?* Murdoch, Western Australia: Murdoch University, School of Management and Governance.

Desai, R., and H. Kharas. 2010. *The Determinants of Aid Volatility.* Global Economy and Development Working Paper 42. Washington, DC: Brookings Institution.

Durlauf, S., J. Johnson, and P. Temple. 2005. "Growth Econometrics." In *Handbook of Economic Growth.* Vol. 1a, edited by P. Agion and S. Durlauf. Amsterdam: Elsevier.

Echeverría, R., and N. Beintema. 2009. *Mobilizing Financial Resources for Agricultural Research in Developing Countries: Trends and Mechanisms.* Rome: Global Forum for Agricultural Research.

Evenson, R., and D. Gollin, eds. 2003. *Crop Variety Improvement and Its Effect on Productivity: The Impact of International Agricultural Research.* Wallingford, UK: CABI Publishing.

Fatás, A., and I. Mihov. 2006. *Fiscal Discipline, Volatility and Growth.* Paris: Institut européen d'administration des affaires; London: Centre for Economic Policy Research.

Fielding, D., and G, Mavrotas. 2008. "Aid Volatility and Donor-Recipient Characteristics in Difficult Partnership Countries." *Economica* 75 (299): 481–494.

Guellec, D., and E. Ioannidis. 1997. "Causes of Fluctuations in R&D Expenditures: A Quantitative Analysis." *OECD Economics Studies* 29 (1997/II). Paris: Organisation for Economic Co-operation and Development.

Hnatkovska, V., and N. Loayza. 2004. *Volatility and Growth.* Policy Research Working Paper 3184. Washington, DC: World Bank.

IAASTD (International Assessment of Agricultural Knowledge, Science and Technology for Development). 2008. *Synthesis Report.* Washington, DC: Island Press.

Johnson, D., and R. Evenson. 2000. *How Far Away Is Africa? Technological Spillovers to Agriculture and Productivity.* Working Paper 2000–01. Wellesley, MA: Wellesley College.

Johnstone, N., I. Haščič, and M. Kalamova. 2011. "Environmental Policy Design Characteristics and Innovation." In *OECD Studies on Environmental Innovation: Invention and Transfer of Environmental Technologies.* Paris: Organisation for Economic Co-operation and Development.

Kharas, H. 2008. *Measuring the Cost of Aid Volatility.* Wolfensohn Center for Development Working Paper 3. Washington, DC: Brookings Institution.

Krebs, T., P. Krishna, and W. Maloney. 2005. *Income Risk and Human Capital in LDCs.* Washington, DC: World Bank.

Perry, G. 2009. *Beyond Lending: How Multilateral Banks Can Help Developing Countries Manage Volatility.* Washington, DC: Center for Global Development.

Servén, L. 1997. *Uncertainty, Instability, and Irreversible Investment: Theory, Evidence, and Lessons for Africa.* Policy Research Working Paper 1722. Washington, DC: World Bank.

Stads, G., and N. Beintema. 2015. "Agricultural R&D Expenditure in Africa: An Analysis of Growth and Volatility." *European Journal of Development Research* 27 (3): 391–406.

Wälde, K., and U. Woitek. 2004. "R&D Expenditure in G7 Countries and the Implications for Endogenous Fluctuations and Growth." *Economics Letters* 82: 91–97.

World Bank. 2007. *World Development Report 2008: Agriculture for Development.* Washington, DC.

———. 2013. *World Development Indicators.* Accessed September 19, 2013. http://databank.worldbank.org/data/views/variableSelection/selectvariables.aspx?source=world-development-indicators.

Chapter 5

WHY DO AFRICAN COUNTRIES UNDERINVEST IN AGRICULTURAL R&D?

Samuel Benin, Linden McBride, and Tewodaj Mogues

Clearly, agricultural research and development (R&D) is among the most important public goods in agriculture and, as such, is a critical component of public agricultural expenditures (PAEs) (Fan 2008; Mogues and Benin 2012; Chapter 3, this volume). The returns to agricultural R&D investments have been shown to be high (Alston et al. 2000; Evenson 2001; Mogues, Fan, and Benin 2015). Thus, it is not surprising that, in its commitment to implementing an agriculture-led development agenda, the African Union's (AU's) New Partnership for Africa's Development (NEPAD) set a target for government spending on agricultural R&D of at least 1 percent of agricultural gross domestic product (GDP.)[1] Given these established high payoffs and the demonstrated political commitment to agricultural R&D in Africa, it is surprising that spending on agricultural R&D in many African countries is lower than would be expected (Chapters 1 and 2, this volume). This contradiction is at the core of the analysis presented in this chapter, which addresses the question, why do African countries underinvest in agricultural R&D?

Whereas underinvestment in agricultural R&D is a global issue (Pardey, Alston, and Piggott 2006), agricultural research intensity ratios in Africa have fallen farther behind those in developed countries (Alston and Pardey 2006).[2] Because agricultural productivity and food security gains in Africa have in a large part derived from agricultural R&D investments funded by or undertaken in a few rich countries (Alston and Pardey 2006), the consequences of continued underinvestment in Africa are likely to escalate, to the extent that international technology spillovers are becoming increasingly less applicable and less accessible as a result of the changing agricultural R&D agenda in developed countries. Policy prescriptions to address underinvestment need to go beyond just

Note that the order of authorship of this chapter is alphabetical.

1 This target derives from the agreement made at the 2007 AU Assembly in Addis Ababa to allocate at least 1 percent of overall GDP to R&D (AU 2007).

2 Intensity ratios are calculated as the share of agricultural GDP a country invests in agricultural research (Chapters 3 and 4, this volume).

increasing the amount of government R&D spending to include equally—if not more—important factors, such as shifting the incentives for nonstate actors to raise their investments, or influencing the type of research conducted (Alston and Pardey 2006). In addition, understanding the budget-allocation process could help improve knowledge of the problem, as well as options for addressing it.

The approach to the analysis presented in this chapter is to examine the political economy of underinvestment in agricultural R&D using a framework of the key drivers of public investment decisionmaking (Mogues 2015): (1) the attributability of investments to key actors, including politicians, bureaucrats, interest groups, and donors; (2) the incentives and constraints of these actors; (3) the budget process; and (4) the political and economic governance environment. The chapter begins by presenting past trends in and future commitments to public agricultural research expenditures compared with other types of PAEs in different parts of Africa. As a means of understanding the resource constraints within which agricultural research expenditures are made, the analysis includes a discussion of trends in total expenditures and PAEs, which also serves as a review of the current progress in meeting the 2003 Maputo Declaration target of investing 10 percent of total national expenditures in the agricultural sector (NEPAD 2005). This discussion is followed by a review of the evidence on the returns to public investment in agricultural R&D compared with other types of public agricultural investments, and a discussion of the political economy drivers of underinvestment in agricultural R&D. Finally, conclusions are drawn in the context of understanding these drivers and their implications for policy dialogue and future research.

Recent Trends in Public Agricultural Research Spending

Meeting the Maputo Declaration Target for Agricultural Expenditures

Although absolute national levels of PAEs have grown faster in many parts of Africa since the 2003 advent of the Comprehensive Africa Agriculture Development Programme (CAADP), the expenditure shares for Africa as a whole declined slightly from 4.0 percent in 2003 to 3.9 percent in 2010 (Figure 5.1).[3] Differences in national agricultural spending are substantial across

3 See Appendix 5A for a description of the data sources and methodology underlying the analysis in this section.

FIGURE 5.1 Agricultural expenditures as a share of total public expenditures by subregion, 2003–2010

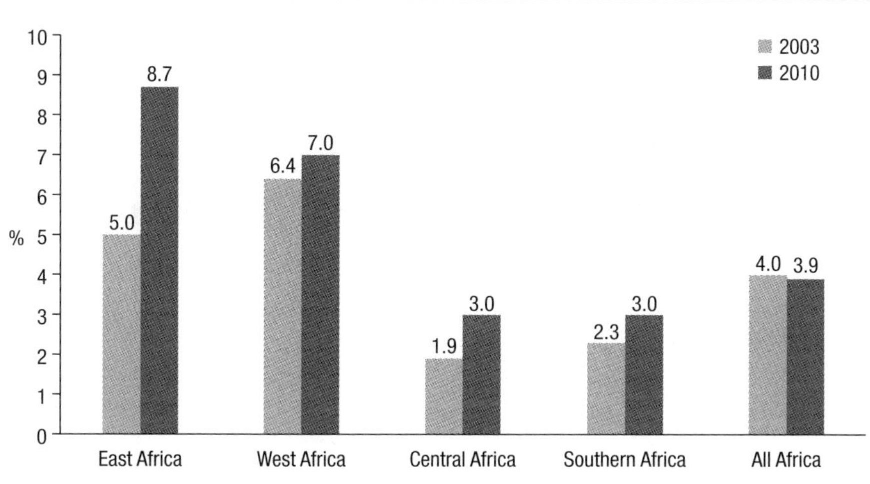

Sources: Calculated by authors based on AUC (2008), Yu (2012), and ReSAKSS (2013).
Note: All Africa includes North Africa.

subregions and countries. None of the five subregions achieved the Maputo Declaration target of spending 10 percent of total government expenditure on agriculture (Figure 5.1). The top performer in terms of a change in the share was East Africa, where spending increased from 5.0 percent in 2003 to 8.7 percent in 2010. West Africa had the highest share in 2003, at 6.4 percent, but it only increased to 7.0 percent in 2010.

Since 2003 (the year of the Maputo Declaration), only 13 countries—Burkina Faso, Burundi, Ethiopia, Ghana, Guinea, Madagascar, Malawi, Mali, Niger, Republic of the Congo, Senegal, Zambia, and Zimbabwe—have surpassed CAADP's 10 percent agricultural expenditure target in any year, and only seven of them—Burkina Faso, Ethiopia, Guinea, Malawi, Mali, Niger, and Senegal—have consistently surpassed the target in most years (Figure 5.2). Within this smaller group, Burkina Faso and Niger are now hovering around the 10 percent threshold, each having reduced its share of PAE in recent years. Several countries show a consistent increase in share of PAE over time: this group includes Burundi, Republic of the Congo, Rwanda, São Tomé and Príncipe, Sudan, Togo, and Zambia. In the remaining countries, expenditure shares have generally declined or stagnated.

CAADP has clearly contributed to raising the profile of agriculture in the development agenda, particularly in West Africa, where CAADP's

FIGURE 5.2 Public agricultural expenditures as a share of total national expenditures by subregion, 2003–2010 yearly average

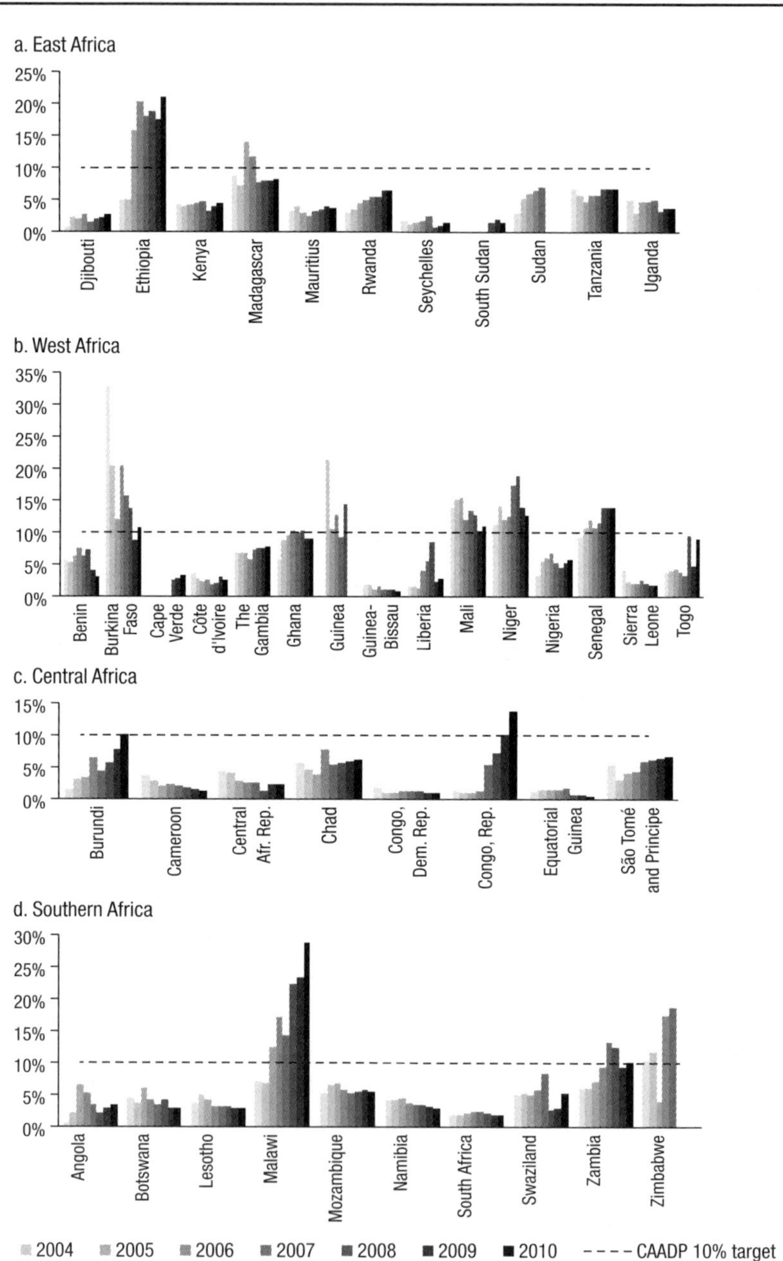

Sources: Calculated by authors based on AUC (2008), Yu (2012), and ReSAKSS (2013).

Notes: Dashed line represents 10%: the 2003 Maputo Declaration target for agricultural expenditures. See appendix Table 5B.3 for details.

implementation is most advanced, and where more countries than in any other subregion have met the target or are moving toward it. All 15 countries in the West African subregion have signed a CAADP Compact and have a national agricultural investment plan (NAIP) in place.

In Southern Africa, which comprises many middle-income countries, agricultural expenditures averaged 5–10 percent of total government expenditures. In fact, as a share of total agricultural value-added (or agricultural GDP), Southern Africa spends more on the sector than does any other subregion (Benin and Yu 2013). Malawi stands out in particular, spending far more than 10 percent of the total government expenditure on agriculture since the start of its farm subsidy program, and particularly since 2005. In most of the other Southern African countries, however, PAE shares have stagnated over time. This could be because these countries have reached an equilibrium, whereby their returns to additional spending on agriculture and nonagriculture are equal—but this, too, needs further investigation.

Against the CAADP's 10 percent expenditure target, Central Africa as a whole has made substantial progress overall, although from a very low base of 1.9 percent in 2003 to 3.0 percent in 2010. This subregion's PAE share rose significantly over time in several countries—particularly Burundi, Republic of the Congo, and São Tomé and Príncipe. In the remaining countries, however, PAE shares were less than 5 percent, with no improvement over the period (Figure 5.2). In East Africa, PAE shares in most countries were between 5 and 10 percent, and increased over time.

Public Agricultural R&D Expenditure Trends in Recent Years

Data obtained from the Food and Agriculture Organization of the United Nations (FAO 2013) on Monitoring African Food and Agricultural Policies (MAFAP)—disaggregated for Burkina Faso, Kenya, Mali, Tanzania, and Uganda only—were used to analyze public agricultural research expenditures compared with other types of PAE. The shares of PAE allocated to agricultural research were moderate during 2006–2010 compared with other types of expenditures (Figure 5.3). For the five countries taken together, the shares spent on agricultural research were about 10–15 percent per year between 2006 and 2010. The share was lowest in Mali (5 percent); moderate in Burkina Faso (10 percent); and larger in Uganda (15 percent), Tanzania (16 percent), and Kenya (17 percent). The three East African countries have larger agriculture budgets (see appendix Table 5B.2 and Table 5B.4) and more developed economies, which perhaps gives policymakers more latitude to focus more resources on research, despite its medium- to long-term impacts.

FIGURE 5.3 Share of national public agricultural expenditures by function for selected African countries, yearly average 2006–2010

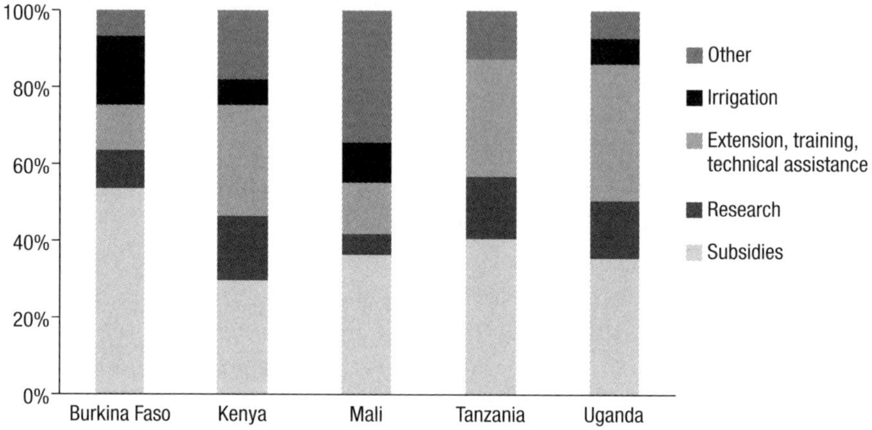

Source: Calculated by authors based on FAO (2013).

Notes: See appendix Table 5B.4 for details. "Other" includes feeder roads and other infrastructure, marketing, storage, public stockholding, inspection, and so on.

A large share of yearly PAE was spent on subsidies, ranging from a yearly average of 30 percent in Kenya, to 54 percent in Burkina Faso. For extension services, training, and other technical assistance, the shares of PAE ranged from 12 to 13 percent in Burkina Faso and Mali, to 30–36 percent in the other countries. Shares of PAE spent on irrigation averaged 6–10 percent, but Burkina Faso's share was much higher (18 percent).

Overall, the functional distribution of PAE seems to be more balanced in Mali compared with the other four countries: the expenditures on subsidies, extension, and research together accounted for 75–88 percent of PAE in the other four countries, compared with only 55 percent in Mali.

Public Agricultural R&D Spending Plans

In analyzing NAIP budgets for the 2010–2015 period, it can be seen that the total budget allocated specifically to research is less than 5 percent in many countries (Figure 5.4), including Malawi (0.4 percent), Niger (1.2 percent), Nigeria (1.5 percent), Rwanda (1.7 percent), The Gambia (2.1 percent), Mali (3.1 percent), Togo (3.3 percent), Ghana (3.4 percent), and Burundi (4.7 percent). Uganda (12.6 percent), Benin (17.0 percent), and Côte d'Ivoire (20.0 percent) recorded the highest shares. The budget shares for Mali (3.1 percent) and Uganda (12.6 percent) are slightly lower than the expenditure shares observed

FIGURE 5.4 Share of national agricultural budget allocations by function, selected countries, and timeframes

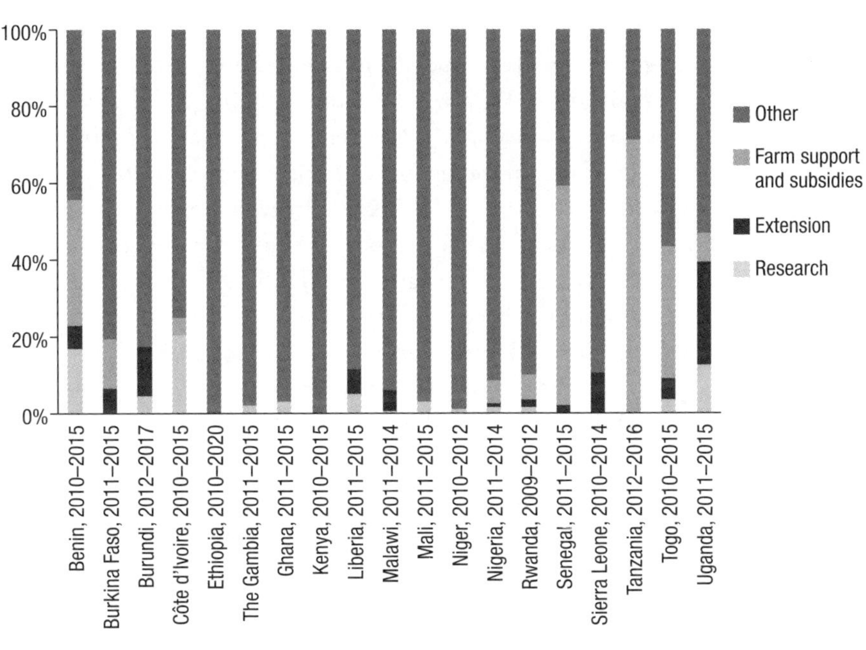

Source: Calculated by authors based on allocations stipulated in the national agricultural investment plans.

Note: This figure only shows the investment plans' budget allocations according to the three agricultural functions indicated; because the budget categorizations differed across plans, not all had allocations for all three agricultural functions. See appendix Table 5B.5 for further details.

from the MAFAP data (5.3 percent for Mali and 15.0 percent for Uganda; see Figure 5.3 and appendix Table 5B.3), suggesting a business-as-usual scenario for the two countries.

Returns to Agricultural R&D Compared with Other Public Investments

High returns to agricultural R&D expenditures have been documented in many parts of the world (Chapter 3, this volume). These returns can be observed in terms of agricultural growth, GDP growth, and poverty reduction, and they are higher than the returns to other types of agricultural and nonagricultural expenditures. In particular, the returns to agricultural R&D investment in terms of poverty and malnutrition objectives are high compared with the returns to nonagricultural R&D expenditures. This section reviews the evidence on the relative returns to agricultural R&D investment in Africa.

Thirtle, Lin, and Piesse (2003) found that agricultural research in Africa has played a substantial role in agricultural growth when compared with other factors contributing to agricultural growth. The mean and weighted mean returns of general agricultural R&D spending—calculated as R&D expenditures per hectare in constant 1995 U.S. dollars and including 44 observations from 22 African countries—are 18 and 22 percent, respectively, for Africa as a whole.[4] Country-level returns range from –12 percent in Lesotho, to 57 percent in Morocco. Agricultural R&D expenditures are one of the larger contributors to yields in Africa, based on a comparison of the impact of agricultural R&D expenditures with other agricultural inputs, such as (1) fertilizer, measured in 100 grams per hectare of arable land; (2) labor, measured in agricultural value-added per agricultural worker (in constant 1995 U.S. dollars); (3) machinery, measured in number of tractors per hectare of arable land; and (4) land quality, measured using an index from Wiebe et al. (2000). Agricultural R&D inputs (with a yield elasticity of 0.36 at the 1 percent level of statistical significance) are second only to labor (with a yield elasticity of 0.63, also at the 1 percent level). Machinery and fertilizer inputs have lower and less statistically significant yield elasticities (0.17 at the 10 percent level and 0.02 at the 5 percent level, respectively).

The relative efficiency of agricultural R&D expenditures seen at the cross-country level is reinforced by country-level case studies that document the comparative impacts of agricultural spending in Tanzania (Fan, Nyange, and Rao 2012) and Uganda (Fan and Zhang 2008), as can been seen from benefit–cost ratios of public spending in Tanzania and Uganda (Figure 5.5). Returns to public spending in Tanzania (captured in terms of per capita income) indicate that for every Tanzanian shilling (TSh) invested in agricultural R&D by the Tanzanian government, income increases by TSh 12.5, on average, compared with TSh 9.0 for education expenditures and TSh 9.1 for road expenditures. Similarly, for every Ugandan shilling (USh) invested in agricultural R&D by the Ugandan government, agricultural labor productivity increases by USh 12.4. Returns to investment in feeder roads are USh 7.2 per USh invested; returns to education lag even further behind, at USh 2.7; and health expenditures only return USh 0.9.

The high relative returns to public spending on agricultural R&D are nuanced by a consideration of the returns to other development objectives, such as poverty reduction. In their cross-country study, Thirtle, Lin, and Piesse (2003) found a reduction of 2.2 million poor people in Africa (as defined by the

4 The regional weighted means allow for country size, whereas the simple means do not.

FIGURE 5.5 Comparing the benefit–cost ratio of agricultural R&D with other investments in Tanzania and Uganda

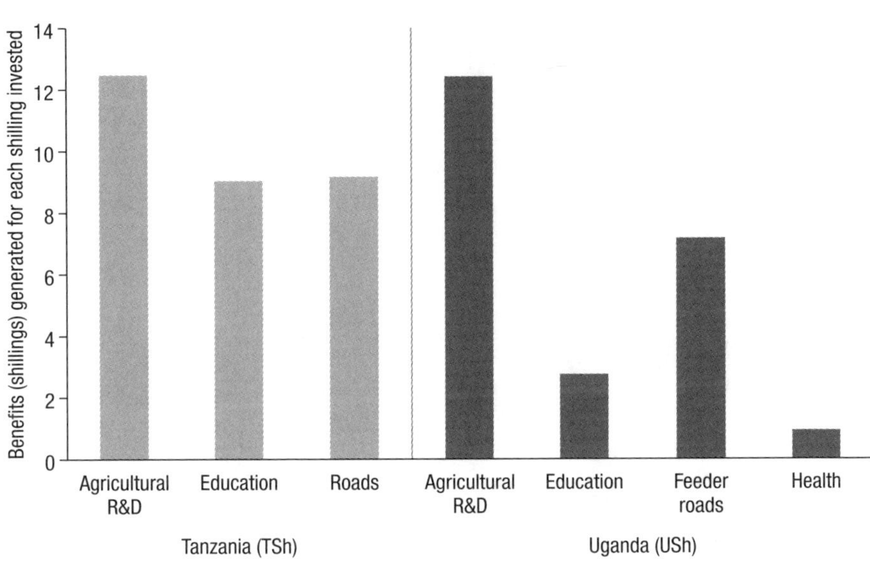

Sources: Data for Tanzania are adapted from Fan, Nyange, and Rao (2012); data for Uganda are adapted from Fan and Zhang (2008).

Notes: For Tanzania, returns are assessed as per capita income in Tanzanian shillings (TSh); the road variable is measured as the distance in kilometers from home to the nearest public transportation; education is measured as the household head's years of education. Ugandan and Tanzanian shillings are not traded one for one. For Uganda, returns are assessed as agricultural labor productivity in Ugandan shillings (USh) per monetary unit invested; education is measured as the share of the population at the community level over the age of 15 that can read and write; the road variable is measured at the community level as the average distance in kilometers from the community to the nearest road; health is measured as the number of days of work lost because of illness in the past 30 days.

dollar per day poverty line) for every 1 percent increase in agricultural value-added resulting from an increase in agricultural R&D expenditures. In the case of Uganda, the poverty alleviation impact of agricultural R&D outpaces those of other interventions, such as road infrastructure and health (Figure 5.6). For every USh 1 million invested in agricultural R&D, 58 poor people cross over the poverty threshold. In Tanzania, however, the poverty alleviation impact of education is slightly higher than that of agricultural R&D.

The Political Economy of Underinvestment in Agricultural R&D

The evidence presented so far in this and other chapters of this book reveals a stark contrast between the significant potential contribution of agricultural

FIGURE 5.6 Comparing the poverty alleviation impact of agricultural R&D with other interventions in Tanzania and Uganda

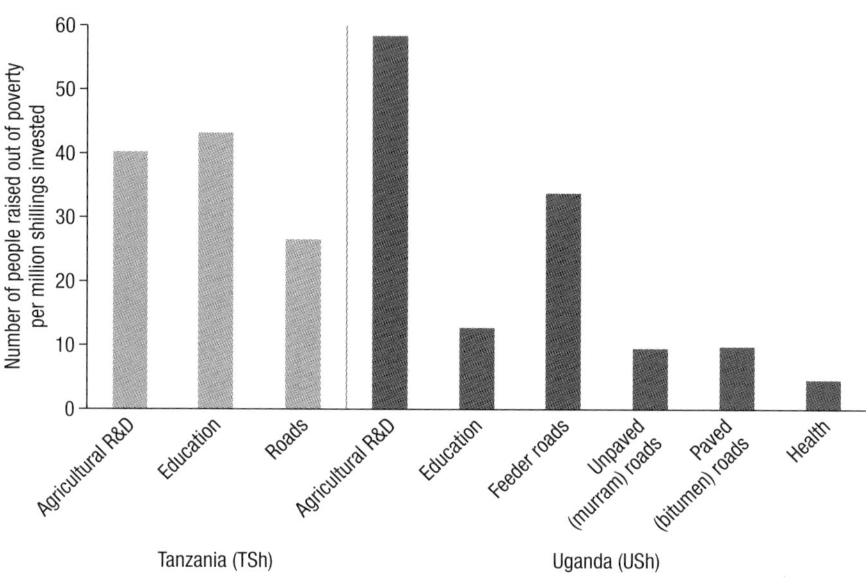

Sources: Data for Tanzania are adapted from Fan, Nyange, and Rao (2012); data for Uganda are adapted from Fan and Zhang (2008).

Notes: Returns are assessed as the number of poor people lifted out of poverty per million shillings of expenditure. For Uganda, poverty is determined at the district level; education is measured as the share of the community-level population over the age of 15 that can read and write; roads are measured in kilometers as the average community-level distance to the nearest road; health is measured as the number of days of work lost because of illness in the past 30 days. For Tanzania, poverty is determined at the national level; roads are measured as the distance in kilometers from home to the nearest public transportation; education is measured in years of education of the household head. Ugandan shillings (USh) and Tanzanian shillings (TSh) are not traded one for one.

R&D to economic and welfare improvements on one hand, and persistent underinvestment in agricultural R&D on the other. Therefore, it is important to understand how budget allocation and expenditure decisions are made, which requires consideration of four key factors, each of which is discussed further in the sections that follow: (1) the time lag between allocating research funding and realizing its outputs and subsequent returns, (2) the manifestation of the collective-action problem in the context of R&D advocacy, (3) the role of international development institutions, and (4) the effect of the rules of the budget process. This framework is motivated by and adapted from the broader theoretical and empirical literature on the drivers of public investments in terms of the political economy in and beyond agriculture (Mogues 2015).

Does the Time Lag in Agricultural Research Contribute to Underinvestment?

A factor that helps explain the presence of underinvestment in R&D is the information asymmetry in the attribution of such investments to the policymakers or politicians who were responsible for them. Citizens can never completely assess a politician's performance based on observed outcomes because they lack the information needed to attribute an outcome to a politician's effort. This information asymmetry dampens policymakers' political incentives to undertake investments for which attribution is a challenge, resulting in a lower prioritization of the requisite investments.

A key characteristic of R&D that affects attributability is the extent of the time lag between allocating resources to provide a service and actually providing it. The longer this gap in time, the more difficult it is to trace the service back to public decisionmakers, who may be national-level government officials or bureaucrats or decisionmakers in international aid agencies. The outcomes of investments in agricultural research typically do not materialize until a long time after the investments are made. For example, developing an improved crop variety takes 7 to 10 years of breeding activity, and this long gestation period must build on past agricultural research investments (Pardey, Alston, and Piggott 2006). Even estimates for agricultural R&D from many decades ago in the United States are only somewhat shorter, taking about six to seven years from the time expenditures are made until the production effect materializes (Evenson 1967).[5] However, another estimate for maize hybridization research in the United States suggests that it took 30 years from the onset of research activities until the product was perfected, and even longer until the returns to this investment were known (Oehmke 1986).

In contrast, public expenditures incurred to subsidize agricultural inputs usually require a time span of months from the time of the investment until the subsidized fertilizer reaches farmers. For example, in Malawi's 2008/09 fertilizer subsidy program, the time span from the expenditures being incurred to the fertilizers being received by farmers ranged from one to six months (Dorward, Chirwa, and Slater 2010); in Ghana's 2008 fertilizer subsidy program, the equivalent time span was about four months (Banful 2011).

Oehmke (1986) formally models how the long time span from investment to outputs and to returns in agricultural research leads to underinvestment,

5 The slightly shorter time lag of Evenson (1967) could be an artificial effect of the model imposed on the data; earlier rate-of-return studies tended to use shorter time lags and benefit streams than do later studies.

by explicitly considering lags in the response of the research funding agency in the face of increases in the demand for agricultural outputs or declines in the cost of doing research. The funding agency bases its decision regarding the quantity of resource allocation to research based on parameters in the recent past, such as past estimated demand for the agricultural commodities in question. In a scenario in which the optimal amount of agricultural research spending increases over time, say, because of the demand or cost factors described above, then the fact that the funder has a backward-looking approach in making allocation decisions—given that returns may lie in the far future and thus may not be easily discernable—means that there will be consistent underinvestment.

This temporal dimension of research investments and outcomes complicates resource allocation. Political decisionmakers may find that the returns to long-term agricultural investments occur too far in the future to be of political benefit to them, reducing their incentive to undertake such investments.[6] Furthermore, a long time span creates opportunities for things to go wrong, such as collapsing commodity prices, thus increasing the riskiness of such investments.

Collective-Action Challenges in Promoting Agricultural Research Spending

Challenges relating to the cost of coordinating societal groups for the purpose of advocating collectively for certain types of public investment or public policies have been well formulated. Becker 1983 and Olson 1985 present highlights of the canonical literature on this topic.) The size in the population of a given interest group; the extent of its spatial concentration and access to transport and communications infrastructure; and the endowment of group members with financial resources, human capital, and information about policies and their consequences can affect how capable the group will be in pressing public decisionmakers to make investments in the group's favor. In particular, small group size reduces the cost of coordination among group members. Having more economic resources enables the group to financially support policymakers who allocate public resources to the group's preferred activities. And well-informed and educated group members are able to more competently assess which policies they should support and which are detrimental to the group's economic interests.

---

6 This can be relevant in both democratic and autocratic systems, as removal from power can take the shape of legitimate elections, coups, and other forms of ouster.

On all of these accounts—group size, dispersion, and economic and human capital endowment—smallholder farmers in African countries are on the unfavorable end of the spectrum. They constitute the mass of the population, are widely dispersed, and are low-resourced economically and in terms of education. Therefore, they face enormous obstacles in engaging in collective action in the political arena. In these respects, they are disadvantaged not only vis-à-vis urban populations, but also vis-à-vis large, commercial farmers: both are smaller in number and have better material and human resources (Olson 1985; de Janvry, Sadoulet, and Fafchamps 1989; Roseboom 2002). And landless agricultural laborers are even more scattered and unorganized than are smallholder farmers, so the types of research investments made (such as labor-saving technologies) may not follow their interests (Lipton 1988). De Janvry, Sadoulet, and Fafchamps (1989) found that, in the interests of maximizing production, the state will allocate resources toward labor-saving (and not land-saving) technologies that benefit large, commercial farmers. Exploring whether the optimal allocation changes with collective action via simulations, de Janvry, Sadoulet, and Fafchamps (1989) found that, where the effectiveness of a collective is inversely proportional to its size (that is, many small farmers versus a few large ones), collective action will only further bias the state toward research investments that benefit large farmers.

These dynamics also play out specifically with regard to agricultural research spending in Africa. A case study of this phenomenon is the striking contrast in government investment in maize seed research between Malawi and Zimbabwe. Smale (1995) illustrates how the different makeup of potential constituencies for such research influenced their ability to advocate for resource allocation for the technological development of maize seeds.

During Zimbabwe's colonial period, areas of the country with high elevation that were relatively free from malaria attracted white settler farmers, who became a nontrivial minority in the population (about 7 percent) in the 1950s, and at their peak presence held about half of the agricultural land. This settler community, a large part of which produced maize commercially, was still small enough to engage in collective action effectively and press the colonial government to undertake significant efforts to develop improved maize varieties (Eicher 1995). In contrast, Malawi's geography was never as conducive for a community of white settler farmers as was Zimbabwe. The white population instead consisted mostly of government functionaries and missionaries. The primary constituency to potentially benefit from maize seed research was millions of dispersed African smallholders who were in no position to organize themselves to appeal for research investments. Accordingly, maize seed development in Malawi remained stagnant.

This is not solely a consequence of the well-known racial legacy of colonial governments paying attention to white rather than African farmers. Even after the end of colonialism in Malawi in 1964, the African government continued to show no interest in ramping up research on the predominant crop for the masses of Malawian farmers. Where the government did dedicate resources to new maize varieties, the collective-action problem made itself felt again. The large number of smallholders, as well as the urban consumers they sold to, preferred as their staple the flinty, white variety of maize, but dent maize was assumed to generate higher yields. Estate owners, who were few in number and thus in a better position to organize themselves to influence public resources, were primarily concerned with their main crop, tobacco, and only grew maize as a secondary crop. To the extent that they cared about government action on maize, the estate owners prioritized maximizing maize yields. Ultimately, government resources heavily prioritized developing improved varieties of dent maize, despite the preferences of the majority of the country's farming population.

The Role of International Development Assistance on R&D Spending

The magnitude of aid within (domestic and international) public spending on agricultural R&D in Africa is substantial. Stads (Chapter 4, this volume) estimates that aid was 30 percent in 2011, the most recent year of available data. Agricultural research activities at the supranational level—mainly CGIAR, but some also funded through the subregional organizations and implemented by national agricultural research systems—are predominantly financed by donor agencies.

Donor institutions exerted significant influence on R&D spending during the 1990s, a period that affected the quantity and composition of total public spending in Africa and in which structural adjustment reforms took root across the continent. A global push to privatize public activities contributed to the funding crisis of agricultural research, despite its evident high payoffs. While the community of scientists quickly increased after the end of colonialism in many African countries, and along with it the ratio of domestic-to-expatriate researchers, with the start of adjustment policies in Africa this trend reversed in some cases and dramatically slowed in others (Byerlee 1998). The share of public spending for agricultural R&D that came from donor or international agencies grew, in part because of declining domestic allocations.

One of the salient examples is Zambia (Elliott and Perrault 2006). During the quarter century of rule by a single president after independence and a heavy state role in economic affairs, including agriculture, Zambia had an extensive system of agricultural research stations. The new rule in 1991 was much more welcoming of structural adjustment measures embraced by international financial institutions, including economic liberalization and reduction in the role of government. In this process, even in legitimate areas the state's role was strongly affected, for example, in the provision of public goods like agricultural research. Declines in the real value of R&D funding led to a secondary exodus of agricultural scientists from the field and from the country, affecting long-term capacity.

Aside from the indirect (but not small) impact of international assistance on R&D through structural reforms that affected the composition and volume of domestic public spending, a direct impact manifested itself through donor funding of R&D. One example is the case of R&D on biochar technology: nongovernmental organizations (NGOs) have in some respects subverted the collective-action problem by making the case for the technical development of biochar among and for smallholder farmers (Fairhead, Leach, and Amonor 2012). Similarly, funding from large international donors and concerted action among local NGOs combined to assist Zimbabwean smallholders in accessing and developing agricultural conservation technologies (Andersson and Giller 2012).

At the national level, donor funding concentrates on operating costs and capital investments; it usually does not cover salaries. In some countries, this has become status quo, so that the national budget hardly provides any funding for operating costs and capital investments associated with R&D. This practice allows donor and other third-party funding to exert an exceptionally large influence over the R&D agenda. Because the implications of this for the quality and direction of research are not clear, this is a useful area for future research.

R&D in the Budget Process

Competing theories have been put forth to describe the workings of the budget process and its implications for budget portfolios. At one extreme, a body of work rejects the notion that interests systematically influence how public expenditures are apportioned. The "garbage can" budgeting model, which comes from the organizational sciences literature, proposes that budget decisions are essentially a random process resulting in unpredictable budget portfolios. At the other extreme, the budgetary model of incrementalism implies high predictability. It describes budget makers as backward looking, and changes in budget allocation as marginal and gradual. The strictest form of the model describes the allocations as increasing or decreasing by the same or very similar proportion each year.

Neither of these concepts leaves room for actors to have much effect by lobbying for changes in allocations or policies. Although the "garbage can" theory has been extended and refined, studies comparing its applicability with alternative models find little support (Davis, Dempster, and Wildavsky 1966; March and Olsen 1976; Weissinger-Baylon 1986; and Reddick 2002, 2003). However, there is some evidence that incrementalism drives the direction of public spending in Africa, typically applied to the recurrent budget (Fölscher 2007b). With the portfolio of the agriculture budget from domestic sources tilted far more toward salaries than that from donor sources (which have much higher shares of capital investment), it is to be expected that incremental budgeting is more pronounced in domestic spending. Incrementalism has been viewed in Africa as a compromise to avoid budget conflicts between agencies, and as the default outcome of input-based budgeting systems. Performance-based resource allocations would force the portfolio away from incrementalism (Fölscher 2007a, 2007c). Administrators of funds may also face greater penalties for errors resulting from large reallocations than for errors resulting from small reallocations (even if the errors from the smaller reallocations are more substantial), so bureaucratic incentive systems are a further potential promoter of incrementalism (Oehmke 1986).

Of course, whether a stable, formulaic approach to research spending is detrimental to allocating funds in line with future benefits depends on the alternative. However, the alternative may not always be optimal spending patterns and could, in fact, be worse for agricultural research than incrementalism.

In South Africa, for example, Liebenberg and Kirsten (2006) document a transition in the budget process affecting public scientific research, from an arrangement that was characterized by relatively high predictability, to a new policy on budget allocation that emphasized innovation. The new budget process for science research, although well meant, ended up hurting the quality and quantity of agricultural R&D. From 1988 to 1994, budget allocation rules for science and research institutes were characterized by a partly formula-based approach, in which resources allocated to a research institute comprised a base amount for core infrastructure and other needs, an amount that depended on the number of research staff employed (in full-time equivalents), and other formulaic components. A 1994 policy did away with most of these features so as to spur research innovation and interagency competition for funds. Core funding was now granted by Parliament on a competitive basis, as were several other sources of funds. This resulted in increased allocations for industrial and medical research, and reduced funds for agricultural sciences. Even where formulas were in place—such as in the allocation of funds to the newly created provincial departments of agriculture—they were voluntary, not mandatory. The departments faltered, as did subnational agricultural research activities (with some provinces ceasing activities altogether). An expected corollary of these policy trends was increased volatility in funding flows to agricultural research.

The evidence on the core features of the budget process in agricultural research in Africa appears more contradictory than it perhaps really is. On face value, it would appear that evidence of high funding volatility would contradict evidence of predictable (even if predictably low) flows implied by the incrementalist model of the budget process in Africa described above (Chapter 4, this volume). However, the findings in the above-mentioned studies point to the fact that this model mostly applies to the recurrent budget (that is, personnel and operating costs). Incrementalism in agricultural R&D allocations on operating costs is, for example, demonstrated in the context of Nigeria in Ayoola and Abdullahi (2011). The most volatile component of agricultural R&D budgets is capital investments (Chapter 4, this volume).

The passage of the final budget—however it is arrived at—is not the end of the budget process; there is still the execution and implementation of the budget, which often do not proceed as planned. Discrepancies between the approved budget and the executed budget in agricultural R&D expenditures can be large, resulting often (but not always) in a downward adjustment of funds for research and other activities. The reasons include leakages; lack of capacity to execute and inability to carry funds over to the following year; poor projections of revenues and, hence, inadequate funds accrued (see Ayoola and Abdullahi 2011 in the case of Nigeria); and changing government or donor priorities (as demonstrated by Lwezaura 2011 for Tanzania) in the middle of the fiscal year.

Implications for Policy Dialogue and Future Research

Analysis of agricultural research expenditures relative to total agricultural expenditures is not fully encouraging. Overall, the data show that the share of total PAEs on agricultural research is very low, which reflects the already low shares of total expenditures allocated to the agricultural sector as a whole. These low shares can be expected to remain so in the foreseeable future, because the shares indicated in the CAADP NAIPs are also low. Very few African countries spend more than 10 percent of their total PAEs on agricultural research. The countries with higher shares (such as Kenya, Tanzania, and Uganda) also tend to have larger agriculture budgets and more developed economies, which, as previously discussed, likely gives policymakers more latitude to channel investments into research. In contrast, countries that spend or allocate low shares to agricultural research face more budget constraints and, therefore, focus spending on more immediate priorities.

These low agricultural R&D investment shares in most African countries are striking in light of the potential high payoffs to agricultural R&D investments. Furthermore, as established in this chapter, returns to public agricultural research expenditures are large not only in absolute terms in Africa, but also in relative terms when compared with returns to other types of agricultural and nonagricultural public spending.

When considering investments in agricultural technology and scientific knowledge, it is important to identify strategies and mechanisms that elevate the political incentives for undertaking such investments. Building on the insights in this chapter, improvements in information symmetry on the relative costs and benefits of different types of agricultural investments (as well as different types of nonagricultural investments) between citizens and politicians will be important. More in-depth research and diagnostic analysis can inform development interventions and policy dialogue. This should include intensive analysis of empirical cases where these incentives and disincentives are very much in place and, perhaps more important, examination of country cases that have managed to sustain a relatively high commitment to agricultural technology development over a prolonged period of time, despite the characteristics of R&D being potentially less conducive to political gain.

Another important factor is the visibility of investments. The effect that CAADP has had on agricultural investments more generally can serve as a useful illustration. A recent study held that the CAADP process has likely had an impact on what was included in African countries' national accounts as "agriculture," which may in fact stem from two drivers (Benin 2012). The first driver relates to the phenomenon concerning incentives to undertake expenditures that allow for greater attributability to policymakers through greater visibility. In other words, the greater visibility afforded agricultural expenditures by the CAADP process, including through the 10 percent Maputo Declaration target against which progress of countries is assessed, may have increased public decisionmakers' motivations to direct more resources to the sector. A second driver of the reported higher expenditures may be a redefinition by countries of agricultural expenditures in the CAADP context. That is, the high-profile CAADP process may have led to increases in reported expenditures through adjustments of what gets counted under this rubric.

Clearly, both explanations for changes in national accounting of agriculture can coexist. An important lesson from the first hypothesis for the particular function of agricultural R&D is that high-visibility initiatives carried out on a continentwide scale—such as a CAADP-like process that is focused on research—could in fact boost agricultural R&D expenditures by directly or indirectly addressing the incentive factor discussed in this chapter.

Comprehension of the budget process also provides important pointers on how public investment decisions on agricultural R&D are made, and useful insights can be gleaned from them for development activity and policy dialogue. Elements of incrementalist behavior in public administration (the practice in many countries) may militate against dramatic portfolio increases, even if evidence on the impacts of agricultural research investments may recommend such shifts. This raises the question of identifying the types of phenomena that may help break through the inertia of incrementalism. Turning again to CAADP, the suggestive evidence in Benin (2012) that funds for agriculture in Africa may have increased as a consequence of CAADP provokes the hypothesis that an initiative like CAADP could provide the necessary jolt—and political license—to move away from incrementalist approaches to agricultural spending, in general, and to agricultural R&D investments in particular. Future research may test this hypothesis in more depth. At the same time, the case of South Africa is an example of a cautionary tale: when the budget process changed to encourage competition and innovation and moved away from stable incrementalism, agricultural R&D lost. Research agencies and leaders in support of agricultural research need to have the wherewithal to make the case for investment in agricultural technology, as analytical evidence of high returns to agricultural research may not necessarily "speak for itself" in the political arena.

Appendix 5A. Methodology and Data Sources

The data on PAE are drawn from five main sources: Statistics on Public Expenditure for Economic Development (SPEED; Yu 2012); the African Union's Agricultural Expenditure Tracking Survey (AETS; AUC 2008); the Monitoring African Food and Agricultural Policies (MAFAP) database (FAO 2013; CAADP NAIP reports); and various national sources, compiled by the Regional Strategic Analysis and Knowledge Support System (ReSAKSS 2013).

The data from SPEED, AETS, and ReSAKSS are used to analyze Africa-wide and subregional trends in PAE, as related to the Maputo Declaration (AUC 2008). To conduct the analysis, total expenditures from 1980 onward were obtained from the SPEED database. Thereafter, data were compiled on the share of PAE in total expenditures based on available data from all the sources cited, with reliance on the more recent source in any case of conflicting data.

The dollar amounts of PAE were then determined by multiplying the shares by total expenditures. Missing values were estimated using extrapolations based on yearly average growth rates in total expenditures and PAE. To adjust for inflation and to allow for comparison across countries, total expenditures and PAEs were converted to constant 2005 purchasing power parity (2005 international PPP dollars), using PPP conversion factors from the World Development Indicators (World Bank 2013). The aggregate value of an indicator was estimated using the weighted sum approach, where the weight for each country is the share of that country's value in the total value for all countries in the subregion/group (see Appendix 5B for the resulting supplementary data tables).

The data from MAFAP cover five countries (Burkina Faso, Kenya, Mali, Tanzania, and Uganda) and are used to analyze public agricultural R&D expenditures compared with other types of PAE and total PAE (details in appendix Table 5B.4). The data from the NAIP reports consist of budget information drawn from 19 country reports; they are used to analyze how the countries are planning to allocate PAE to agricultural R&D compared with other types of PAE.[7]

---

7 The NAIPs reviewed include Ethiopia, Kenya, Rwanda, Tanzania, and Uganda in East Africa; Benin, Burkina Faso, Côte d'Ivoire, The Gambia, Ghana, Liberia, Mali, Niger, Nigeria, Senegal, Sierra Leone, and Togo in West Africa; Burundi in Central Africa; and Malawi in Southern Africa. See Appendix Table 5B.5 for details of the plans, their duration, and their total budgets.

Appendix 5B. Supplementary Tables

TABLE 5B.1 Total government expenditures, 2003–2010

Country	2003	2004	2005	2006	2007	2008	2009	2010
	2005 PPP dollars (billions)							
Angola	25.963	13.441	19.883	28.495	30.875	35.095	51.814	57.865
Benin	2.058	2.007	2.232	2.090	2.507	2.520	3.119	2.645
Botswana	8.174	7.850	7.283	6.839	7.792	9.405	11.181	11.418
Burkina Faso	2.466	2.863	3.190	3.762	4.098	3.514	4.208	4.836
Burundi	0.689	0.790	0.895	1.055	1.161	1.142	1.282	1.501
Cameroon	5.720	5.500	5.880	5.864	6.052	5.955	6.416	6.488
Cape Verde	0.413	0.470	0.538	0.516	0.587	0.645	0.744	0.800
Central Afr. Rep.	0.335	0.327	0.547	0.625	0.626	0.448	0.552	0.590
Chad	0.490	0.554	0.614	0.274	0.676	0.647	0.766	0.732
Comoros	na	na	na	na	na	na	na	na
Congo, Dem. Rep.	2.753	3.191	5.477	5.180	5.445	6.166	6.646	6.296
Congo, Rep.	2.732	3.280	2.623	2.581	2.977	2.777	4.080	3.942
Côte d'Ivoire	5.846	5.998	5.885	5.701	6.223	6.566	6.818	7.249
Djibouti	0.529	0.529	0.556	0.577	0.644	0.639	0.683	0.710
Equatorial Guinea	8.460	6.957	4.682	3.922	3.795	4.376	10.518	12.731
Eritrea	1.722	1.430	1.537	1.090	1.073	1.021	0.771	0.892
Ethiopia	10.384	9.930	11.495	12.142	12.482	12.432	12.346	14.967
Gabon	na	na	na	na	na	na	na	na
The Gambia	0.208	0.190	0.195	0.204	0.208	0.217	0.219	0.224
Ghana	6.643	8.034	8.251	6.037	7.238	7.884	8.092	9.532
Guinea	na	na	na	na	na	na	na	na
Guinea-Bissau	0.127	0.147	0.207	0.194	0.197	0.199	0.219	0.240
Kenya	8.711	10.289	9.488	11.176	12.308	13.879	14.691	16.248
Lesotho	1.190	1.169	1.246	1.406	1.547	1.735	1.951	2.210
Liberia	0.003	0.003	0.003	0.002	0.004	0.006	0.005	0.006
Madagascar	2.292	2.972	2.165	2.978	6.825	8.864	11.563	14.931
Malawi	1.933	1.923	2.424	1.979	2.087	3.223	3.004	3.446
Mali	0.024	0.027	0.029	0.032	0.033	0.028	0.036	0.033
Mauritius	2.861	3.005	2.945	3.091	2.916	3.016	3.607	4.064
Mozambique	2.980	3.033	3.665	3.872	4.543	4.626	5.356	6.340
Namibia	3.084	3.076	3.069	3.169	3.373	3.230	3.374	3.634
Niger	1.317	1.410	1.582	1.656	1.882	2.246	2.384	2.418
Nigeria	29.429	28.359	32.052	25.651	32.767	34.384	38.929	37.885
Rwanda	1.242	1.406	1.653	1.938	2.333	2.627	2.996	3.433
São Tomé & Príncipe	0.092	0.082	0.076	0.083	0.105	0.097	0.101	0.106
Senegal	3.502	4.051	4.283	4.975	5.297	5.335	5.520	5.875
Seychelles	0.506	0.642	0.624	0.742	0.746	0.558	0.580	0.663

(continued)

Table 5B.1 (continued)

Country	2003	2004	2005	2006	2007	2008	2009	2010
	2005 PPP dollars (billions)							
Sierra Leone	0.001	0.001	0.001	0.001	0.001	0.001	0.001	0.001
Somalia	na	na	na	na	na	na	na	na
South Africa	95.221	100.791	107.614	113.993	119.757	131.724	143.768	143.625
South Sudan	na	na	na	na	na	na	na	na
Sudan	na	na	na	na	na	na	na	na
Swaziland	1.441	1.486	2.366	2.730	3.856	4.774	6.761	8.929
Tanzania	6.359	5.903	7.866	11.020	13.369	14.436	17.727	21.615
Togo	0.710	0.743	0.914	1.024	0.946	0.903	1.142	1.177
Uganda	5.753	4.780	5.189	5.496	5.810	5.961	5.963	7.466
Zambia	2.672	2.823	3.919	2.676	3.899	3.465	3.487	3.810
Zimbabwe	na	na	na	na	na	na	na	na
East Africa	38.638	39.457	41.983	49.159	57.433	62.413	70.156	84.097
West Africa	52.336	53.834	58.824	51.330	61.400	63.803	70.692	72.120
Central Africa	21.272	20.681	20.793	19.583	20.837	21.609	30.360	32.386
Southern Africa	142.657	135.592	151.470	165.159	177.728	197.277	230.697	241.276
All Africa (incl. North Africa)	446.293	446.471	475.751	515.806	561.660	632.378	718.662	743.276

Sources: Calculated by authors based on Yu (2012), AUC (2008), and ReSAKSS (2013).

Notes: PPP = purchasing power parity; na = data were not available.

TABLE 5B.2 Public agricultural expenditures, 2003–2010

Country	2003	2004	2005	2006	2007	2008	2009	2010
	2005 PPP dollars (billions)							
Angola	0.167	0.301	1.286	1.507	1.096	0.797	1.456	2.013
Benin	0.114	0.107	0.143	0.158	0.158	0.184	0.126	0.079
Botswana	0.370	0.288	0.432	0.282	0.272	0.401	0.336	0.325
Burkina Faso	0.807	0.586	0.386	0.766	0.648	0.483	0.367	0.524
Burundi	0.010	0.024	0.031	0.068	0.050	0.066	0.099	0.154
Cameroon	0.205	0.160	0.128	0.139	0.123	0.104	0.096	0.084
Cape Verde	0.167	0.301	1.286	1.507	1.096	0.797	1.456	2.013
Central Afr. Rep.	0.014	0.014	0.016	0.016	0.017	0.006	0.012	0.014
Chad	0.028	0.026	0.024	0.021	0.037	0.037	0.045	0.045
Comoros	na	na	na	na	na	na	na	na
Congo, Dem. Rep.	0.051	0.033	0.050	0.062	0.065	0.071	0.068	0.071
Congo, Rep.	0.032	0.035	0.025	0.035	0.162	0.205	0.411	0.541
Côte d'Ivoire	0.211	0.171	0.135	0.144	0.112	0.141	0.210	0.182
Djibouti	0.004	0.011	0.011	0.016	0.010	0.012	0.016	0.020
Equatorial Guinea	0.113	0.099	0.071	0.064	0.066	0.035	0.084	0.069
Eritrea	na	na	na	na	na	na	na	na

(continued)

Table 5B.2 (continued)

Country	2003	2004	2005	2006	2007	2008	2009	2010
				2005 PPP dollars (billions)				
Ethiopia	0.517	0.493	1.831	2.466	2.251	2.352	2.159	3.167
Gabon	na	na	na	na	na	na	na	na
The Gambia	0.014	0.013	0.013	0.012	0.015	0.016	0.017	0.017
Ghana	0.379	0.710	0.792	0.622	0.719	0.805	0.730	0.866
Guinea	na	na	na	na	na	na	na	na
Guinea-Bissau	0.002	0.003	0.002	0.003	0.002	0.002	0.002	0.002
Kenya	0.371	0.426	0.414	0.502	0.600	0.441	0.574	0.750
Lesotho	0.043	0.059	0.052	0.044	0.051	0.056	0.059	0.063
Liberia	0.000	0.000	0.000	0.000	0.000	0.000	0.000	0.000
Madagascar	0.199	0.215	0.303	0.348	0.528	0.703	0.940	1.244
Malawi	0.139	0.131	0.305	0.338	0.299	0.724	0.698	0.994
Mali	0.003	0.004	0.004	0.004	0.004	0.004	0.004	0.004
Mauritius	0.096	0.119	0.086	0.079	0.092	0.106	0.143	0.153
Mozambique	0.160	0.197	0.247	0.219	0.235	0.250	0.313	0.351
Namibia	0.127	0.129	0.140	0.114	0.118	0.108	0.107	0.110
Niger	0.148	0.200	0.189	0.207	0.328	0.425	0.332	0.306
Nigeria	1.011	1.608	1.955	1.772	1.712	1.562	2.079	2.176
Rwanda	0.038	0.051	0.071	0.099	0.129	0.148	0.193	0.226
São Tomé & Príncipe	0.005	0.003	0.003	0.004	0.006	0.006	0.007	0.007
Senegal	0.328	0.440	0.514	0.533	0.615	0.742	0.767	0.817
Seychelles	0.009	0.008	0.009	0.014	0.018	0.004	0.006	0.009
Sierra Leone	0.000	0.000	0.000	0.000	0.000	0.000	0.000	0.000
Somalia	na	na	na	na	na	na	na	na
South Africa	1.862	1.949	2.214	2.655	2.873	2.888	2.644	2.609
South Sudan	na	na	na	na	na	na	na	na
Sudan	na	na	na	na	na	na	na	na
Swaziland	0.073	0.080	0.120	0.160	0.318	0.127	0.195	0.473
Tanzania	0.432	0.336	0.371	0.637	0.773	0.989	1.188	1.477
Togo	0.027	0.030	0.039	0.038	0.032	0.086	0.055	0.107
Uganda	0.283	0.146	0.245	0.261	0.290	0.188	0.229	0.290
Zambia	0.164	0.173	0.280	0.250	0.514	0.434	0.323	0.388
Zimbabwe	na	na	na	na	na	na	na	na
East Africa	1.949	1.807	3.341	4.421	4.691	4.944	5.448	7.337
West Africa	3.045	3.872	4.174	4.258	4.346	4.450	4.688	5.080
Central Africa	0.459	0.394	0.348	0.409	0.526	0.529	0.822	0.986
Southern Africa	3.105	3.307	5.075	5.570	5.777	5.785	6.133	7.326
All Africa (incl. North Africa)	17.295	17.819	21.154	22.851	22.262	25.445	25.646	29.112

Sources: Calculated by authors based on Yu (2012), AUC (2008), and ReSAKSS (2013).

Notes: PPP = purchasing power parity; na = data were not available.

TABLE 5B.3 Public agricultural expenditures as a share of total public expenditures, 2003–2010

Country	2003	2004	2005	2006	2007	2008	2009	2010
				Share (%)				
Angola	0.6	2.2	6.5	5.3	3.6	2.3	2.8	3.5
Benin	5.5	5.3	6.4	7.5	6.3	7.3	4.0	3.0
Botswana	4.5	3.7	5.9	4.1	3.5	4.3	3.0	2.8
Burkina Faso	32.7	20.5	12.1	20.4	15.8	13.8	8.7	10.8
Burundi	1.5	3.1	3.5	6.5	4.3	5.8	7.7	10.3
Cameroon	3.6	2.9	2.2	2.4	2.0	1.7	1.5	1.3
Cape Verde	na	na	na	na	na	2.6	2.8	3.3
Central Afr. Rep.	4.3	4.3	2.8	2.6	2.6	1.3	2.2	2.3
Chad	5.7	4.7	3.9	7.8	5.5	5.7	5.9	6.2
Comoros	na	na	na	na	na	na	na	na
Congo, Dem. Rep.	1.9	1.0	0.9	1.2	1.2	1.1	1.0	1.1
Congo, Rep.	1.2	1.1	0.9	1.3	5.4	7.4	10.1	13.7
Côte d'Ivoire	3.6	2.9	2.3	2.5	1.8	2.2	3.1	2.5
Djibouti	0.7	2.2	2.0	2.8	1.6	1.9	2.3	2.8
Equatorial Guinea	1.3	1.4	1.5	1.6	1.7	0.8	0.8	0.5
Eritrea	na	na	na	na	na	na	na	na
Ethiopia	5.0	5.0	15.9	20.3	18.0	18.9	17.5	21.2
Gabon	na	na	na	na	na	na	na	na
The Gambia	6.9	6.7	6.9	5.7	7.3	7.4	7.6	7.8
Ghana	5.7	8.8	9.6	10.3	9.9	10.2	9.0	9.1
Guinea	na	21.4	10.5	12.7	9.3	14.5	na	na
Guinea-Bissau	1.9	1.8	1.2	1.5	1.2	1.1	1.0	0.9
Kenya	4.3	4.1	4.4	4.5	4.9	3.2	3.9	4.6
Lesotho	3.6	5.1	4.1	3.1	3.3	3.2	3.0	2.9
Liberia	1.7	1.5	1.3	4.0	5.5	8.6	2.3	2.9
Madagascar	8.7	7.2	14.0	11.7	7.7	7.9	8.1	8.3
Malawi	7.2	6.8	12.6	17.1	14.4	22.4	23.2	28.9
Mali	14.0	15.1	15.5	12.1	13.4	12.7	10.2	11.1
Mauritius	3.4	4.0	2.9	2.6	3.2	3.5	4.0	3.8
Mozambique	5.4	6.5	6.7	5.7	5.2	5.4	5.8	5.5
Namibia	4.1	4.2	4.5	3.6	3.5	3.3	3.2	3.0
Niger	11.2	14.2	11.9	12.5	17.4	18.9	13.9	12.7
Nigeria	3.4	5.7	6.1	6.9	5.2	4.5	5.3	5.7
Rwanda	2.9	3.6	4.5	5.1	5.5	5.6	6.4	6.6
São Tomé & Príncipe	5.4	3.1	4.0	4.4	5.9	6.2	6.5	6.9
Senegal	9.4	10.9	12.0	10.7	11.6	13.9	13.9	13.9
Seychelles	1.8	1.2	1.5	1.8	2.5	0.7	1.0	1.4
Sierra Leone	4.1	2.4	2.1	2.1	2.5	2.2	2.0	1.7
Somalia	na	na	na	na	na	na	na	na

(continued)

Table 5B.3 (continued)

Country	2003	2004	2005	2006	2007	2008	2009	2010
				Share (%)				
South Africa	2.0	1.9	2.1	2.3	2.4	2.2	1.8	1.8
South Sudan	na	na	na	na	na	1.4	1.9	1.4
Sudan	3.1	5.4	5.9	6.5	7.0	na	na	na
Swaziland	5.0	5.4	5.1	5.9	8.2	2.7	2.9	5.3
Tanzania	6.8	5.7	4.7	5.8	5.8	6.9	6.7	6.8
Togo	3.9	4.1	4.2	3.7	3.4	9.6	4.8	9.1
Uganda	4.9	3.1	4.7	4.7	5.0	3.2	3.8	3.9
Zambia	6.1	6.1	7.2	9.3	13.2	12.5	9.3	10.2
Zimbabwe	10.4	11.7	4.0	17.3	18.8	22.0	25.8	30.2
East Africa	5.0	4.6	8.0	9.0	8.2	7.9	7.8	8.7
West Africa	5.8	7.2	7.1	8.3	7.1	7.0	6.6	7.0
Central Africa	2.2	1.9	1.7	2.1	2.5	2.5	2.7	3.0
Southern Africa	2.2	2.4	3.4	3.4	3.3	2.9	2.7	3.0
All Africa (incl. North Africa)	3.9	4.0	4.4	4.4	4.0	4.0	3.6	3.9

Sources: Calculated by authors based on Yu (2012), AUC (2008), and ReSAKSS (2013).

Note: na = data were not available.

TABLE 5B.4 Public agricultural spending by function in selected countries, yearly average 2006–2010

Function	Burkina Faso	Kenya	Mali	Uganda	Tanzania
			Share (%)		
Subsidies	53.5	29.6	36.5	35.4	40.5
Research	10.0	16.9	5.3	15.1	16.3
Extension, training, and technical assistance	11.9	28.7	13.7	35.9	30.9
Irrigation	18.2	7.0	10.1	6.4	0.0
Feeder roads and other infrastructure	1.4	3.7	13.5	4.0	0.0
Marketing, storage, and public stock-holding	1.9	9.2	14.2	1.9	4.9
Inspection	1.4	3.0	4.1	1.4	0.5
Other	1.8	1.9	2.7	0.0	6.8
Total 2005 PPP dollars (millions)	75.4	727.3	67.8	280.1	407.9

Source: Calculated by authors based on FAO (2013).

Note: PPP = purchasing power parity.

TABLE 5B.5 National agricultural investment plans, reviewed budget allocations, selected countries

Country/name of plan/timeframe	Unit	Total budget	Share allocated to agricultural R&D (%)
Benin: Agricultural Investment Plan, 2010–2015	Billion FCFA	491.25	17.1
Burkina Faso: Global Agriculture and Food Security Program, 2011–2015	Billion FCFA	26.78	na
Burundi: National Agricultural Investment Plan, 2012–2017	Billion FBU	1,452.30	4.7
Côte d'Ivoire: National Agriculture Investment Plan, 2010–2015	Billion FCFA	660.18	20.3
Ethiopia: Agricultural Sector Policy and Investment Framework, 2010–2020	Billion USD	15.50	na
Gambia National Agricultural Investment Plan, 2011–2015	Billion USD	296.58	2.1
Ghana: Medium Term Agriculture Sector Investment Plan, 2011–2015	Million GHS	1,532.40	3.3
Kenya: Agricultural Development Sector Strategy Medium-Term Investment Plan, 2010–2015	Billion KSh	247.01	na
Liberia: Agriculture Sector Investment Program, 2011–2015	Million USD	772.30	5.2
Malawi: Agriculture Sector Wide Approach, 2011–2014	Million USD	1,752.00	0.4
Mali: National Priority Investment Plan in Agriculture, 2011–2015	Billion FCFA	358.85	3.1
Niger: National Agricultural Investment Plan, 2010–2012	Billion FCFA	547.31	1.2
Nigeria: National Agriculture Investment Plan, 2011–2014	Billion Naira	235.09	1.5
Rwanda: Agriculture Sector Investment Plan, 2009–2012	Million USD	848.12	1.7
Senegal: National Agricultural Investment Plan, 2011–2015	Billion francs	1,346.01	na
Sierra Leone: Smallholder Commercialization Program Investment Plan, 2010–2014	Million USD	402.60	na
Tanzania: Agriculture and Food Security Investment Plan, 2011/12–2015/16	Billion TSh	8,752.33	na
Togo: National Agriculture and Food Security Investment Plan, 2010–2015	Billion FCFA	569.14	3.3
Uganda: Agriculture Sector Development Strategy and Investment Plan, 2010/11–2014/15	Billion USh	2,731.30	12.6

Source: Calculated by authors based on national agricultural investment plans, available at www.resakss.org.

Notes: FBU = franc of Burundi; FCFA = franc of the Communauté Financière Africaine; GHS = Ghanaian cedi; KSh = Kenyan shilling; na = data were not available (because of a lack of sufficient detail to identify research and development [R&D] spending); TSh = Tanzanian shilling; USh = Ugandan shilling; USD = US dollar.

References

Alston, J., M. Marra, P. Pardey, and T. Wyatt. 2000. "Research Returns Redux: A Meta-analysis of the Returns to Agricultural R&D." *Australian Journal of Agricultural and Resource Economics* 44 (2): 185–215.

Alston, J., and P. Pardey. 2006. "Developing-Country Perspectives on Agricultural R&D: New Pressures for Self-Reliance." In *Agricultural R&D in the Developing World: Too Little Too Late?* Edited by P. Pardey, J. Alston, and R. Piggott. Washington, DC: International Food Policy Research Institute.

Andersson, J., and K. Giller. 2012. "On Heretics and God's Blanket Salesmen: Contested Claims for Conservation Agriculture and the Politics of Its Promotion in African Smallholder Farming." In *Contested Agronomy: Agricultural Research in a Changing World*, edited by J. Sumberg and J. Thompson. New York: Routledge.

AU (African Union). 2007. *Assembly/AU/Decl.5 (VIII): Addis Ababa Declaration on Science Technology and Scientific Research for Development.* Assembly of the African Union Eighth Ordinary Session, 29–30 January 2007, Addis Ababa, Ethiopia. Accessed April 2016. unesdoc.unesco.org/images/0015/001502/150216e.pdf.

AUC (African Union Commission). 2008. *National Compliance with 2003 African Union–Maputo Declaration to Allocate at Least 10% of National Budget to Agriculture Development.* Addis Ababa, Ethiopia: Department of Rural Economy and Agriculture.

Ayoola, G., and A. Abdullahi. 2011. *Nationally Financed Agricultural Research: A Case Study on Nigeria.* Background paper prepared for the ASTI/IFPRI–FARA Conference "Agricultural R&D: Investing in Africa's Future: Analyzing Trends, Challenges, and Opportunities," Accra, Ghana. December 5–7. Accessed June 2013. www.asti.cgiar.org/pdf/conference/Theme1/CaseStudies/Ayoola.pdf.

Banful, A. 2011. "Old Problems in the New Solutions? Politically Motivated Allocation of Program Benefits and the 'New' Fertilizer Subsidies." *World Development* 39 (7): 1166–1176.

Becker, G. 1983. "A Theory of Competition among Pressure Groups for Political Influence." *Quarterly Journal of Economics* 98 (3): 371–400.

Benin, S. 2012. *Are African Governments Serious about Agriculture? Recent Trends and Patterns of Investments.* Ghana Strategy Support Programme Conference Brief. Washington, DC: International Food Policy Research Institute.

Benin, S., and B. Yu. 2013. *Complying with the Maputo Declaration Target: Trends in Public Agricultural Expenditures and Implications for Pursuit of Optimal Allocation of Public Agricultural Spending.* ReSAKSS Annual Trends and Outlook Report 2012. Washington, DC: International Food Policy Research Institute.

Byerlee, D. 1998. "The Search for a New Paradigm for Development of National Agricultural Research Systems." *World Development* 26 (6): 1049–1055.

Davis, O., M. Dempster, and A. Wildavsky. 1966. "A Theory of the Budgetary Process." *American Political Science Review* 60 (3): 529– 547.

de Janvry, A., E. Sadoulet, and M. Fafchamps. 1989. "Agrarian Structure, Technological Innovation and the State." In *The Economic Theory of Agrarian Institutions*, edited by P. Bardhan. Oxford: Oxford University Press.

Dorward, A., E. Chirwa, and R. Slater. 2010. *Evaluation of the 2008/9 Agricultural Input Subsidy Programme, Malawi. Report on Programme Implementation.* Accessed April 23, 2014. Updated paper available at http://eprints.soas.ac.uk/9598/.

Eicher, C. 1995. "Zimbabwe's Maize-based Green Revolution: Preconditions for Replication." *World Development* 23 (5): 805–818.

Elliott, H., and P. Perrault. 2006. "Zambia: A Quiet Crisis in African Research and Development." In *Agricultural R&D in the Developing World: Too Little Too Late?* Edited by P. Pardey, J. Alston, and R. Piggott. Washington, DC: International Food Policy Research Institute.

Evenson, R. 1967. "The Contribution of Agricultural Research to Production." *Journal of Farm Economics*. 49 (5): 1415–1425.

———. 2001. "Economic Impacts of Agricultural Research and Extension." In *Handbook of Agricultural Economics*, edited by B. Gardner and G. Rausser. Amsterdam: Elsevier Science.

Fairhead, J., M. Leach, and K. Amanor. 2012. "Anthropogenic Dark Earths and Africa: A Political Agronomy of Research Disjunctures." In *Contested Agronomy: Agricultural Research in a Changing World*, edited by J. Sumberg and J. Thompson. New York: Routledge.

Fan, S. 2008. *Public Expenditures, Growth, and Poverty: Lessons from Developing Countries.* Baltimore: Johns Hopkins University Press.

Fan, S., D. Nyange, and N. Rao. 2012. "Public Investment and Poverty Reduction in Tanzania: Evidence from Household Survey Data." In *Public Expenditures for Agricultural and Rural Development in Africa*, edited by T. Mogues and S. Benin. London: Routledge/Taylor and Francis.

Fan, S., and X. Zhang. 2008. "Public Expenditure, Growth and Poverty Reduction in Rural Uganda." *African Development Review* 20 (3): 466–496.

FAO (Food and Agriculture Organization of the United Nations). 2013. "Monitoring and Analysing Food and Agricultural Policies (MAFAP)." Accessed August 30, 2013. www.fao.org/mafap/database/public-expenditure/en/.

Fölscher, A. 2007a. "Budget Methods and Practices." Chapter 4 in *Budgeting and Budgetary Institutions*, edited by A. Shah. Public Sector Governance and Accountability Series. Washington, DC: World Bank.

———. 2007b. "Country Case Study: Kenya." Chapter 14 in *Budgeting and Budgetary Institutions*, edited by A. Shah. Public Sector Governance and Accountability Series. Washington, DC: World Bank.

———. 2007c. "Country Case Study: South Africa." Chapter 15 in *Budgeting and Budgetary Institutions*, edited by A. Shah. Public Sector Governance and Accountability Series. Washington, DC: World Bank.

Liebenberg, K., and J. Kirsten. 2006. "South Africa: Coping with Structural Changes." In *Agricultural R&D in the Developing World: Too Little Too Late?* Edited by P. Pardey, J. Alston, and R. Piggott. Washington, DC: International Food Policy Research Institute.

Lipton, M. 1988. "The Place of Agricultural Research in the Development of Sub-Saharan Africa." *World Development* 16 (10): 1231–1257.

Lwezaura, D. 2011. *Government Funding for Agricultural R&D: A Case Study on the Tanzanian Division of Research and Development.* Background paper prepared for the ASTI/IFPRI–FARA Conference "Agricultural R&D: Investing in Africa's Future: Analyzing Trends, Challenges, and Opportunities," Accra, Ghana. December 5–7. Accessed June 2013. www.asti.cgiar.org/pdf/conference/Theme1/CaseStudies/Lwezaura.pdf.

March, J., and J. Olsen. 1976. *Ambiguity and Choice in Organizations*. Bergen, Norway: Universitetsforlaget.

Mogues, T. 2015. "Political Economy Determinants of Public Spending Allocations: A Review of Theories, and Implications for Agricultural Public Investment." *European Journal of Development Research* 27 (3): 452–473.

Mogues, T., and S. Benin, eds. 2012. *Public Expenditures for Agricultural and Rural Development in Africa*. London: Routledge/Taylor and Francis.

Mogues, T., S. Fan, and S. Benin. 2015. "Public Investments in and for Agriculture." *European Journal of Development Research* 27 (3): 337–352.

NEPAD (New Partnership for Africa's Development). 2005. *Guidance Note for Agriculture Expenditure Tracking System in African Countries*. Midrand, South Africa.

Oehmke, J. 1986. "Persistent Underinvestment in Public Agricultural Research." *Agricultural Economics* 1 (1): 53–65.

Olson, M. 1985. "Space, Agriculture and Organization." *American Journal of Agricultural Economics* 67 (5): 928–937.

Pardey, P., J. Alston, and R. Piggott, eds. 2006. *Agricultural R&D in the Developing World: Too Little Too Late?* Washington, DC: International Food Policy Research Institute.

Reddick, C. 2002. "Testing Rival Decision-Making Theories on Budget Outputs: Theories and Comparative Evidence." *Public Budgeting and Finance* 22 (3): 1–25.

———. 2003. "Budgetary Decision Making in the Twentieth Century: Theories and Evidence." *Journal of Public Budgeting, Accounting and Financial Management* 15 (2): 251–274.

ReSAKSS (Regional Strategic Analysis and Knowledge Support System). 2013. "CAADP M&E Indicators: Share of Public Agriculture Expenditure (PAE) in Total Expenditure." Accessed September 15, 2013. www.resakss.org/sites/default/files/pdfs/ReSAKSS_AgExp_2013_website.pdf.

Roseboom, J. 2002. "Essays on Agricultural Research Investment." PhD dissertation, Wageningen University, The Netherlands.

Smale, M. 1995. "'Maize is Life': Malawi's Delayed Green Revolution." *World Development* 23 (5): 819–831.

Thirtle, C., L. Lin, and J. Piesse. 2003. "The Impact of Research-Led Agricultural Productivity Growth on Poverty Reduction in Africa, Asia, and Latin America." *World Development* 31 (12): 1959–1975.

Weissinger-Baylon, R. 1986. "Garbage Can Decision Processes in Naval Warfare." In *Ambiguity and Command*, edited by J. March and R. Weissinger-Baylon. Marshfield, MA: Pitman.

Wiebe, K., M. Soule, C. Narrod, and V. Breneman. 2000. "Resource Quality and Agricultural Productivity: A Multi-Country Comparison." Selected paper presented at the Annual Meeting of the American Agricultural Economics Association, Tampa, Florida, July 31.

World Bank. 2013. *World Development Indicators.* Accessed June, 2013. http://data.worldbank.org/data-catalog/world-development-indicators.

Yu, B. 2012. "SPEED Database: Statistics on Public Expenditure for Economic Development." In *2011 Global Food Policy Report.* Washington, DC: International Food Policy Research Institute.

Chapter 6

CHANGING DONOR TRENDS IN ASSISTANCE TO AGRICULTURAL RESEARCH AND DEVELOPMENT IN AFRICA SOUTH OF THE SAHARA

Prabhu Pingali, David Spielman, and Fatima Zaidi

The challenge of feeding 2.4 billion people in Africa south of the Sahara (SSA) by the middle of the current century is a topic set squarely on the development agenda of the global donor community. But this community is also a loose collection of actors—developing-country governments, multilateral agencies, and charitable foundations—with varying levels of commitment, coordination, and interaction. If solutions to Africa's challenges require long-term investments in agricultural research for development (AR4D), as demonstrated by many of the chapters in this volume, and if few governments in the region have adequate public resources to commit to research, then donor assistance will likely play a key role in financing. While recent trends suggest that development assistance flows are on an upswing for many agricultural sector activities, including research, it remains to be seen whether the donor community and its many stakeholders will stay the course and continue investing in AR4D in Africa over the long haul.

This chapter examines one facet of this resourcing challenge—the role and contribution of the donor community and official development assistance (ODA) to AR4D. Specifically, it examines historical and emerging trends in the international donor community in terms of who is investing, how much they are investing, and where their investments are targeted. The chapter goes on to explore the potential impacts of these donor trends on future human and agroecological landscapes, with a particular focus on SSA.

The authors thank Mumukshu Patel, Kathleen Flaherty, Nienke Beintema, and Mary Jane Banks; participants of the ASTI/IFPRI-FARA Conference "Agricultural R&D: Investing in Africa's Future—Analyzing Trends, Challenges, and Opportunities," held in Accra, Ghana, December 5–7, 2011; and participants of several other workshops and conferences for their comments on earlier versions of this chapter. Any and all errors are the sole responsibility of the authors.

Past Trends in Donor Assistance to Agricultural Development

Since the 1960s, donor assistance to agriculture and rural development has been a largely successful investment. Development assistance allocated to agricultural research, rural infrastructure, human capital development, and agricultural policy reforms has demonstrated the important contribution of agricultural development to poverty reduction and economic growth (Staatz and Eicher 1990; World Bank 2007a). In general, the returns to agricultural development assistance have been positive, despite occasional failures resulting from poorly designed projects and policies. And within the broad category of agricultural development, agricultural research is often cited as the single-best investment in terms of increasing productivity and reducing poverty (Fan and Pardey 1997; Fan 2000; Fan, Hazell, and Thorat 2000).

Among many investments made in agricultural research during the past five decades, South Asia's Green Revolution—the doubling of the yields and output of South Asia's major food staples between 1965 and 1985—is one of the most cited examples of this high payoff (Hazell 2010; Pingali 2012b). But similar successes have also been achieved in Africa at different scales and with different crops and technologies (Spielman and Pandya-Lorch 2009; Haggblade and Hazell 2010). A shared characteristic of many of these high-return investments was the contribution of modern science, particularly plant breeding and cultivar improvement, which was supported by the donor community (Evenson and Gollin 2003; Raitzer and Kelley 2008; Renkow and Byerlee 2010). Long-term donor commitments paid handsomely relative to many alternative public investment opportunities (Chapters 3, 5, 6, and 11, this volume).

The public and private donors who financed many of these investments were considered visionaries of their time (Lele and Nabi 1991). The Ford Foundation and Rockefeller Foundation were the drivers behind the creation of an international agricultural research system focusing on major staple food crops (rice, wheat, and maize), while the World Bank and other members of the bilateral and multilateral donor community invested in the creation of a broader research network under the CGIAR umbrella.

Yet despite the enthusiasm for agricultural development through the 1970s—and despite the many development successes in Asia, Africa, and Latin America—donor assistance declined dramatically in the mid-1980s (Pingali 2010). ODA trend figures point to several "lost decades" in donor support to AR4D, including research in and for African agriculture (Chapter 1, this volume). The lost decades in funding began from a peak in 1983–1986 when yearly ODA disbursements to agriculture, forestry, and

FIGURE 6.1 Official development assistance to agriculture and related sectors, 1967–2011

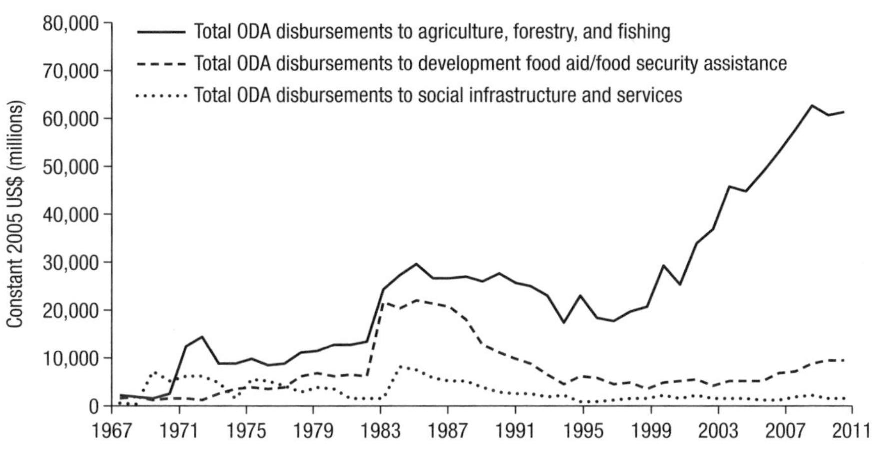

Source: Calculated by authors based on data from OECD (2013a) and World Bank (2014a).
Note: ODA = official development assistance.

fishing averaged $21.4 billion[1] (in constant 2005 terms) and declined to just $4.5 billion by 1997–2000, according to data from the Organisation for Economic Co-operation and Development (OECD 2013a) data. The period of stagnation continued through to 2004–2006 (Figure 6.1).[2] As a proportion of net ODA disbursements, agriculture, forestry, and fishing dropped from almost 14.7 percent in 1983–1986 to 1.9 percent in 1997–2000. And while it can be argued that figures derived from OECD are fraught with data-quality concerns, there is sufficient acknowledgment in the donor community itself that the so-called lost decades are real (see, for example, World Bank 2007a).

The causes are fairly well documented. Donor support for agricultural development and agricultural research and development (R&D) began to dwindle in the mid-1980s in response to low and declining real food prices worldwide, particularly between 1985 and 2005; growth in the number of aid recipients competing for a relatively fixed pool of funding; and concerns about persistent policy biases against agriculture, inefficient bureaucracies, poor project management, short funding cycles, long delays in completion, and

1 All currency is in US dollars, unless specifically noted otherwise.

2 Note that Figure 6.1 shows a sharp increase in ODA disbursements to agriculture, forestry, and fishing between 1982 and 1983; OECD/Development Assistance Committee data do not provide a sufficient level of disaggregation by country, donor, or sectoral activity to fully explain this trend in the data.

FIGURE 6.2 **Official development assistance, 1967–2011**

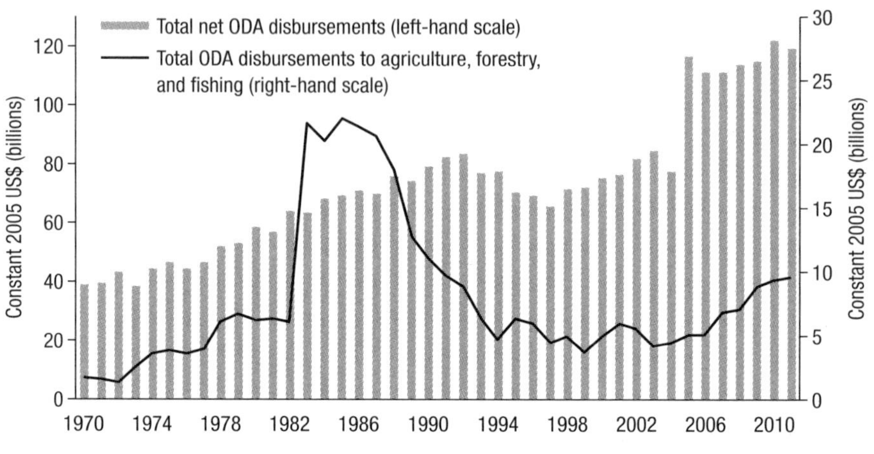

Source: Calculated by authors based on data from OECD (2013a) and World Bank (2014a).
Note: ODA = official development assistance.

lags between research and impact (Byerlee 1998; Islam 2011). More generally, the donor withdrawal from agriculture was part of a larger withdrawal from many sectors receiving development assistance in SSA during the mid-1980s (Figure 6.2). This was a period of widespread complacency about agriculture resulting from high global food surpluses; low commodity prices; competing commitments to health, education, and other social-sector investments; and structural adjustment programs designed to reduce public expenditures on agriculture and other sectors of the economy (Christensen 1994; Islam 2011). However, this withdrawal from agriculture persisted even as ODA disbursements began to recover in the mid- to late 1990s (Figure 6.2), particularly for social infrastructure and services (Figure 6.1).[3]

Only during the past decade have there been signs of a recovery in ODA for agricultural research. Recent signs of a recovery in ODA to agriculture, forestry, and fishing emerged after funding levels hit bottom in 1997–2000: following a stagnant period in 2004–2006, average yearly ODA disbursements increased to $10.38 billion in 2009–2011, although they remained at just 2.3 percent of total ODA during that period.

3 Note that this chapter does not address the related issues of aid effectiveness, tied aid, or the extent to which ODA earmarked for developing countries actually flows back to donor countries in the procurement of goods and services associated with development assistance, or the issue of ODA's impact or the attribution of impact.

The groundwork for this tentative upswing in donor funding was laid during a succession of high-profile global, regional, and national political commitments that began in 2000. The obvious watershed event was the United Nation's Millennium Development Goals (MDGs), a compact under which 189 countries committed to new and lofty medium-term development outcomes. The first and most visible goal—halving extreme poverty and hunger by 2015—contributed to a renewed global interest in food security; agricultural development; and, more implicitly, agricultural research. On the global stage, this priority was further taken up by the Group of Eight (G8) nations at their 2005 Gleneagles Summit and in subsequent forums.

In Africa, the MDGs were followed by the Comprehensive Africa Agriculture Development Programme (CAADP) (Chapter 1, this volume). African ownership and leadership of CAADP played a key role in putting AR4D squarely on the regional and national development agendas, and the 2006 Framework for African Agricultural Productivity (FAAP), organized with support from the Forum for Agricultural Research in Africa (FARA), provided a roadmap for improving agricultural productivity through applications of modern science, better donor coordination, and stakeholder engagement (FARA 2006).

These regional efforts gained further footing when the World Bank capped off a renewed global commitment to agriculture with its 2008 World Development Report titled *Agriculture for Development* (World Bank 2007a). This was the first time agriculture was featured in the organization's highly influential annual report since 1982, and helped frame several World Bank initiatives on agricultural research in the years that followed. For example, one of the World Bank's hallmarks of its renewed commitment to agricultural research has been the introduction of a regionalized approach to programs aimed at enhancing agricultural productivity in East, West, and Southern Africa (Chapter 2, this volume). These programs allocate upward of $80 million over six years in each subregion to strengthen scientific and technical capacity in agriculture, encourage technology transfers and knowledge sharing, and establish subregional centers of scientific excellence in specific crops and systems (World Bank 2007b, 2009, 2014b).

But do these subregional and country-level efforts actually account for the recent upswing in donor funding to agricultural research? While CAADP and the FAAP roadmap focus attention on AR4D, and while the World Bank's regional agricultural productivity programs bring new funds to the task, their overall contributions need to be seen in terms of how effective they are in leveraging funds from other sources, including African governments themselves. Expenditure reviews suggest that budgetary support to agriculture is not meeting the CAADP target in most countries (Chapter 5, this volume). A number of issues ranging from poor data availability and quality, to the lack of an agreed-upon measurement system covering agriculture, to low stakeholder participation in the CAADP process may partly explain why these targets have not been met by so many countries (Morton 2010; ONE 2013; Randall 2011).

Furthermore, any new resources garnered from subregional and country-level efforts bring with them some of the same challenges as experienced in the past. For example, a new subregional approach to funding agricultural research may not change the traditional financing mechanism by which the World Bank and other donors direct funding to specific national agricultural research institutes, programs, and projects. This may benefit some research activities, but there is sufficient evidence suggesting that short-term commitments leave national agricultural research systems to struggle with the sharp loss of funding when the funding ends (Chapter 4, this volume). Moreover, other areas of agricultural research may be left behind when certain commodities take precedence. Whether these programs are constructive engagements that leverage regional spillover effects, overcome small-country constraints, and sustainably improve capacity and resource deficits of national agricultural research institutes (NARIs) remains to be seen (Chapter 14, this volume).

So is there a different way to understand the recent upswing in donor funding to agricultural research? Arguably, the single-most significant driver of this upswing was the global food price crisis of 2007–2008, when the international prices of major food cereals rapidly increased (Headey and Fan 2010). In response to the price crisis, G8 countries assembled to pledge $20 billion to agricultural development at their 2009 L'Aquila Summit—pledges that increased to $22 billion at the Pittsburg G20 summit with commitments from additional countries. The 2009 G8 L'Aquila Food Security Initiative (AFSI) pledges and the wider commitment to global food security, agricultural research, and productivity growth have since been the subject of much follow-up discussion at subsequent gatherings of the G8 and G20 groups of nations (Coppard 2010). This included an increase in disbursements to CGIAR, which invests more than 50 percent of its resources in Africa—on the order of $922 million between January 2008 and July 2009—to raise agricultural productivity through scientific research (G8 2009).[4]

4 Note, however, that some partner countries have considered their ODA loans for agriculture development as financial disbursements under the G8 commitment.

A key driver behind this upward shift in assistance to agricultural research is the renewed commitment of traditional bilateral donors, such as the European Union (EU), United Kingdom, and United States, as well as the traditional multilateral agencies, such as the World Bank. The EU pledged $3.8 billion to agricultural development in response to AFSI in 2009. Similarly, the US "Feed the Future" initiative committed $3.5 billion in 2010 over three years to results-driven programming in agricultural development and food security that targets some of the world's poorest and most vulnerable countries and communities (Feed the Future 2011). The World Bank reentered the field by expanding its investments in agricultural development lending and grants, while also assuming trusteeship of the Global Agriculture and Food Security Program (GAFSP), an initiative begun in 2010 that made available some $521 million (of $925 million pledged by most of the world's major donors) to support strategic investment plans for national and regional agriculture and food security through both public- and private-sector financing (GAFSP 2011).[5] All of these donors—working in concert with other bilateral donors, multilateral agencies, and charitable foundations to encourage far-reaching governance and structural reforms—boosted CGIAR funding from $500 million in 2008 to $1 billion in 2013, raising hopes that the world's largest agricultural research partnership will play an expanded role in tackling the world's major development challenges (CGIAR 2013).

These major donors—alongside many other bilateral donors and multilateral agencies—have committed significant resources to agriculture since 2009, and a significant portion of this commitment has been further allocated to agricultural research. There are also signs of even greater commitment to agricultural research in the near future. For example, the Sustainable Development Goals (SDGs)—currently under development and expected to be issued in 2016 as a follow-on to the MDGs—are expected to include both agricultural productivity and environmental stewardship as key

5 GAFSP is the first global fund designed exclusively for smallholder productivity growth in least developed countries. It is a competitive grants program that was set up to support public- as well as private-sector efforts for enhancing smallholder productivity. While the GAFSP fund is administered by the World Bank, decisions about grants and performance tracking are made by a steering committee comprising major GAFSP donors, national partners, and civil society representatives. National proposals are reviewed by an independent technical committee, then the steering committee makes a final selection. In the case of SSA, grants are explicitly tied to CAADP country plans and priorities. Since the amount of funding provided by GAFSP is relatively small, its success depends on its ability to leverage additional resources from ongoing ODA and domestic efforts in each country. GAFSP may also play an important role in helping to coordinate disparate national efforts targeting the same smallholder populations. It is too early to say whether GAFSP will be successful in reinvigorating smallholder agriculture in the least developed world, but it is an experiment worth monitoring.

FIGURE 6.3 Estimated funding allocated to agricultural research by the United States Agency for International Development, 1950–2011

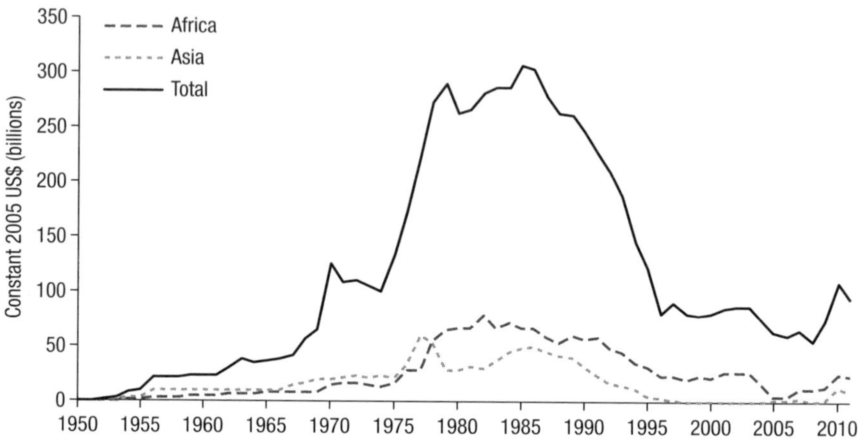

Source: Alex (2013).

goals, potentially encouraging new global, regional, and national commitments toward agricultural research to eradicate hunger and poverty by 2030 (Farming First 2013; UN 2013).

Nevertheless, it remains to be seen whether these new resources will lead to significant, long-term, and steady increases in assistance to international and national research institutes, programs, and activities. For example, US assistance to agricultural research since 2009 (Figure 6.3) is far less in real terms than what was provided in the 1970s, suggesting that the US renewed commitment is still a strategic "work-in-progress" at best (Anderson and Roseboom 2013). Also, it is not entirely clear whether bilateral funds are ultimately destined for national systems, or instead earmarked for CGIAR and regional organizations—a strategy that does little to reverse the depreciation that occurred in national systems during the lost decades. Ultimately, this may suggest that only multilateral donors—primarily the World Bank—will constitute a renewed source of funding for national systems, although data from past trends do not indicate what may be in store from the World Bank for the future (Pardey et al. 2006). Further, it is unclear whether assistance to agricultural development and agricultural research is being better coordinated, targeted, and evaluated in light of the Paris Declaration on Aid Effectiveness (2005), the Accra Agenda for Action (2008), and the Busan High Level Forum on Aid Effectiveness (2011), among other events. Most donors would argue yes. But critics—for example, ONE (2013) and Coppard (2010)—remain skeptical. Others, such as Benin (2014), who focuses more specifically on AFSI, suggest that, while there are positive signs, it is still too early to tell.

Emerging Trends in Donor Assistance to Agricultural Development

Regardless of what the major donors are doing, another important aspect of the tentative increase in assistance to agricultural research is the entry of new donors to the landscape (Lele 2009; Pingali 2010). Unfortunately, statistics compiled by OECD, the primary data source for most analyses of ODA trends, shed little light on these new donors, because few report their development spending to OECD, and among those that do report, few do so in a manner that is comparable with other sources of assistance or in a way that separates funding for agricultural research. As of 2011, only three non-OECD countries and one charitable foundation reported their assistance to agricultural development to OECD's Creditor Reporting System (OECD-CRS).[6]

But even with this underreporting and measurement problem, OECD statistics suggest several emerging trends. First, OECD reports that nonbilateral/multilateral channels for ODA disbursement are diversifying: between 2004 and 2011, the share of ODA disbursements through nongovernmental organizations (NGOs), civil society, and public–private partnerships increased from 1.3 percent of the total to 13.7 percent (Figure 6.4). Second, OECD reports that the share of total assistance provided by new donors to agricultural development grew from zero in 2008 to 5 percent in 2011 (Figure 6.5). And from among those reporting developing assistance spending to OECD, the Bill & Melinda Gates Foundation accounted for 92 percent of the yearly funding attributable to sources other than traditional bilateral and multilateral donors during 2009–2011.

Necessarily, the analytical usefulness of OECD's statistics on ODA to agricultural development is limited by several known shortcomings (Coppard 2010). First, as already mentioned, many nonmembers of OECD's Development Assistance Committee (OECD-DAC) and nonofficial donors simply do not report to OECD-CRS, including, among many others, the Rockefeller Foundation and the Ford Foundation—two of the earliest and most influential donors to agricultural development. Second, the actual types of assistance

6 Only the Czech Republic, Kuwait, the United Arab Emirates, and the Bill & Melinda Gates Foundation reported their assistance to agricultural development to OECD's Creditor Reporting System. For these donors, the earliest year for which assistance was reported is 2009.

FIGURE 6.4 Official development assistance disbursements by type of channel, Africa South of the Sahara, 2004–2011

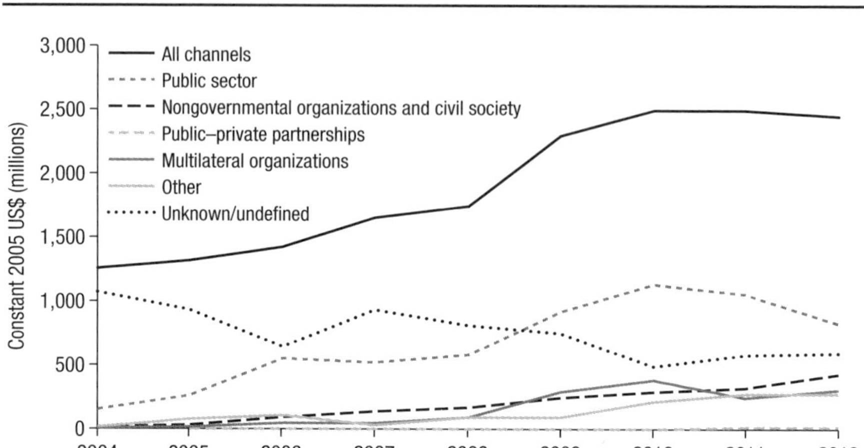

Source: Calculated by authors using data from OECD (2013b).

Note: Other channels include, for example, universities, colleges, or other teaching institutions, and research institutes or think tanks; unknown channels are listed as such by OECD (2013b).

reported cover a range of funding mechanisms that vary in their degree of relevance to development or research: categories, such as ODA grants, ODA-like grants, ODA loans, private grants, and equity investments, are all open to interpretation. Third, complete data series on agricultural development begin only at 2002, making it difficult to match trends by individual donors with the more aggregated trends dating back to the mid-1960s.

As a result, OECD-CRS data overlook the rising proportion of assistance to agricultural development from other avenues, chiefly (1) new, high-profile bilateral donors, such as Brazil, Russia, India, and China; (2) new, smaller, low-profile bilateral donors, such as Argentina, the Czech Republic, Israel, Poland, Korea, Kuwait, Qatar, Saudi Arabia, Slovakia, Taiwan, Thailand, and Turkey; and (3) many charitable foundations, private philanthropies, and corporations, such as the Tata Trusts, Gatsby Charitable Foundation, and McKnight Foundation (Table 6.1).

Smith, Yamashiro, and Zimmermann (2010) attempt to remedy this with a snapshot of ODA circa 2007–2009 from new and emerging bilateral donors. Their analysis draws primarily on net disbursement data from OECD data for those countries reporting to OECD and official government sources for Brazil, China, India, Russia, and South Africa. They report a total of $11.8 billion in ODA, 80 percent of which is attributable to developing-country donors, and

FIGURE 6.5 Development assistance to agriculture, Africa south of the Sahara, 2002–2011

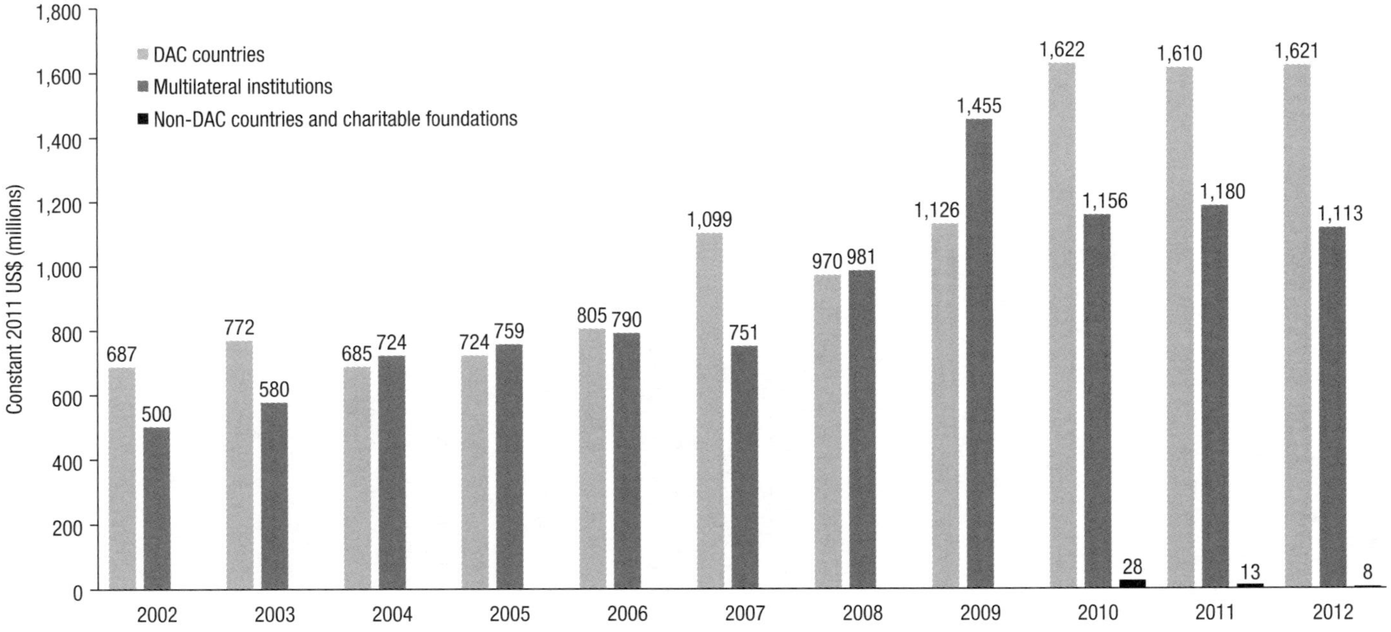

Source: Calculated by authors based on data from OECD (2013b).

Note: DAC countries = countries that are members of the Organisation for Economic Co-operation and Development's Development Assistance Committee.

TABLE 6.1 Official development assistance from new and emerging donors to developing countries, 2007–2009

Region/country	Estimated official development assistance (US$ millions)
Emerging European Union donors	1,140
Other emerging developed-country donors	1,190
Brazil	437
China	1,800
India	610
Kuwait	283
Saudi Arabia	5,564
South Africa	109
Taiwan	435
Thailand	179
United Arab Emirates	88
Total (all new/emerging bilateral donors)	11,835
Total (all new/emerging bilateral developing-country donors)	9,505

Sources: Calculated by authors based on Smith, Yamashiro, and Zimmermann (2010) citing official government publications, OECD statistics, and various other sources.

Notes: Emerging European Union donors refers to Cyprus, Czech Republic, Estonia, Hungary, Latvia, Lithuania, Poland, Romania, Slovak Republic, and Slovenia; other emerging developed-country donors refers to Iceland, Israel, Liechtenstein, Russia, and Turkey. The figure for China is based on the lower estimate from Smith, Yamashiro, and Zimmermann (2010).

58 percent of which is attributable to a single donor—Saudi Arabia. Although these figures do not provide a complete sense of how resources are allocated between Africa and other developing regions, they do provide some evidence of assistance from new donors.

Unfortunately, these estimates are not disaggregated by sector, making it difficult to determine exactly how much is directed to agricultural development or, within this area, to agricultural research. However, other sources do shed some light on the magnitude and nature of these donors' contributions to agriculture. These contributions are explored below, focusing on Brazil, China, India, and the "new philanthropists."

Brazil

Following President Luiz Inácio Lula da Silva's first visit to Africa in 2003 to advance his country's engagement in Africa's development,[7] Brazil has committed

7 Between 2003 and 2010, President Lula made 12 separate trips to Africa, visiting 27 countries—more than all of his predecessors combined (see www.bbc.co.uk/news/world-latin-america-11717757).

new resources to the task. ODA from Brazil totaled $90 million between 2003 and 2009, of which $45 million was allocated to Africa, according to Patriota and Pierri (2013), who draw on figures from the Brazilian Cooperation Agency.

Brazil's most notable engagement comes in the form of extending its experience in agricultural intensification to Africa. This has included vocational training and technology transfer projects in more than a dozen countries (Patriota and Pierri 2013). A notable focus of this engagement revolves around the possibilities of transforming the Guinea savanna into a highly productive agriculture and livestock area, as Brazil did with its Cerrado savanna.

Although technical cooperation represents just 3 percent of Brazil's ODA (excluding debt relief and export credits), agriculture is the key focal point of its cooperation activities. Between 2003 and 2010, agriculture accounted for 22 percent of the country's technical cooperation portfolio and 26 percent of its Africa portfolio (Cabral and Shankland 2013). In addition, Brazil has committed resources under the Brazil–Africa Dialogue on Food Security, Fighting Hunger, and Rural Development to support initiatives modeled on recent national programs implemented in Brazil. Examples include the 2008 More Food Africa Program (Programa Mais Alimentos África), aimed at increasing productivity through technology transfers and mechanization, and the 2010 Food Acquisition Programme (Programa de Aquisição de Alimentos), aimed at providing social protection services through local food procurement (Barka 2011; Cabral and Shankland 2013).

Not unlike China, Brazil has received some criticism for its assistance to agriculture in Africa. Concerns include the bias toward large-scale mechanized farming of cash crops, such as cotton, soybeans, and tobacco—a strategy that may have performed well in Brazil but is criticized by some as inappropriate to many smallholder systems in Africa (Chichava et al. 2013; Mukwereza 2013; Rada 2013). Another criticism is leveled at the apparent lack of coordination across the 20-plus Brazilian agencies working on agricultural development issues in Africa (Cabral and Shankland 2013).

Despite these criticisms, Brazil's encounter with African agriculture seems to have gathered momentum (Cabral and Shankland 2013 reported Brazilian initiatives in 38 countries). The main vehicle for Brazil's engagement in Africa has been the Brazilian Cooperation Agency. For agriculture, however, the opening of an international office of the Brazilian Agricultural Research Corporation (Embrapa) in Accra, Ghana, in 2006 is heralded as one of the critical drivers in strengthening research linkages and partnerships with Brazil. This office, though later scaled down to cover only Ghana, pioneered a number of technology transfer partnerships across the region that opened the door for dozens of new projects.

China

Aid flows from China vary widely, depending on the nature and measurement of the flow, but they generally fell in the neighborhood of $1–$3 billion between 2007 and 2009 (Bräutigam 2011; Buckley 2013). China's approach to ODA, first outlined in a white paper published in 2011 (People's Republic of China 2011) suggests that China's sectoral allocations are quite similar to allocations by most OECD-DAC member countries: allocations to economic infrastructure development account for 61 percent of the total, whereas agriculture accounts for just 4.3 percent (People's Republic of China 2011).

Although China's ODA primarily focuses on neighboring Asian countries, significant funding is also flowing to Africa. Estimates reported by Pingali (2012a) indicate that between 2001 and 2009, China's assistance to Africa consistently captured 38–44 percent of the total development assistance budget. This funding follows directly from the 2000 Forum on China–Africa Cooperation that elaborated China's assistance strategy for the region, and the ambitious Program for China–Africa Cooperation in Economic and Social Development (FOCAC 2009; AATF 2010). Beyond its commitments to canceling debts, reducing trade barriers, and increasing development assistance for African countries, the program committed resources to training African agricultural scientists and establishing agricultural technology demonstration centers with the support of Chinese expertise.

A significant part of China's commitment to African agricultural development was contained in a donation of $30 million to the Food and Agriculture Organization of the United Nations (FAO) in 2009. The aim of this donation was to expand China's contribution to FAO-led efforts to eradicate hunger and poverty by assisting developing countries in improving agriculture and food production under the umbrella of FAO's Special Program for Food Security (SPFS). China, along with other developing countries, uses SPFS as a vehicle to provide experts, technicians, and technical support to national and regional food security activities (FAO 2010). Technology transfers and support to NARIs are an implicit outcome of the program design, with the potential to parallel other donor initiatives and programs, such as the Alliance for a Green Revolution in Africa (AGRA) funded by the Rockefeller Foundation and Bill & Melinda Gates Foundation (Bräutigam 2009).

Strange et al. (2013) examine Chinese assistance to Africa between 2000 and 2011 and identify 71 projects totaling $981 million similarly accounting for approximately 4.6 percent of total ODA to Africa (Figure 6.6). A looser definition of ODA that includes projects related more to investment than to assistance suggests 107 projects during the same timeframe totaling more

FIGURE 6.6 Official development assistance to Africa from China for agriculture, forestry, and fishing, 2000–2011

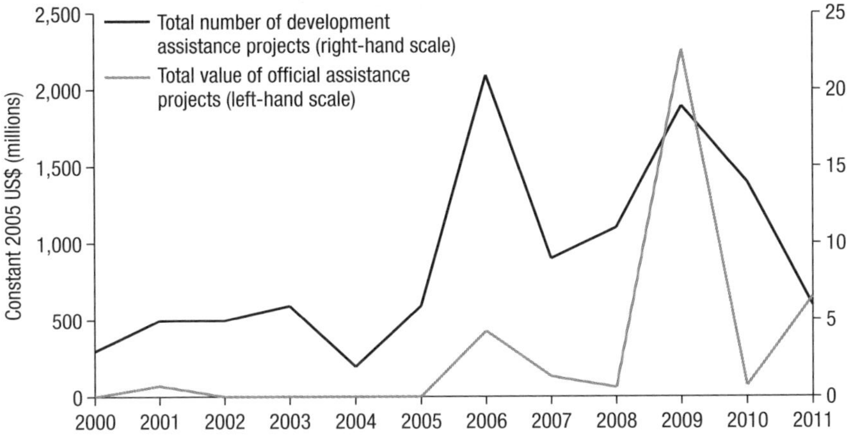

Sources: Calculated by authors based on data from AidData (2013) and Strange et al. (2013).

than $4 billion and still accounting for about 5.4 percent of all flows to Africa. Notably, many of these projects tend to be a combination of public and private Chinese investment, as is often characteristic of recent trends in Chinese assistance to agriculture (Freemantle and Stevens 2013).

China's support to African agriculture has attracted significant attention in the global development community. Some critics have focused on the potentially controversial aspects of the political "noninterference" principle underlying China's ODA policy. Thus, Chinese development assistance supported Zimbabwe with technology demonstration centers, inputs and equipment for tobacco and cotton farming, and other activities during the height of Zimbabwe's international political isolation in the mid-2000s (Mukwereza 2013). Others have focused on many of the same historical issues raised about assistance from OECD-DAC member countries: aid tied to procurement of goods and services from the donor country, aid focused on supporting extractive industries and nonrenewable resource exploitation, and aid associated with weak growth linkages or poor labor standards or land appropriation.

Evidence both for and against these criticisms is scant, but the debate is not likely to dissipate any time soon.[8] And given the central importance of agriculture to many developing countries, particularly in Africa, it is likely that

8 See, for example, Bräutigam (2011), Buckley (2013), Freemantle and Stevens (2013), and Strange et al. (2013), among many others.

this debate will intensify as China explores new opportunities for agricultural technology transfers, food staple production, and large-scale mechanized farming in the region—many of which are areas of interest shared by other OECD-DAC donors and African governments.

India

Similar to Brazil, India's presence in Africa has been a topic of note in recent years. India has many long-standing political and economic ties with African countries, as well as the successes of the Green Revolution during the 1960s and 1970s to inspire partnerships supporting agricultural research. There is, however, very little documentation on the size, scope, or nature of Indian ODA to the region. Primarily, attention has been given to Indian corporate investment in land deals related to agricultural ventures (von Braun and Meinzen-Dick 2009; Rahmato 2013; Rowden 2013), or to private-sector technology transfers embodied in farm machinery and irrigation equipment (Modi 2013).

Nonetheless, there is some evidence that India is driving toward greater research collaboration with African countries. The first India–Africa Forum Summit held in New Delhi in 2008 marked a symbolic shift in this direction (Singh 2013). Official cooperation supporting agricultural research—formally organized under the India–Africa Framework of Cooperation—has included study tours, technology transfer programs, higher-education scholarships, participation in international projects, and extension training, with future plans that include rural technology parks, farm science centers, and other public and private investments (AU 2011; Singh 2013). Further exploration of the relationship between Africa and India is warranted.

The "New Philanthropists"

The past decade has seen a rapid expansion of philanthropic and charitable organizations ranging from privately funded foundations to religious charities to corporate initiatives focused on social responsibility and social entrepreneurship. Although precise definitions, sources, and uses of philanthropic spending in developing countries vary, and data collection efforts are limited, trend data strongly indicate an acceleration of development assistance from philanthropic donors. For example, the Hudson Institute (2012)—using its own figures and data—estimated total philanthropic flows in 2010 from DAC member countries at $56 billion, a figure that compares favorably with the $128 billion in ODA flows from the same countries (Figure 6.7). Again, while these figures are not disaggregated between Africa and other developing regions, they nonetheless point to trends that may affect funding for agricultural research in SSA.

FIGURE 6.7 Philanthropic and official development assistance flows to developing countries, 1991–2010

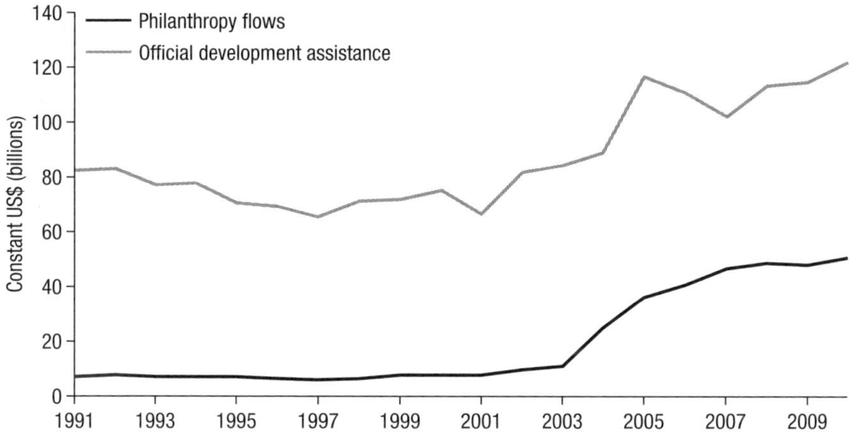

Philanthropy flows
Official development assistance
Constant US$ (billions)
140
120
100
80
60
40
20
0
1991 1993 1995 1997 1999 2001 2003 2005 2007 2009

Source: Hudson Institute (2012); Pingali (2012a); OECD (2013a).

Philanthropic assistance to agricultural development is directed through a wide range of organizations, including foundations where a related corporate entity is directly associated, sometimes (but not exclusively) as part of their corporate social responsibility activities. The Syngenta Foundation for Sustainable Agriculture and Monsanto Fund are examples emerging from the global crop and bioscience industry. The Barwale Foundation is a similar example from India, and although its R&D funding and activities are not directly focused on African agriculture, its work on such topics as marker-assisted selection for crop improvement and hybrid seed production, as well as its close linkages to both private seed companies and the international agricultural research system, suggests potential for positive research spillovers for SSA (Barwale Foundation 2012).

The landscape also includes foundations that are effectively separate from their corporate parent or drawn from a corporate parent with little or no direct association with agriculture or agricultural development. This was the case with the Ford Foundation and Rockefeller Foundation when they forayed into agricultural development in the 1950s, the Tata Trusts in 2003, and the Bill & Melinda Gates Foundation in 2006.

Also included in this landscape are industry organizations that represent a group of corporate entities with direct interests in agricultural products and services. Examples include Croplife International, which draws its membership from the six largest multinational crop science companies. Other philanthropies include social entrepreneurship and venture capital funds with explicit investments in agricultural development activities, such as the Pearl Capital Partners (PCP) group, an investment initiative targeting small and medium-sized agribusinesses in East Africa with funding from the Gatsby Charitable Foundation. PCP has made several investments in seed companies that host their own breeding programs, including $1 million in the Kenyan company Western Seed in 2008 and $350,000 in Uganda's Nalweyo Seed Company in 2006 (Fletcher 2011). Still, other philanthropies are nonprofit, nongovernmental, direct-giving, faith-based, and charitable organizations that provide financial support to agricultural development activities.

Although many philanthropic organizations are engaged in agricultural development around the world, the entry of the Bill & Melinda Gates Foundation (BMGF) into agriculture is often cited as a catalytic event in changing the assistance trend (Pingali 2012c). Between November 2006 and September 2013, the Foundation awarded 461 grants totaling approximately $2.523 billion for agricultural development globally.[9] More specifically, between 2009 and 2011, the Foundation invested $50.8 million in agriculture policies, $134.8 million in R&D, and $145.0 million in access and markets. (Figure 6.8), a significant share of which is allocated to programs that focus directly or indirectly on SSA. In a few short years, the BMGF has invested more than half of its sizable African agricultural development portfolio in agricultural research.

Initially, the BMGF's approach to making grants involved identifying market and government failures and addressing those gaps through technology R&D, as well as strategic advocacy investments. It brought high-profile tranches of new funding to the research agenda, and invested in both the established field of cultivar improvement and the more novel field of biofortification. This approach has also helped convene several agricultural research initiatives that have successfully leveraged funding from other donors willing to buy into BMGF's strategic priorities. These investments—alongside other large-scale initiatives on livestock improvement, nutrition, data and statistics, and organizational reform—represent a significant injection of new life into the global agricultural research system.

9 In fact, while the Bill & Melinda Gates Foundation's program on agricultural development did not begin until 2005, its first grant in the field of agriculture was given in September 2003 to support the activities of the HarvestPlus Challenge Program to reduce micronutrient deficiencies in developing countries by breeding higher levels of essential micronutrients into staple crops. The grant was given to the International Food Policy Research Institute (IFPRI) in the amount of $25 million over four years. The second grant was awarded in November 2005 in the amount of $0.6 million to the International Center for Tropical Agriculture (CIAT) to demonstrate that biofortified crops can be delivered effectively to farmers and consumers.

FIGURE 6.8 **Early grant disbursements for agricultural development from the Bill & Melinda Gates Foundation, 2006–2011**

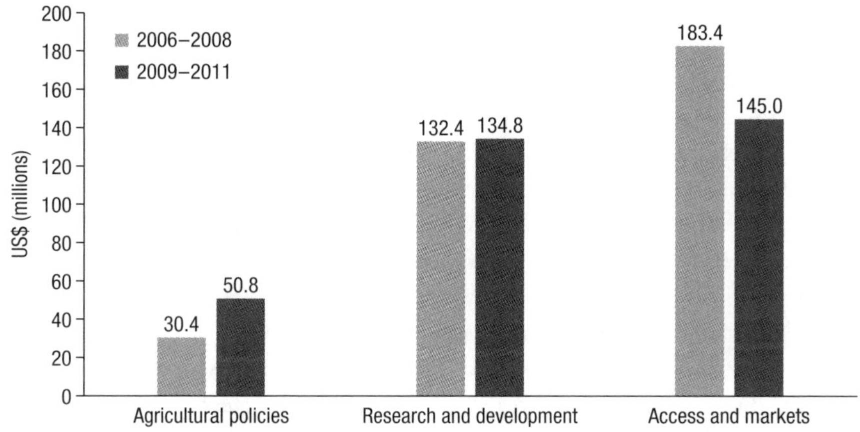

US$ (millions)
2006–2008
2009–2011
Agricultural policies: 30.4, 50.8
Research and development: 132.4, 134.8
Access and markets: 183.4, 145.0

Source: Adapted from Pingali (2012a).

BMGF's influence on the global agricultural development agenda has been more than proportional to its financial contribution over the past five years. The foundation has helped the donor community "rediscover" agricultural development by advocating for greater attention to agriculture, setting examples through its own strategic investments, and engaging in multilateral initiatives and global networks on agriculture (Pingali 2012c).

As part of this effort, BMGF maintains a dialogue with major bilateral donors, such as the United States Agency for International Development and the United Kingdom's Department for International Development, while also supporting the commitments to agricultural development set by the G20 group of nations and multilateral efforts, such as GAFSP. BMGF formally joined CGIAR in 2010, and has been a strong proponent for CGIAR's return to its historical strength in crop improvement. The reversal in declining trends in funding for crop genetic improvement over the past few years show the foundation's impact on CGIAR priorities through its own investments, as well as its advocacy with the other donors to the system.

Creating an effective "hand-off" from global public-good R&D to technology dissemination at the regional and national levels requires identifying and strengthening partnerships with actors along the commodity value chain. BMGF has been experimenting with institutional innovations at the regional level, such as through AGRA, and at the national level, such as with the Ethiopian government's Agricultural Transformation Agency (ATA). AGRA, in particular, is intended to provide a bridge from global innovation to local technology adoption in Africa. The AGRA program on African Seed Systems has already released and disseminated more than 150 new varieties of the major staple crops across SSA, working with local private-sector entities. AGRA has also become an influential advocate for agricultural R&D in the region, closely collaborating with CAADP and other regional bodies.

BMGF has faced major challenges in converting its significant global public good investments into impact on the lives of smallholders, particularly in SSA. Poor infrastructure investments and weak policy incentives continue to constrain the rapid uptake of improved varieties and technologies. For example, current seed policies across SSA do not promote the free movement of improved seed across borders, and sometimes even across states within a country. BMGF has also been constrained in building effective partnerships at the national level because of poor technical, policy, and managerial capacity. Broad-based capacity-building efforts, though crucial, are beyond the scope of a single donor and require sustained commitment from a larger coalition of bilateral and multilateral funders. Developing countries themselves need to step up their own commitment in this regard.

Philanthropists are not without critics. Herdt (2012) shares several insights based on experiences of the Ford Foundation and Rockefeller Foundation in light of the large scale at which BMGF has entered the field of agricultural development. His most salient concern is the tentative nature of charitable support to agricultural development—symbolized by the eventual exit of both the Ford Foundation and the Rockefeller Foundation from the field—and the need for national organizations and policies to replace development assistance. He further argues that today's foundations may be pursuing strategies that are too narrowly focused on technology solutions without commensurate focus on developing the national capacity that would eventually "work themselves out of their jobs," particularly in SSA. That said, BMGF does have a fairly broad portfolio of education and training activities focused on building national capacity to support agricultural development and R&D through programs that are administered through regional initiatives in SSA.[10]

10 Such initiatives include degree and other training (in such disciplines as plant breeding, seed systems, and agricultural economics) run with support from AGRA under the auspices of such programs as the Regional Universities Forum for Capacity Building in Agriculture; the African Centre for Crop Improvement (ACCI) at the University of Kwazulu-Natal, South Africa; the West Africa Centre for Crop Improvement at the University of Ghana; and the Collaborative Masters of Agricultural and Applied Economics.

New Priorities in Agricultural Development

The re-emergence of traditional donors to agricultural development and agricultural research, and a growing role for emerging donors in these areas, suggest a new landscape with the potential for improving the prospects for rapid agricultural productivity growth and poverty reduction in developing countries. New sources of funding—from such large emerging economies as Brazil and China, to such middle-income and resource-rich countries as Saudi Arabia, to philanthropic organizations—may rewrite the rules of development assistance in terms of whom, how, and what they fund.

Greater diversification of funding sources can insulate recipient developing countries and their research systems from the shocks and volatility associated with dependence on a small number of donors and donor-funded projects. It can also spread the benefits of assistance from a singular focus on public agricultural development and research projects to a wider set of engagements targeting the private and civil society sectors. Ultimately, such diversification would also bring with it a variety of innovative ideas and approaches that could be tailored to fit the development needs of a given country or community.

However, this new landscape may also come with high transaction costs if it is not accompanied by a coherent vision and an improved cooperation strategy in support of agricultural research. Improved reporting mechanisms are an immediate priority, so that donor assistance to agriculture and agricultural research can be accurately characterized and coordinated. Although the OECD-CRS reporting tool (OECD 2013b) is not the only data warehouse for public and private donors to record their contributions, it is probably the foremost platform for comparing assistance levels and trends by type, source, sector, and destination. Thus, greater effort by donors and OECD to encourage accurate reporting and to improve the coding of development and research activities is crucial to future analysis. Another priority is better coordination of the assistance to agricultural development and research. New bilateral donors, such as China and Saudi Arabia, could do more on this front by engaging with the international community to adhere to and promote the aforementioned Paris Principles on Aid Effectiveness, the Accra Agenda for Action, and the Busan High Level Forum on Aid Effectiveness.

Last, but equally important—keeping in mind that agricultural development is ultimately the domain of sovereign nations—is the need to ensure that donor assistance is closely aligned with national strategies and policies. This is not as simple or obvious as it may seem. Many developing countries continue to pursue ill-advised policy regimes that are biased against agricultural growth, smallholder farming systems, and sustainable use of scarce natural resources. Donors can do much more to understand underlying political processes that lead to such policy regimes (Future Agricultures 2012), and to use this understanding to support evidence-based policy research that demonstrates the importance of formulating suitable and realistic national agricultural policies, and the potential role of agriculture as an engine of broader economic growth and poverty alleviation. Donors can also do much by reducing the influence of their own political and economic interests on official assistance programs. While the surge in philanthropic engagement in agricultural development may dampen the promotion of vested interests by bilateral donors, philanthropies are no less charitable when it comes to promoting their own agendas. Stronger advocacy and push-back from developing countries will likely be critical to improvement on this front.

There are signs that some regions and countries have recently begun to operate on a more equal footing with bilateral and multilateral donors as a result of diversification in assistance sources. In Africa, the CAADP process and regional organizations, such as the Common Market for Eastern and Southern Africa (COMESA) and the Economic Community of West African States (ECOWAS), have made concerted efforts with member countries to put priority setting firmly in the hands of African governments and their stakeholders. However, challenges remain.

The first and more obvious challenge may simply be poor national capacity. While overall ODA for agriculture has risen in recent years, the effectiveness of this assistance is still in question because of poor national-level capacity to absorb and use the funds effectively. Inadequate human capacity at the technical and managerial levels may be a consistently significant barrier to creating sustainable rural change in the least developed countries. Poor capacity may also limit the ability of many countries to establish their own priorities and to effectively coordinate the funds and agendas of their numerous foreign donors. Indeed, the demand for talented nationals at the country offices of the donor agencies may have further added to the human capacity strain faced by several country governments.

National capacity limitations are just a part of the story, and the emphasis here is more on the donors themselves. Donors face a broad range of challenges, an important one of which is their ability to understand and address the local context. There is widespread recognition that donor priorities and programs do not always reflect the needs and priorities of the country concerned. This is partly because of poor national capacity to articulate needs and priorities, but also because of capacity constraints within donor agencies. Often the problem lies in donors' limited understanding of ground-level needs and realities, and—in the case of aid targeted toward agricultural R&D—an understanding of complex tropical farming systems is particularly limited. Strengthening in-house capacity by building a cadre of tropical agricultural specialists and experienced development practitioners should be a high-priority investment for most donor agencies.

This inability to understand local context often leads to poorly aligned or irrelevant strategies espoused by both traditional and emerging donors. Each donor that engages in Africa's agriculture sector seems to have a slightly different view of agriculture's role in wider economic growth and poverty reduction. For some, such as BMGF, economic growth hinges on agricultural development, which is driven by productivity growth in smallholder farming. For other donors, the agricultural sector is less of a strategic focus and takes a second seat to humanitarian relief, social-sector interventions, and other priorities. Other donors espouse strategies that seek to replicate their own successes—whether in terms of China's achievements in rural development, India's Green Revolution, or Brazil's success in the Cerrado—through technology transfers and infrastructure development. The question may not be whether these strategies are well reasoned or otherwise convincing, but whether donors can (1) find the common ground to support African agriculture with new expertise and resources, (2) pause long enough from their own strategic rhetoric to understand the nuances of the region's challenges, and (3) dispense with cookie-cutter approaches taken in the past and instead cooperate with each other to find "best-fit" solutions for African agriculture.

Even where strategies may be aligned, the control and management of resources earmarked for ODA still pose a challenge for agricultural development in the region. International intermediaries still control funds and manage programs, such that a large share of the increased funding noted earlier flows through outside intermediaries, rather than being given directly to developing-country governments or local organizations. Traditional intermediaries include CGIAR and international NGOs engaged in agricultural development activities. Newer intermediaries include GAFSP, which has been administering its grants through multilateral donors and United Nations agencies. Funds flowing to countries through these channels tend to be tightly controlled for specific projects and are often managed by technical and administrative staff from the international intermediaries. Country ownership of such efforts tends to be tenuous and to accentuate the problem of building local capacity, discussed above.

A less obvious challenge lurking in the background relates to the credibility and influence of new and emerging donors. Traditional bilateral and multilateral donors operate in the region through a well-established sense of credibility that has allowed them to lend support and exert influence on decisions taken by sovereign governments. But it is worth asking whether their preeminence in the donor landscape is coming to an end with the entry of new donors with similar levels of credibility, such as Brazil, China, and India—many of which are symbolically aligned with the region's political and historical struggles. Meanwhile, BMGF and other philanthropists—though lacking in credibility as nonstate actors—seem to have exerted influence in such countries as Ethiopia, where the combination of a strong and independent national development strategy has allowed similarly independent donors to engage directly with the government.

Even where these emerging donors have both credibility and influence, the long-standing issue of donor coordination poses a final challenge worth consideration. Least developed countries have long struggled with the problem of donor coordination to ensure that support fits within the country's overall framework of development priorities. Unfortunately, many of the least developed countries lack the capacity to set priorities or establish and enforce a coordination mechanism. Hence, donor coordination mechanisms have been established by the donors themselves. However, these mechanisms tend to exclude the new and emerging donors. Private philanthropies, in particular, have generally been excluded from the donor coordination groups, while emerging donors, such as China, have avoided them.

A few SSA countries, such as Ethiopia and Ghana, have emerged as exceptions. Both have taken strong ownership of the development agenda and have established country-driven donor coordination mechanisms. Ethiopia's previously mentioned ATA was established with the specific intention of building strong local capacity for strategic planning, priority setting, and overall coordination of development resources targeted toward agriculture. Many other countries in Africa are interested in emulating the ATA model; however, human capacity constraints may hold them back.

There have been signs of greater donor coordination in recent years, with CAADP serving as a strong mechanism in support of donor coordination. Similarly, the CGIAR reform process has emerged as a strong mechanism for donor coordination, specifically on the agricultural R&D front (Chapter 15, this volume). However, the links between CGIAR priorities and those of specific countries have been historically weak, and there is little evidence to indicate that the reforms have changed that situation. This strongly suggests the need for the new aid structure to create mechanisms to improve linkages between international R&D and national agriculture development strategies (Pingali 2010). Even within a country, the process for identifying technology needs and prioritizing them for budget support is uncertain. As a result, R&D activities continue to be undervalued in national strategies and donor priorities.

More than two decades ago, Vernon Ruttan suggested the formation of "National Research Support Groups" that would assess and prioritize research demands and champion their supply at the national level (Ruttan 1987). Such groups could also be conduits between national R&D with international research pipelines, and between client demand at the farm level and other actors in commodity-specific value chains or the wider innovation system (Ekboir and Rajalahti 2012; Lynam 2012). Better data generation and analysis could strengthen the ability of such national research groups to identify high-priority problems and potential solutions available from the global research community, and coordinate their adaptation and dissemination at the national level. Most important, these research groups could help strengthen subregional and regional voices by working collectively with regional groups, such as the Southern African Development Community and ECOWAS, and with global alliances, such as the Global Forum for Agriculture Research.

Conclusion

There are many innovative approaches to the challenge of bringing the best science to bear on feeding an expected population of nine billion by 2050, the most significant portion of whom will be in SSA. Considerable financial resources will likely be needed to underwrite those approaches, and donor assistance will be a likely source of funding for agricultural research in the region. What remains to be seen is whether public policymakers and their constituents in donor countries are willing to invest in agricultural research for developing countries.

Despite science-led successes in agricultural development between the 1960s and 1980s, the international donor community largely turned its back on agriculture beginning in the mid-1980s. The consequences for national and international research capacity are still evident today, and recovery has been slow. Yet several watershed events in recent years have encouraged the donor community to re-engage with global efforts in support of agricultural development and research. There are growing expectations of a long-term renewal in commitments to agricultural research by both the leading bilateral and multilateral donors and the growing group of new players in the international donor community. And although these new participants provide just a sliver of the development assistance channeled toward agricultural development, they bring new resources, expertise, and perspectives to the table. ODA from Brazil and China, coupled with strategic investments from BMGF, are among the most influential contributions amassing in this area.

But if the past is any predictor of the future, caution is warranted. Quite simply, it is too soon to determine whether this is a momentary blip in the longer downward trendline. Several efforts to improve the transparency and efficacy of donor assistance to agricultural research may help address this concern. Going forward, greater transparency is needed to ensure accurate and constructive analysis of development assistance levels, trends, priorities, destinations, and uses—especially for new funding sources that are not reporting to the OECD-CRS. Continued monitoring of both new donors and the more traditional bilateral and multilateral donors is still needed to continuously improve coordination and effectiveness in compliance with the Paris Principles on Aid Effectiveness and the Accra Agenda for Action.

Greater ownership of the development assistance agenda in support of agricultural R&D by leaders, agricultural ministries, research organizations, farmers' associations, and other constituencies in SSA is also needed. Experiences from other developing countries—most notably China, India, and Brazil—suggest that more assertive priority setting, resource allocation, and R&D management at both national and subnational levels are feasible and desirable. However, this can only occur if SSA countries open the door to a wider discussion on the opportunities and limits of development assistance to agricultural R&D.

Efforts must also be made to monitor the performance of the recipients of development assistance for agricultural research—in terms of both developing-country governments and international agricultural research organizations. Close monitoring is needed to determine whether recipients are chalking out their own, independent priorities and strategies; making necessary changes to the governance, organization, and management of agricultural research efforts to increase research impact; and monitoring the returns on their investments of scarce public resources for research. It may be too early to tell, but the signs are both positive and worrisome—for policymakers, for donors, and for food-insecure households throughout the developing world.

References

AATF (African Agricultural Technology Foundation). 2010. *A Study on the Relevance of Chinese Agricultural Technologies to Smallholder Farmers in Africa.* Nairobi, Kenya.

AidData. 2013. *Chinese Official Finance to Africa Dataset, Version 1.0. Excel file.* Accessed June 2013. http://china.aiddata.org/datasets/1.0, Washington, DC: Center for Global Development.

Alex, G. 2013. Unpublished data files. United States Agency for International Development, Washington, DC.

Anderson, J., and J. Roseboom. 2013. "Towards USAID Re-engaging in NARS Development: An Issues Paper and Literature Compilation." Mimeo, Weidemann Associates Inc. for the United States Agency for International Development, Washington, DC.

AU (African Union). 2011. "2nd Africa–India Forum Summit—Enhancing Partnership: Shared Vision." Accessed March 27, 2014. http://summits.au.int/en/AfricaIndia.

Barka, H. 2011. *Brazil's Economic Engagement with Africa.* Africa Economic Brief 2 (5). Tunis, Tunisia: African Development Bank Group. Accessed November 4, 2011. www.afdb.org/fileadmin/uploads/afdb/Documents/Publications/Brazil%27s_Economic_Engagement_with_Africa_rev.pdf.

Barwale Foundation. 2012. "The Barwale Foundation: Activities." Accessed March 20, 2014. www.barwalefoundation.org/html/activities.htm.

Benin, S. 2014. *Aid Effectiveness: How Is the L'Aquila Food Security Initiative Doing?* IFPRI Discussion Paper 1329. Washington, DC: International Food Policy Research Institute.

Bräutigam, D. 2009. *The Dragon's Gift: the Real Story of China in Africa.* Oxford: Oxford University Press.

———. 2011. "Aid 'with Chinese Characteristics': Chinese Foreign Aid and Development Finance Meet the OECD-DAC Aid Regime." *Journal of International Development* 23 (5): 752–764.

Buckley, L. 2013. *Narratives of China–Africa Cooperation for Agricultural Development: New Paradigms?* Future Agricultures Consortium Working Paper 52. Sussex, UK: Future Agricultures Consortium.

Byerlee, D. 1998. "The Search for a New Paradigm for the Development of National Agricultural Research Systems." *World Development* 26: 1049–1055.

Cabral, L., and A. Shankland. 2013. *Narratives of Brazil–Africa Cooperation for Agricultural Development: New Paradigms?* Future Agricultures Consortium's China and Brazil in African Agriculture (CBAA) Project Working Paper Series 48. Brighton, UK: Institute of Development Studies.

CGIAR. 2013. "Media: Press Releases: CGIAR Doubles Funding to $1 Billion in Five Years." Accessed March 7, 2014. www.cgiar.org/media/hopes-rise-for-global-food-security-as-worlds-largest-agricultural-research-partnership-marks-major-milestone-doubles-funding-to-1-billion-in-five-years/.

Chichava, S., J. Duran, L. Cabral, A. Shankland, L. Buckley, T. Lixia, and Z. Yue. 2013. *Chinese and Brazilian Cooperation with African Agriculture: The Case of Mozambique.* Future Agricultures Consortium's China and Brazil in African Agriculture (CBAA) Project Working Paper Series 48. Brighton, UK: Institute of Development Studies.

Christensen, C. 1994. *Agricultural Research in Africa: A Review of USAID Strategies and Experience.* Technical Paper 3. Washington, DC: United States Department of Agriculture, Economic Research Service.

Coppard, D. 2010. *Agricultural Development Assistance: A Summary Review of Trends and the Challenges of Monitoring Progress.* London: Development Initiatives/ONE.

Ekboir, J., and R. Rajalahti. 2012. "Coordination and Collective Action for Agricultural Innovation." In *Agricultural Innovation Systems: An Investment Sourcebook.* Washington, DC: World Bank.

Evenson, R., and D. Gollin. 2003. "Assessing the Impact of the Green Revolution, 1960 to 2000." *Science* 300 (5620): 758–762.

Fan, S. 2000. "Research Investment and the Economic Returns to Chinese Agricultural Research." *Journal of Productivity Analysis* 14 (2): 163–182.

Fan, S., P. Hazell, and S. Thorat. 2000. "Government Spending, Growth and Poverty in Rural India." *American Journal of Agricultural Economics* 82 (4): 1038–1051.

Fan, S., and P. Pardey. 1997. "Research, Productivity, and Output Growth in Chinese Agriculture." *Journal of Development Economics* 53: 115–137.

FAO (Food and Agriculture Organization of the United Nations). 2009. *The State of Food Insecurity in the World: Economic Crises: Impacts and Lessons Learned.* Rome.

———. 2010. "Food Security Programmes Support." Accessed March 25, 2014. www.fao.org/spfs/south-south-spfs/ssc-spfs/en/.

FARA (Forum for Agricultural Research in Africa). 2006. *Framework for African Agricultural Productivity.* Accra, Ghana.

Farming First. 2013. "Event: Eradicating Hunger and Malnutrition in Our Lifetime." Accessed March 10, 2014. www.farmingfirst.org/tag/sdgs/.

Feed the Future. 2011. "Feed the Future." Accessed November 28, 2011. www.feedthefuture.gov.

Fletcher, P., ed. 2011. "Impact Investment: Understanding Financial and Social Impact of Investments in East African Agricultural Businesses." The Gatsby Charitable Foundation." April 2016. www.gatsby.org.uk/uploads/africa/reports/pdf/gatsby-africa-aac-impact-investment-study-2011.pdf.

FOCAC (Forum on China-Africa Cooperation). 2009. "Sharm El Sheikh Action Plan, 2010–2012." Accessed March 25, 2014. www.focac.org/eng/ltda/dsjbzjhy/hywj/t626387.htm.

Freemantle, S., and J. Stevens. 2013. "China's Food Security Challenge: What Role for Africa?" In *Agricultural Development and Food Security in Africa: The Impacts of Chinese, Indian and Brazilian Investments,* edited by F. Cheru and R. Modi. London: Zed Books.

Future Agricultures Consortium. 2012. "Political Economy of Agricultural Policy in Africa (PEAPA)." Accessed March 25, 2014. www.future-agricultures.org/research/policy-processes/592-political-economy-framework#.UzHPjvldWAg.

G8 (Group of Eight). 2009. "G-8 Efforts towards Global Food Security." Accessed March 7, 2014. www.g8italia2009.it/static/G8_Allegato/G8_Report_Global_Food_Security,2.pdf.

GAFSP (Global Agriculture and Food Security Program). 2011. GAFSP website. Accessed August 30, 2011. www.gafspfund.org.

Haggblade, S., and P. Hazell, ed. 2010. *Successes in African Agriculture: Lessons for the Future.* Baltimore: Johns Hopkins University Press.

Hazell, P. 2010. "The Asian Green Revolution." In *Proven Successes in Agricultural Development: A Technical Compendium to Millions Fed,* edited by D. Spielman and R. Pandya-Lorch. Washington, DC: International Food Policy Research Institute.

Headey, D., and S. Fan. 2010. *Reflections on the Global Food Crisis: How Did It Happen? How Has It Hurt? And How Can We Prevent the Next One?* Research Monograph 165. Washington, DC: International Food Policy Research Institute.

Herdt, R. 2012. "People, Institutions, and Technology: A Personal View of the Role of Foundations in International Agricultural Research and Development 1960–2010." *Food Security* 37 (2): 179–190.

Hudson Institute. 2012. *The Index of Global Philanthropy and Remittances 2012.* Washington, DC: Hudson Institute Center for Global Prosperity.

Islam, N. 2011. *Foreign Aid to Agriculture: Review of Facts and Analysis.* IFPRI Discussion Paper 1053. Washington, DC: International Food Policy Research Institute.

Lele, U. 2009. "Global Food and Agriculture Institutions: Cosmology of International Development Assistance." *Development Policy Review* 27 (6): 771–784.

Lele, U., and I. Nabi, eds. 1991. *Transitions in Development: The Role of Aid and Commercial Flows.* San Francisco: ICS Press.

Lynam, J. 2012. "Agricultural Research within an Agricultural Innovation System." In *Agricultural Innovation Systems: An Investment Sourcebook.* Washington, DC: World Bank.

Modi, R. 2013. "India's Strategy for African Agriculture: Assessing the Technology, Knowledge, and Finance Platforms." In *Agricultural Development and Food Security in Africa: The Impacts of Chinese, Indian and Brazilian Investments,* edited by F. Cheru and R. Modi. London: Zed Books.

Morton, J. 2010. *Analysis of Donor Support to CAADP Pillar 4–Phase 1.* Brussels: European Initiative for Agricultural Research for Development.

Mukwereza, L. 2013. *Chinese and Brazilian Cooperation with African Agriculture: The Case of Zimbabwe.* Future Agricultures Consortium's China and Brazil in African Agriculture (CBAA) project Working Paper Series 48. Brighton, UK: Institute of Development Studies.

OECD (Organisation for Economic Co-operation and Development). 2013a. "OECD. StatExtracts: Aid (ODA) by sector and donor [DAC5]." Accessed March 25, 2014. http://stats.oecd.org/Index.aspx?DataSetCode=TABLE5.

———. 2013b. "OECD. StatExtracts: Creditor Reporting System." Accessed March 25, 2014. http://stats.oecd.org/Index.aspx?DataSetCode=CRS1.

ONE. 2013. "Food Aid Reform." ONE 2013 Data Report." Accessed September 2013. http://one-org.s3.amazonaws.com/us/wp-contentuploads/2013/10/131007_ONE_FoodAidReform_FINAL.pdf.

Pardey, P., N. Beintema, S. Dehmer, and S. Wood. 2006. *Agricultural Research: A Growing Global Divide?* Washington, DC: International Food Policy Research Institute.

Patriota, T., and F. Pierri. 2013. "Brazil's Cooperation in African Agricultural Development and Food Security." In *Agricultural Development and Food Security in Africa: The Impacts of Chinese, Indian and Brazilian Investments,* edited by F. Cheru and R. Modi. London: Zed Books.

People's Republic of China. 2011. *China's Foreign Aid.* Beijing: Information Office of the State Council.

Pingali, P. 2010. "Global Agricultural R&D and the Changing Aid Architecture." *Agricultural Economics* 41 (S1): 145–153.

———. 2012a. "Changing Trends in the Demand and Supply of Aid for Agricultural Development." Paper presented at the International Agricultural Economics Association meetings, Foz de Iguazu, Brazil, August 20.

———. 2012b. "Green Revolution: Impacts, Limits and the Path Ahead." *Proceedings of the National Academies of Science* 109 (31): 12302–12308.

———. 2012c. *The Bill & Melinda Gates Foundation: Catalyzing Agricultural Innovation. 2020 Policy Brief 16, Focus 19, Scaling Up in Agriculture, Rural Development, and Nutrition*, edited by J. Linn. Washington, DC: International Food Policy Research Institute.

Rada, N. 2013. "Assessing Brazil's Cerrado Agricultural Miracle." *Food Policy* 38: 146–155.

Rahmato, D. 2013. "Up for Grabs: The Case of Large Indian Investments in Ethiopian Agriculture." In *Agricultural Development and Food Security in Africa: The Impacts of Chinese, Indian and Brazilian Investments,* edited by F. Cheru and R. Modi. London: Zed Books.

Raitzer, D., and T. Kelley. 2008. "Benefit–Cost Meta-Analysis of Investment in the International Agricultural Research Centers of the CGIAR." *Agricultural Systems* 96: 108–123.

Randall, I. 2011. "Guidelines for Non State Actor Participation in CAADP Processes." CAADP Working Group on Non State Actor Participation. Accessed April 2016. www.nepad-caadp.net/download/file/fid/269.

Renkow, M., and D. Byerlee. 2010. "The Impacts of CGIAR Research: A Review of Recent Evidence." *Food Policy* 35 (5): 391–402.

Rowden, R. 2013. "Indian Agricultural Companies, 'Land Grabbing in Africa and Activists' Responses." In *Agricultural Development and Food Security in Africa: The Impacts of Chinese, Indian and Brazilian Investments,* edited by F. Cheru and R. Modi. London: Zed Books.

Ruttan, V. 1987. *Agriculture Research Policy.* Minneapolis: University of Minnesota Press.

Singh, G. 2013. "India and Africa: New Trends in Sustainable Agricultural Development." In *Agricultural Development and Food Security in Africa: The Impacts of Chinese, Indian and Brazilian Investments,* edited by F. Cheru and R. Modi. London: Zed Books.

Smith, K., T. Yamashiro, and F. Zimmermann. 2010. *Beyond the DAC: The Welcome Role of Other Providers of Development Co-operation.* DCD Issues Brief. Paris: OECD Development Co-operation Directorate.

Spielman, D., and R. Pandya-Lorch, eds. 2009. *Millions Fed: Proven Successes in Agricultural Development.* Washington, DC: International Food Policy Research Institute.

Staatz, J., and C. Eicher. 1990. "Agricultural Development Ideas in Historical Perspective." In *Agricultural Development in the Third World,* edited by C. Eicher and J. Staatz. Baltimore: John Hopkins University Press.

Strange, A., B. Parks, M. Tierney, A. Fuchs, A. Deher, and V. Ramachandran. 2013. *China's Development Finance to Africa: A Media-Based Approach to Data Collection.* Center for Global Development Working Paper 323. Washington, DC: Center for Global Development.

UN (United Nations). 2013. "Sustainable Development Knowledge Platform." Accessed March 10, 2014. http://sustainabledevelopment.un.org/index.php?menu=1300.

von Braun, J., and R. Meinzen-Dick. 2009. *"Land Grabbing" by Foreign Investors in Developing Countries: Risks and Opportunities.* IFPRI Policy Brief 13. Washington, DC: International Food Policy Research Institute.

World Bank. 2007a. *Agriculture for Development.* World Development Report. Washington, DC.

———. 2007b. "West Africa Agricultural Productivity Program (WAAPP)." Accessed March 20, 2014. http://web.worldbank.org/external/projects/main?pagePK=64283627&piPK=73230&theSitePK=40941&menuPK=228424&Projectid=P094084.

———. 2009. "East Africa Agricultural Productivity Program (EAPP)." Accessed March 20, 2014. http://web.worldbank.org/external/projects/main?menuPK=228424&pagePK=64283627&piPK=73230&theSitePK=40941&Projectid=P112688.

———. 2014a. "Data: Inflation, GDP deflator (annual %)." Accessed March 7, 2014. http://data.worldbank.org/indicator/NY.GDP.DEFL.KD.ZG.

———. 2014b. "AFCC2/RI. Agricultural Productivity Program for Southern Africa (APPSA)." Accessed March 20, 2014. www.worldbank.org/projects/P094183/agricultural-productivity-program-southern-africa-appsa?lang=en.

Chapter 7

PRIVATE-SECTOR INVESTMENT IN AFRICAN AGRICULTURAL RESEARCH

Carl Pray, Derek Byerlee, and Latha Nagarajan

Africa has entered a phase of rapid commercialization of its food and agricultural system that provides major new opportunities for privately conducted research and development (R&D). Indeed, unless both public and private investment in agricultural R&D is stepped up sharply, lack of competitiveness will prevent Africa from seizing these opportunities. Rapid growth of domestic and regional markets offers the most attractive opportunities for African commercial agriculture. Assuming that Africa meets a 6 percent growth rate (which many countries are already doing), rising consumer incomes and the projected doubling of the urban population in Africa by 2030 imply that urban food markets will quadruple in the next 20 years (World Bank 2013).

Rising market opportunities can be met by domestic production or by food imports. Given its substantial land and water resources, Africa has a comparative advantage in most food products (with important exceptions, such as wheat). Nevertheless, food import shares have been rising in recent years, such that Africa has converted from a significant net agricultural exporter in the 1970s, to a significant net agricultural importer in the 2000s, reflecting poor competitiveness in many products, such as rice—the fastest-growing import.

Similarly, opportunities are provided by booming export markets where world agricultural trade approximately tripled in nominal value terms from 1993 to 2008. Middle-income developing countries successfully captured the bulk of this market growth (with Brazil, Argentina, Indonesia, Thailand, and Malaysia occupying five of the top six places), but African countries have

Derek Byerlee gratefully acknowledges the assistance of Jonathan Nzuma and S. Doumbia for the case studies of Kenyan tea research and Côte d'Ivoire research, respectively. Carl Pray and Latha Nagarajan gratefully acknowledge the financial support of the Bill & Melinda Gates Foundation and the International Food Policy Research Institute; they also thank their country study collaborators David Gisselquist, Anwar Naseem, David Spielman, Gert-Jan Stads, Louis Sène, Johann Kirsten, Ruan Stander, Choolwe Haankuku, Hannington Odame, Mick Mwala, Harun Ar Rashid, Elsie Kangai, Oscar Okumu, and Rose Matiko Ubwe.

ranked very poorly. While Africa contributes 12.1 percent of world population and 5.3 percent of agricultural gross domestic product (AgGDP), its share of global agricultural exports fell to 2.0 percent in 2009, compared with 7.6 percent in the early 1960s.

Declining competitiveness is in large part due to low and stagnant productivity. Fuglie and Rada (Chapter 3, this volume) indicate that although total factor productivity (TFP) growth in Africa south of the Sahara (SSA) has reversed its negative trend since the mid-1980s, it is still increasing at only half of the rate of developing countries as a whole. They attribute poor productivity performance to low and stagnating investment in R&D, along with poor macroeconomic policy up to 1990. Although Fuglie and Rada (this volume) only analyzed public-sector R&D and CGIAR investments, the lack of private-sector research is also likely to be a factor undermining Africa's performance. As one example, Africa was the leading producer of oil palm up to 1975, a crop that originated in Africa; however, in the 1960s, Malaysia mounted a strong industry-led effort to fund and conduct R&D on oil palm, which quickly made the country the world's leading palm oil exporter. Much of this research spilled over to Indonesia, and exports of palm oil by these two countries now exceed the value of *all* agricultural exports from SSA.

Today, with a more conducive policy environment, commercial agriculture could be a major source of growth in many SSA countries, following the recent path of Brazil, Thailand, and other emerging economies over the past 20 years. The challenge is to invest more in R&D to ensure that African countries can compete in growing markets. Already, a more open policy environment in the 2000s is stimulating strong private investor interest in Africa that could spur private R&D. Agriculture and associated industries are now among the favored sectors for foreign direct investments to the tune of about US$1 billion in 2007 (Miller et al. 2010).

The potential for growth in R&D by private firms and industry associations in Africa is great. Research by private input firms outside of South Africa is just getting started and is still limited to maize, some vegetables, and export crops. Applied research in some large food-processing firms in beer, horticulture, and sugar, as well as research by industry associations funded by levies on export crops, was started in the colonial period. Private research, however, remains constrained by small markets, weak public-sector research programs, a shortage of scientists and technicians, and a difficult business environment, including competition with government corporations and weak intellectual property rights (IPR). Levy-funded industry associations lack internal support for greater funding and government incentives to increase private R&D funding.

This chapter provides an overview of private agricultural R&D in SSA. The approach is not comprehensive, but is analyzed through a series of case studies. The next section provides an overview of three types of private agricultural R&D and their main drivers as a preamble to a more detailed case study review centered on the typology. Thereafter, the discussion focuses on policies needed to realize the potential of private African agricultural R&D into the future, before providing final conclusions.

Types of Private Agricultural R&D

It is well known that the private sector underinvests in research, in large part because of the nonexclusive and nonrival nature of the products of research (Ruttan 2001). In some cases, such as hybrid seed, profits from specialized private input firms that carry out R&D can be appropriated, although these are infrequent in the African context and are mainly confined to hybrid maize. Stronger IPR could assist in overcoming this barrier. A less stated reason for private underinvestment is that much research involves significant economies of size resulting from high fixed costs relative to market size. In agriculture—outside of a few cases, such as plantation crops and commercial horticulture—farms and other firms in the industry are generally too small to efficiently undertake R&D for their own use beyond very simple adaptive research for testing new technologies.

Private agricultural research is generally funded by three groups of industries: agricultural input industries, production agriculture, and the agricultural processing industry. These firms may finance and conduct their research in-house, outsource research to research organizations or universities, or finance research collectively with other firms. In Africa, the three main combinations of industry and research organizations are in-house research conducted by input firms, in-house research conducted by large plantations or agricultural processing firms, and collective research undertaken by producers and processors. Each of these is discussed below in turn.

Intramural R&D by Agricultural Input Industries

Agricultural input firms invest in research to develop new inputs that will increase the productivity of farmers or increase the quality of agricultural output. Input firms profit by increasing their sales to farmers. The firms must decide whether to invest in agricultural R&D based on its cost, the probability of its success, the size of the potential market for the improved input, the cost of producing and marketing the improved input, and the firms' ability to keep it from being copied by competitors. Moreover, the decision to conduct R&D has to be weighed against options to license public or private technology from within or outside the country.

Research to develop new chemicals, new biotech traits, and new tractors can require large investments in experiment stations, laboratories, computers, and engineering facilities, as well as large numbers of well-qualified scientists, engineers, and technicians. Often only a few multinational companies make sufficient turnover worldwide to fund the research that generates these new technologies. Monsanto, for example, invests more than US$1.5 billion per year in R&D (Monsanto 2013). Some types of applied research, however—such as plant breeding or engineering new agricultural implements—do not require such massive investments. This is particularly the case when African firms can build on research conducted by government institutes, universities, or international agricultural research centers. An important factor that drives up private firms' research costs in Africa and reduces the probability that research will be successful is the shortage of scientists and weak public-sector research in many countries.

In small countries, of which Africa has many, the small size of both the agricultural sector—in terms of AgGDP—and of modern input markets significantly limits opportunities for private firms to profit from investing in agricultural research. The level of adoption of improved inputs in SSA is far below other regions of the world. For example, the 2008 World Development Report showed that the adoption of improved crop varieties in Africa in 2000–2005 was considerably lower than in Asia and Latin America for all major staple crops except cassava (Walker et al. 2014). Government policies that tax agriculture rather than staying neutral or subsidizing it are in part to blame for holding back the demand for modern inputs in many African countries. The nominal rate of assistance to agriculture shows that Senegal, Tanzania, and Zambia taxed agriculture in the period 2000–2004, and that South Africa was neutral. Only Kenya subsidized agriculture (Anderson and Valenzuela 2008). Another factor that, until recently, limited the incentive for African firms to invest in modern inputs was the large role that parastatal corporations played in the provision of inputs. In the seed sector, Tanzania had the parastatal Tanseeds, Zambia had Zamseeds, and Kenya had the Kenya Seed Company (KSC). Only South Africa had a fully privatized seed production industry before 2000.

The ability of agricultural input firms to capture some of the economic surplus created by new technologies they develop may also be limited by weak

IPR. Park's (2008) index of the strength of IPR ranges from 0 to 5. With a score of 4.25, South Africa has the strongest IPR of the sample countries; scores for the other African countries in the study range from 1.94 to 3.22. In some cases, firms can protect their investments in innovation by offering technologies that are difficult to copy for technical reasons and, hence, preserve trade secrets (such as hybrid cultivars whose parentage is not divulged, or pesticides produced using complicated chemistry).

Intramural R&D Undertaken by Large Plantation or Processing Companies

Agricultural processing and plantation firms invest in research and innovation to reduce their production costs or improve the quality of the agricultural products they produce or purchase.[1] Like input firms, they will have to make substantial investments in experimental fields, laboratories, and scientists, and compare the costs for research with the costs and effectiveness of importing technologies.

For plantations, the profits from these investments in R&D or imported technologies depend on the size of the plantation and how much the innovation reduces the cost of production or increases the market value of their commodity. If the innovation quickly spreads to competing firms and pushes down commodity prices, the R&D investment will yield little profit. Only a few plantation companies, such as those producing pineapples in Kenya and tea in several East African countries, are big enough to fund their own research.

For agricultural processing firms, the returns to R&D investments are a function of the degree to which agricultural costs can be reduced, based on reduced prices for commodities purchased combined with the quantity of products purchased—both of which can be strengthened by monopsonistic power in the commodity market. Some sugar mills in Africa own large sugarcane plantations or have sufficient market power to profit from investing in their own sugarcane research. Their returns depend on the degree to which the research can reduce production costs, while maintaining quality. Tobacco

1 Note that these industries are involved in research targeting agricultural production and agricultural processing. However, in the context of this discussion, the focus is on research related to agricultural production. Processing and plantation firms are interested in having a steady supply of good-quality agricultural produce at low prices. For these firms to invest in research on agricultural production, the market structure of the processing industry needs to be monopolistic or oligopolistic.

processors in several countries invest in tobacco research to keep the costs of tobacco production down and improve quality.

When individual plantation and processing firms need to solve key problems but are not large enough to afford their own research infrastructure, they often contract with public-sector providers to conduct targeted research, for example, on specific diseases or pests or breeding-related issues. Alternatively, they may organize the industry to support levy-based R&D (described below). Like input industries, private plantation and processing industries have been limited by government policies and ownership. Parastatals in the processing industry and marketing boards for export commodities limited the role of the private sector. The sugar industry, for example, was controlled by parastatals in Senegal and Tanzania up until 1995 and 1998, respectively. Cotton processing and exports and groundnut oil production and exports were also government owned in Senegal until the mid-2000s. Privatization and breaking up processing-industry monopolies does not necessarily lead to more research, however, because the new companies may have less ability to capture the gains from their research: their markets may be smaller, and they may lose their ability to extract gains from research that cuts agricultural production costs.

Levy-Based R&D Undertaken through Collective Action by Agricultural Producers and Processors

This type of private R&D is almost always backed by legislation to impose a small tax—commonly known as a "levy" or "cess"—on production in order to fund collective goods or services that are in turn made accountable to industry representatives. This funding arrangement has two major objectives. First, and most obviously, it aims to increase the total funding of R&D in a specific industry (Bingen and Brinkerhoff 2000), given the overwhelming evidence of underfunding of R&D nearly everywhere. Second, if constructed as truly collective action by users rather than a dictate by government, it aims to empower users in setting the research agenda and making research organizations accountable to them (Klerkx and Leeuwis 2008).

Given potential incentives for free riders, this kind of collective action depends on the industry being able to make the cost of research mandatory for all firms, which in turn requires a sufficient number of firms acting through the political process to make the case for the necessary legislation. In a diverse industry with firms of different sizes, R&D costs can be shared according to a measure of size to achieve some level of equity. A common way to do this is to set a levy as a share of production value. In addition, agriculture is agroclimatically and structurally diverse, so some farms will inevitably gain more, and some may even lose from technological change. These disparities are further accentuated if larger and more politically powerful members have disproportionate influence in a collective process to set the research agenda. At the extreme, very large firms able to efficiently conduct their own R&D may have little incentive to join in collective action on R&D.

Beyond these conceptual issues, a number of practical issues have also been identified with using levies to fund research (for example, Brennan and Mullen 2003; Kangasniemi 2003). The most obvious limitations are the feasibility and cost of collecting levies in smallholder agriculture. In general, collecting levies is only cost-effective for commercial crop and livestock products that pass through a small number of processing or marketing points. This is obvious for most export-oriented products, but many opportunities also exist within domestic markets, especially where production is largely commercial and geographically concentrated. Examples in Africa include irrigated rice, wheat, and sugarcane and, in many cases, some partly commercial products, such as groundnuts, poultry, and dairy.

There are also practical issues related to the objective of making R&D more accountable to industry in terms of priorities and delivering results. Levies therefore require strong industry governance and accountability mechanisms, with appropriate means to aggregate demand from different segments of the industry and from different geographical subregions. An additional complication is that the case for collective action goes beyond R&D to include other industry-related public goods or services, such as market promotion, extension, and control of pests. Clearly, efficiencies exist in having one levy cover a variety of activities, but the allocation of the funds among research and other uses becomes a further decision point. Long-term risky activities, such as R&D, are likely to be penalized in this process (as are R&D expenditures in public budgets).

Finally, there are a variety of institutional design issues for undertaking the research generated by such funds. Funds may be managed by a dedicated funding body (common in Australia) that outsources research competitively or through other means to existing, largely public, research organizations. In other cases, a dedicated commodity research institute under the control of the industry is funded through the levy, although this may reduce economies of size and scope. A levy may also be applied across commodity subsectors, with a single governing body that either outsources research (for example, Côte d'Ivoire) or funds a multicommodity research institute (for example, Uruguay). This complicates governance, because allocation across subsectors is a further decision point.

Evidence of Private Intramural R&D by Input Firms, Plantations, and Processing Firms

The evidence presented in this section was collected through a series of case studies in five African countries: Kenya, Senegal, South Africa, Tanzania, and Zambia. Collaborative research teams, including scientists or economists from each country and collaborators from the International Food Policy Research Institute, McGill University, and Rutgers University, conducted country studies during 2009–2011 involving a survey on innovations, research expenditures, and personnel from a sample of private organizations from all segments of agribusiness.[2]

Private research in SSA, in terms of expenditures and the number of scientists, is limited. For the five study countries, total 2008 expenditures on agricultural R&D were at least $75 million purchasing power parity (PPP) dollars (in 2005 prices), and 331 scientists were privately employed (Table 7.1). The survey data, however, do not capture all of the private research undertaken, because a number of firms known to conduct R&D either did not respond or responded without providing expenditure data. As a result, country teams estimated *actual* private R&D expenditure in 2008 to be about $100 million based on their knowledge of the firms that did not provide R&D data (last row of Table 7.1).

In 2008 South Africa accounted for 72–78 percent percent of private agricultural R&D expenditures in the sample (Table 7.1). R&D related to the seed industry was the largest component, followed by research on sugarcane and citrus fruit, which is performed by private organizations paid for by these industries. Senegal had the next-largest private R&D expenditures and number of scientists. Much of Senegal's private research is by recently privatized corporations processing groundnuts and cotton, and a sugar mill that conducts research on sugarcane, sugar milling, and biofuels. Kenya recorded the third-highest private R&D expenditures in 2008. A number of companies in Kenya invest in plant breeding, and a few invest in R&D for fertilizers and processing. Whereas private sugar mills and tea and coffee plantations in some other African countries conduct research for these commercial crops, in Kenya, research on these commodities is conducted by parastatal or nonprofit institutes, and is paid for by a combination of funds derived through commodity levies (discussed further later in this chapter) and from government contributions. Tanzania had the next-highest R&D expenditures, with research concentrated in seed and sugar. In the study sample, private firms in

---

2 Private agribusiness was defined to include agricultural input firms, farms and plantations, and industries that process agricultural products and that were at least 51 percent privately owned. Research foundations and trusts funded through commodity taxes and managed by the government were excluded.

TABLE 7.1 Private R&D, research staff, and research budgets in study countries, 2008

	South Africa		Kenya		Senegal		Tanzania		Zambia	
Industry	Researchers	R&D spending 2005 PPP dollars (thousands)	Researchers	R&D spending 2005 PPP dollars (thousands)	Researchers	R&D spending 2005 PPP dollars (thousands)	Researchers	R&D spending 2005 PPP dollars (thousands)	Researchers	R&D spending 2005 PPP dollars (thousands)
Input supply										
Seed	95	27,143	8	1,600	19	NR	16	740	7	1,340
Fertilizer			2	NR	3	NR	7			
Pesticide	6	4,286			5	NR			2	
Machinery							1	NR	1	NR
Livestock and fisheries inputs	9	2,857	0	NR	7	NR			5	
Plantation processing					9	NR	4	NR		
Crop	91	22,857	2	NR	5	NR	4	477	10	980
Livestock										
Fish					13					
Total for surveyed organizations	201	58,571	12	4,000	61	7,200	32	3,000	25	2,600
Estimated actual total		71,429		8,000		9,400		6,000		5,000

Source: Pray, Gisselquist, and Nagarajan (2011).

Note: NR indicates that data are not reported to protect the organizations' confidentiality, given that only one company in each of these categories reported research. Data on researchers are for individuals, not full-time equivalents; some organizations reported that research staff may also have part-time nonresearch duties. The estimated actual data for South Africa (the last row of the table) include estimates of private research spending by companies that were not contacted or that did not return questionnaires. For example, in the case of South Africa, major research programs, such as Pioneer, Illovo Sugar, Sappi Limited, and Mondi, did not respond to the questionnaire, so as much as an estimated 20 percent of the country's private research could have been omitted. Also, as much as half the country's private research could have been omitted based on lack of data for Del Monte and floriculture firms, such as Oserian. Similarly, as much as half of Kenya's private research, primarily on commercial crops like tobacco and sugarcane, could have been omitted. PPP = purchasing power parity.

TABLE 7.2 Private agricultural R&D intensity, 2008

Measures	South Africa	Kenya	Senegal	Tanzania	Zambia	Bangladesh	India
AgGDP (billions of 2005 US dollars)	8.3	6.3	2.0	6.2	2.8	15	218
Private agricultural R&D (millions of 2005 US dollars)	41–50	1.6–3.2	3.6–4.7	0.9–1.8	1.3–2.5	10–20	251
Private agricultural R&D as a share of AgGDP (%)	0.49–0.60	0.025–0.05	0.18–0.24	0.015–0.03	0.05–0.09	0.07–0.13	0.115

Source: IFPRI–McGill–Rutgers (2010/11); AgGDP data were calculated from World Bank (2011).

Zambia spent the least on R&D, although the country employs more private scientists than Kenya.

The two industries that attracted the most R&D investment are the seed and processing industries, a pattern that is common across all five case study countries (Table 7.1). Research on livestock inputs and pesticides (primarily trials for registration) is important in South Africa, Senegal, and Zambia. Research on sugarcane is important in Senegal, South Africa, and Zambia, as is research on tea and coffee in Tanzania. At 0.6 percent, South Africa's research intensity (private agricultural R&D expenditure as a share of AgGDP) is the highest among the study countries, followed by Senegal (Table 7.2). Zambia has higher research intensity than Kenya and Tanzania, but this is partly because it has a small agricultural sector. Kenya and Tanzania, which have small R&D expenditures and large agricultural sectors, recorded the lowest R&D intensities of the study countries. Bangladesh and India, which were studied at the same time, are included for comparison. They have research intensity levels below South Africa, but about the same as the other African countries.

About half of the research recorded in the 2008 survey was conducted by African firms, some of which are regional multinational corporations. The other half of the R&D was conducted by multinational corporations headquartered in Europe and the United States. In South Africa, US–based firms conducted about half the seed and biotech research and most of the pesticide research; other research is conducted by South Africa–based firms, some of which are themselves multinational corporations—for example, Pannar (seeds), Illovo Sugar, and South Africa Breweries.[3] In Kenya, both local com-

3 Pannar was purchased by DuPont in 2013 and is gradually being integrated into the DuPont Pioneer subsidiary; 51 percent of Illovo was purchased by Associated British Foods in 2006, but it continues to have its headquarters in South Africa and is listed on the Johannesburg Stock Exchange.

panies (such as East African Seed and Western Seed) and Africa-based multinational corporations (such as Pannar and Zimbabwe's Seed Co), as well as Pioneer and Monsanto, have small maize-breeding programs. Multinational corporations are also active in research on tobacco, pineapples, sugarcane, and tea in East Africa. In Senegal the major groundnut, cotton, sugarcane, and horticulture firms that conduct much of the private research are now controlled by French or Swiss firms (Stads and Sène 2011).

The Senegal, South Africa, and Zambia studies contained data on private R&D growth. Between 2001 and 2008, expenditures in South Africa grew by 22 percent in constant US dollars. Eighty percent of the growth was accounted for by seed companies and some smaller companies that are both seed and fertilizer businesses (Kirsten, Stander, and Haankuku 2011). In Senegal, R&D expenditures rose by 40 percent during the same period; all of the increase was accounted for by vegetable seed or vegetable and fruit processing firms (Stads and Sène 2011). In Zambia, the number of scientists who worked at least part time on research grew by 56 percent during 2001–2008[4]; most of this growth was in the processing industry, followed by the seed industry (Mwala and Gisselquist 2012). The interviews conducted in the five countries indicated that plant breeding also grew in Kenya and Tanzania. In Kenya, livestock-related research grew, but data on other industries in Kenya and Tanzania are insufficient to indicate whether private research actually increased in the aggregate during 2001–2008.

The Seed Industry

Maize dominates the seed market in East Africa and Southern Africa. Recent studies of the East Africa market showed that 87 percent of seed sales in Kenya, 71 percent in Tanzania, and 75 percent in Uganda were maize seed sales (Erenstein, Kassie, and Mwangi 2011), whereas vegetables were a small but growing component of the market. Maize is not as important in West Africa, particularly in Senegal, where vegetables constitute a much larger share of the commercial seed market. As of 2007, global multinational companies accounted for 18 percent of maize sales in SSA, whereas regional multinationals, such as Pannar and Seed Co, accounted for 46 percent, and national companies accounted for 36 percent (Langyintuo et al. 2010).

Growth in privately performed R&D in the seed industry developed in response to the liberalization and subsequent growth of the seed industry in

4 More companies reported scientists than R&D trends, so researcher numbers are considered a more accurate indicator of trends (Mwala and Gisselquist 2012).

all study countries, ending government monopolies on the sale of improved seed in combination with varying degrees of privatization. The seed industries of Tanzania and Zambia followed the path of first allowing competition and then privatizing the government seed companies, Zamseed and Tanseed, in the 1990s. Cargill was the first private company to register a maize hybrid in Tanzania, followed by Pannar, Monsanto, and Pioneer later in the decade. In Zambia, Pioneer was the first company to bring in maize hybrids in 1992; other private companies registering one or more maize hybrids in the 1990s included Carnia (from South Africa), Cargill, Pannar, Seed Co, and the local company Maize Research Institute.

Like Zambia and Tanzania, Kenya allowed competition in the 1990s, but unlike those countries, it did not privatize its seed parastatal, KSC. KSC, which produced seed of the Kenya Agricultural Research Institute's (KARI's) new cultivars, had a monopoly on the distribution of certified maize seed until 1993. A key change allowing private companies to enter the market was a new willingness on the part of seed regulators to register cultivars from private companies. A number of small Kenyan companies, such as Western Seed Company and East Africa Seed Co., entered or expanded their seed business in the 1990s. Pannar and Cargill also entered the seed business in that period. Monsanto entered the Kenyan market by buying Cargill's international seed business, and then registering its first maize hybrids in Kenya in 2000. Seed Co's first maize hybrid was approved for sale in 2003. In 2003–2004, however, KSC still accounted for 86.5 percent of the total volume of maize seed produced by the formal seed industry in Kenya, according to Ministry of Agriculture estimates (Odame, Kangai, and Spielman 2012). The partial opening of seed markets in East Africa to maize hybrids from other countries and to seed imports has allowed national, regional, and other multinational seed companies to develop and introduce new cultivars, focusing on crops for which seed markets are large (especially in terms of hybrid maize) and for which the introduction of new cultivars is unregulated (vegetables and forage crops in Kenya and Zambia).[5]

5 Zimbabwe's Seed Co is a useful example. It began in 1940 as a private farmers' cooperative producing maize seed; in 1973 the cooperative purchased a breeding station; and in 1996 it was re-registered as a publicly owned company selling shares on the Zimbabwe Stock Exchange. Seed Co has developed into a regional multinational, expanding into Ethiopia, Kenya, Malawi, Zimbabwe, and other countries as their markets opened. As the business climate in Zimbabwe deteriorated after 2000, Seed Co moved some of its research to Zambia. The company also has a technology agreement with Syngenta that provides Seed Co with access to Syngenta's technologies from elsewhere in the world, and Syngenta with access to Seed Co's white maize hybrids and soybean lines. Recently, the French company Limagrain purchased 25 percent of Seed Co's shares, which will help Seed Co expand its operations in Zambia and Malawi (Mapakame 2013).

Maize research has also increased because of the growth in maize production, which increased in response to the quantity of the maize seed sown. Data on maize seed use from the Food and Agriculture Organization of the United Nations show that in East Africa, maize seed increased from about 300,000 tons in 2000 to 400,000 tons in 2011. Seed use in West Africa (primarily Nigeria) also increased rapidly. The only subregion where maize seed use declined was Southern Africa, which includes South Africa and Zimbabwe. Further driving R&D on maize, most of the maize sold is hybrid maize, giving developers the ability to capture a substantial part of the benefits from new hybrids, as they must be purchased every year to maintain their yield performance.

South Africa's seed industry research dwarfs research in the other sample countries (more than $19 million in seed and biotech R&D in 2008, compared with less than $2 million for the other four countries combined). The size of the market for innovation in maize and soybeans in South Africa increased dramatically during 2001–2008 because of the spread of genetically modified (GM) maize from 57,000 hectares in 2000/01 to 1.89 million hectares in 2009/10 (Gouse 2013). Even in 2001, almost all of the maize seed was high-quality hybrids that companies sold at high prices. Adding GM insect resistance and herbicide tolerance allowed firms to charge a technology fee on top of the hybrid price. Demand for conventional hybrids from South Africa also increased as markets in Southern and East Africa opened up, but where the release of GM seeds has not been approved. IPR on new plant varieties, hybrids, and transgenic traits are as strong in South Africa as they are in the United States. With the end of apartheid, foreign firms brought in new capital to establish subsidiaries, purchase local seed companies, and finance research.

The investments in seed industry research in South Africa were encouraged by several other factors. The dominance of large commercial farms that regularly purchase the latest variety or hybrid has provided a ready market for companies. The availability of scientists in South Africa was greater than other African countries because of a relatively large public agricultural research system and a strong agricultural university system. South Africa is also a comparatively comfortable place for foreign scientists to live, so private firms can attract scientists from elsewhere in Africa and the rest of the world. Strong human resource capacity and IPR, together with an efficient biosafety regulatory system, allow the private and the public sectors to use the latest biotechnology research tools and GM traits to produce improved crop varieties faster. South Africa's capacity to use biotechnology has been helped by biotechnology research funded by the Rockefeller Foundation, the Bill &

Melinda Gates Foundation, and the United States Agency for International Development.

The Plantation and Processing Industries

The basis for private research by firms in the agricultural processing industry in Africa was also the elimination of government monopolies and the privatization of government processing industries. The impact of nationalization and later liberalization on private technology transfer and research varies considerably among countries, products, and time periods. After 2000, Senegal privatized the companies that controlled the processing of two major cash crops: cotton and groundnuts. The government sold 51 percent of SODEFITEX, the cotton monopoly, to a French company in 2003. And Suneor, the government groundnut processing company, was privatized in 2005. Both of these companies had their own research programs before privatization; the privatization of these programs led to much of the growth in private research in Senegal (although not necessarily to more R&D for the country in total).

In East Africa, nationalization followed by liberalization and privatization had mixed impacts on R&D in the plantation and processing subsectors. During colonial times, research on plantation crops—such as coffee and tea in Kenya and Tanzania, sugarcane in Kenya, and such cash crops as cotton in Zambia—was originally financed by commodity levies. Independent governments nationalized some of the monopolies and extended government control over formerly autonomous research institutes. In Kenya in the 1970s, monopolists controlled pineapple processing and plantations (Del Monte), barley (Kenya Breweries Ltd.), and tobacco (BAT Kenya Ltd.), paying for and managing research on these crops. Liberalization did not affect the pineapple monopoly, but South African Breweries entered the beer industry in 1998, and a local company, Mastermind Tobacco, and another multinational, R. J. Reynolds, entered the tobacco industry. A study by Ndii and Byerlee in 2004 suggested that private research on plantation crops and processing in Kenya declined with increased competition after liberalization.

The end of apartheid in South Africa and liberalization of trade and regulations on foreign investment in the rest of Africa have encouraged South African firms to expand into African regional markets. An important example of the impact of liberalization on agricultural research is Illovo Sugar. Its expansion outside South Africa began in 1996, when it bought 50 percent of a Mozambique sugar mill. In 1997, it bought Lonrho Sugar Corporation, which had sugar mills and land in Malawi, Mauritius, South

Africa, and Swaziland. In 1998, Illovo bought 55 percent of the Tanzanian government's sugar company at Kilombero, and in 2001 it bought the Zambian sugar company that had been a parastatal. Illovo Sugar is now Africa's biggest sugar producer. In 2009/10, the estates it managed produced 6.1 million tons of cane, while contract growers supplied about 8 million tons of cane. About 40 percent of its production is in South Africa. Illovo Sugar accounts for 32 percent of sugar production in Mozambique, 30 percent in South Africa, 35 percent in Swaziland, 46 percent in Tanzania, and 94 percent in Zambia (Illovo Sugar 2011). In 2011, the firm spent $3.5 million on research for all of its African operations, most of which was spent in South Africa (Illovo Sugar 2011).

While privatization and liberalization of rules on foreign investment made the expansion of these companies possible, demand for sugar is what made the investments in expansion into new countries profitable. In turn, this demand provided the incentive to invest in importing technology, such as sugarcane varieties from Mauritius and South Africa, and in the adaptive research that was required to make those technologies productive. International demand for sugar from such countries as India has expanded rapidly in recent years. Preferential trade agreements offered by the European Union to low-income countries have helped some African countries to improve their position in supplying European demand for sugar.

Commodity Levies to Fund R&D

The Current Use of Levies and Their Potential

Agricultural Science and Technology Indicators (ASTI) data and reports provide an overview of current use of commodity levies (or cesses). A very rough overall estimate indicates that, in 2008, about $93 million in 2005 PPP dollars, or about $45 million in 2005 US dollars at official exchange rates, was provided by industry for research in SSA, most commonly for coffee, cocoa, tea, sugar, and tobacco (Table 7.3). This amounted to less than 6 percent of total agricultural R&D spending. Only 9 of 26 ASTI countries appear to use any levy, and where they do, only a couple of commodities are covered. Additionally, more than half of these funds were generated in just two countries: Ghana for cocoa (33 million 2005 PPP dollars) and South Africa for sugar (19 million 2005 PPP dollars) (Nieuwoudt and Nieuwoudt 2004). Côte d'Ivoire, Kenya, Malawi, Mauritius, and Tanzania each generated 5–10 million 2005 PPP dollars in levies.

TABLE 7.3 Overview of industry funding of crop-specific research institutes in Africa, 2008

Country	Commodity	R&D expenditure	Levy	Industry value	Specific crop R&D intensity (%)[a]	Public R&D intensity (%)[a]
			2005 PPP dollars (millions)			
Ghana	Cocoa	33.3	32.7	757	4.4	0.9
Kenya	Tea	3.1	1.7	369	0.8	1.3
Kenya	Coffee	5.9	4.4	45	12.8	1.3
Malawi/ Zimbabwe	Tea	2.4	1.4	65	3.7	0.7[b]
Mauritius	Sugar	9.9	9.4	149	6.6	3.9
South Africa	Sugar	18.6	na	673	2.8	2.0
Tanzania	Coffee	3.4	0.4	46	7.4	0.5
Tanzania	Tobacco	0.1	na	81	0.1	0.5
Tanzania	Tea	4.1	na	37	11.1	0.5
Uganda	Coffee	4.8	1.5[c]	228	2.1	1.2

Sources: Calculated by authors based on ASTI data; industry value data are estimated values of production in purchasing power parity (PPP) dollars 2004–2006 based on FAO (various years).

Notes: Research intensity is research expenditures relative to the value of sales expressed as a percentage. na = data were not available. [a]Specific crop R&D intensity refers to levy-funded R&D as a share of industry value of production; public R&D intensity refers to investment in R&D for all public agricultural research as a share of agricultural GDP. Public R&D estimates include levied crop research for Ghana, Tanzania, and Uganda. [b]2001 data. [c]Uncertain estimate. ASTI = Agricultural Science and Technology Indicators; FAO = Food and Agriculture Organization of the United Nations; GDP = gross domestic product; R&D = research and development.

As previously discussed, levies are mainly legislated, often at the request of the industry. However, in most cases, the revenues raised are not allocated exclusively to research; research institutes have to compete with other uses, such as extension and market promotion. In a few cases, the industry may make ad hoc contributions to funding. This is most evident for the Cocoa Research Institute of Ghana, which receives a yearly budget directly from the earnings of the Cocoa Board of Ghana. This is the largest industry-funded research effort in Africa, although it is not strictly a levy.

The institutional arrangements under which the industry funds are allocated to R&D also vary considerably.

- Most common is a legally required levy that is managed by an industry council or board with official status, which then allocates a portion of the funds to a dedicated nonprofit research institute affiliated with that board (for example, tea and sugar in most countries). The influence of producer organizations in these boards may be quite variable. A variant is to have a subregional research institute funded by a levy applied in several neighboring countries. The only current example is the Tea Research Foundation of Central Africa, funded by a levy on tea production in Malawi and Zimbabwe and governed by representatives of both countries.

- A second mechanism is a legally required levy allocated to fund research on a commodity at a public research institute, with varying degrees of industry input into the research program. Cotton in both Mozambique and Tanzania seems to be in this category. In both cases, there have been difficulties in establishing an institutional structure that provides the industry with a sufficient sense of ownership of the funds they provide (Boughton and Poulton 2011).
- Côte d'Ivoire, discussed further below, is a special case, where a council of several producer organizations organizes the collection of the levy and then allocates it to the public research institutes.

Comparing crop-specific research expenditures relative to crop production values (assuming no research is conducted on these crops outside of these institutes) provides an estimate of crop-specific research intensity that can be compared with the research intensity for all public research in each country. In eight cases, crop-specific research intensity supported by industry funding is higher than for overall research intensity. In some cases, research intensity is very high by global standards, although some are small industries (tea in Malawi and Tanzania), or the industry is declining (coffee in Kenya and sugar in Mauritius). In two cases, tea in Kenya (a large industry) and tobacco in Tanzania (a relatively small industry), crop-specific research intensity is lower than the national average. In industry-managed research institutes, there is also some evidence that spending per scientist is higher than in public institutes (Byerlee 2011).

Currently, collected commodity levies only represent a small share of the potential for levy funding in SSA. To estimate potential, it was assumed that all export crops could be levied, since they pass through one or very few ports, and are relatively easy to implement. IFurthermore, in some cases the nature of production or processing could facilitate the collection of levies on commercially oriented production for domestic markets. Sugarcane, oil palm, some other oil crops, and wheat are in this category because they mostly pass through a few fairly large-scale mills. In addition, crops extensively produced under irrigation, such as rice in some countries, are largely commercial and would be easy to levy. Some commercial livestock, such as dairy and poultry production, could also be levied.

In addition, a minimum threshold industry size is needed to introduce a levy because the levy generated has to be large enough to cover the costs of collecting and managing it. Setting an arbitrary threshold industry size of $100 million per country, a 1 percent levy would generate at least $1 million for R&D, which is sufficient to fund a small research institute. Above this threshold (assuming all large export commodities are levied and about half of the large commercially oriented commodities for domestic markets are levied), the potential total levy would be about $250 million—some five times the amount currently levied for R&D. The potential of levy funding could be greater if regional collective action that could fund R&D on commercial crops in small countries were included. Regional collective action, as already employed for rice in Latin America, would be a logical extension of the current move toward formal regional collaboration in food crop research in Africa. These estimates show the great potential to increase R&D funding through collective action for commercial crops and livestock in SSA.

Case Studies of Levy Funding in Africa

Two cases of levy funding in Africa were chosen for further analysis: (1) multiple products in Côte d'Ivoire and (2) tea in Kenya. Two additional cases from Latin America—the National Agricultural Research Institute (INIA) in Uruguay and coffee and oil palm research in Colombia—were reviewed for comparative purposes; details are provided in Byerlee (2011). The reviews focused on four issues: institutional structure, levels of and trends in funding, evidence of the effectiveness of the research funded, and the role of producer associations in enhancing effectiveness. The two African cases are summarized below, followed by a synthesis of the four cases. Full information on all four cases can be found in Byerlee (2011).

PROFESSIONAL FUND FOR AGRICULTURAL RESEARCH AND EXTENSION, CÔTE D'IVOIRE

The Interprofessional Fund for Agricultural Research and Advisory Services (FIRCA) was established in 2003 as a professional agency operating under private company law to fund research and extension.[6] FIRCA is a federation of 14 industry associations (as of 2012), including producers and processors. The associations have a majority vote (73 percent) both in the General Assembly and in FIRCA's Executive Board, which is appointed by the assembly. Only 5 percent of the seats are allocated to government officials.

Funds are provided through levies on exports (cocoa or rubber) or processed products (palm oil), or in one case on imports (rice), per agreement of the member organizations. A total of about $15 million in levies was collected in 2008, amounting to 0.26 percent of the value of agricultural production, and 0.34 percent of the value of exports in that year. However, four

6 This section is based on FIRCA (various years).

commodities—cocoa, coffee, palm oil, and rubber—provided 92 percent of the funding. This in part relates to the difficulty of collecting the levy for some commodities (such as poultry and swine for domestic markets) and to the exemption of most food crops, although FIRCA does fund research on these commodities. The levies are all volume based and fixed by the industry association. The actual levy as a share of production value varies substantially, from 0.6 percent for cocoa/coffee, to 1.6 percent for rubber, and 3.1 percent for cotton. FIRCA also receives about 10 percent of its funding from the state.

Industry associations determine the projects to be funded for that subsector; however, research tends to receive a small share of the allocation, depending on the industry, relative to extension. Over the period 2004–2008, FIRCA allocated only 18 percent of its budget to research; 59 percent was allocated to extension, and most of the rest covered administrative costs. The relatively modest amount of the levy and the low share allocated to research mean that research intensity is often low. In the case of cocoa, for example—by far the largest and most important sector—it was only about 0.2 percent of industry value in 2008, compared with 4.4 percent in Ghana.[7]

FIRCA contracts most research to the National Center for Agricultural Research (CNRA). Originally, the government was to provide 40 percent of CNRA's funding; however, in practice, this share has been much lower and was only 15 percent in 2008. Presumably, government funding was intended to cover research gaps, especially for noncommercial food crops. FIRCA has attempted to fill this gap through a solidarity fund for food crops but, even so, research on food crops is seriously underfunded.

FIRCA is an interesting example of funding commercial research in Africa. In industries where there is strong buy-in from producers and other industry stakeholders, such as rubber, industry associations are clearly in the driver's seat in setting the research and extension agenda. However, other industries, such as sugar, remain unconvinced and have not joined FIRCA. Moreover, research investment in key sectors is still low as a share of production value—and is extremely low for the largest sector, cocoa. The associations have tended to prefer short-term gains from extension over longer-term, riskier gains from research. Administrative costs, at 18 percent of the total budget, also seem very high relative to international norms.[8]

---

7 The figures for Côte d'Ivoire include coffee, which is about 10 percent of the value of cocoa.

8 As a general rule of thumb, a research funding agency that does not undertake research should be able to administer the funds with less than 5 percent of the total research expenditures.

TEA RESEARCH FOUNDATION OF KENYA

Kenya is the world's largest exporter of tea, and tea is Kenya's largest export—amounting to more than $1 billion per year, or 26 percent of agricultural export earnings.[9] Originally, tea was mostly produced on large estates, many in the hands of multinational companies. Over time, with support from government and donors, smallholders have increased their share to now account for 62 percent of national tea production. An estimated 630,000 smallholders have an average of 0.25 hectares, and 63 tea-processing factories are owned by smallholders and managed on a fee basis by the Kenya Tea Development Agency (KTDA), a private company owned by smallholder tea producers. KTDA also provides inputs and advisory services to smallholders. Significantly, the yield gap between smallholders and estates has fallen from 68 percent in 1980 to only 18 percent in 2007 (Mitchell 2012).

The Tea Research Foundation of Kenya (TRFK) was established as a parastatal in 1980 incorporated under the Companies Act to conduct tea research in Kenya. TRFK has a board of 13 members representing various tea organizations, although the majority—including the chair and chief executive officer—are appointed by the government. Smallholder interests on the board are represented by KTDA. TRFK is a small organization with only 13 scientific staff and 124 support staff. TRFK receives more than 80 percent of its funding from a volume-based levy, and the remainder from other sources, including self-generated income. The levy funding as a share of output value has been low and declining—only about 0.1 percent of the production value, which is half of what it was in 2000. Some tea research is conducted by large companies, but combined research intensity on tea is still likely to be well below research intensity estimates for public research in Kenya of 1.3 percent (Flaherty et al. 2010). In addition, TRFK's expenditures have fallen by more than half in real terms since 2000.

TRFK works closely with the industry in setting its research program and disseminating results; however, smallholders largely participate in governance through KTDA. This has generated criticism that TRFK does not adequately respond to the needs of its clients, and the composition of the board was under review in 2011.

TRFK is generally regarded as being an effective research organization. Salaries are competitive, and scientists have access to a reasonable operating budget, equivalent to the budget for salaries. However, the capital budget has been very small, at less than 2 percent of the 2010 budget.

---

9 This section is based on Nzuma (2011).

As already discussed, Kenyan tea is generally regarded as a success story in terms of yields and competitiveness. TRFK products, such as improved clones, agronomic practices, and innovative processing methods, have been widely adopted, especially by smallholders. Kangasniemi (2003) characterized tea research in Kenya as effective but underfunded, and that applies even more today. The very low funding for tea research was recognized as a problem, leading to changes in legislation in 2010 to replace the volume-based levy with a 1 percent *ad valorem* levy, 40 percent of which was to be allocated to TRFK. Once implemented, this will represent a significant increase in the tea research budget over 2010 levels, but the intensity of tea research would still be below Kenya's average research intensity for all commodities. The legislation also called for a review of the governance structure of TRFK to allow more direct influence by producers.

IMPLICATIONS OF THE CASE STUDIES

What do the cases, including those from Latin America, collectively reveal about the two objectives of implementing produce levies for R&D—increased funding and more accountability? On the first objective, while the levies have increased funding allocated to R&D overall, the contribution has mostly been modest. INIA in Uruguay is the major exception, where the levy is dedicated to R&D, and where, by law, the government must match industry contributions (Byerlee 2011). In other cases, a surprising finding is that little evidence exists to indicate that R&D on commodities funded by levies is better funded than for commodities where R&D is publicly funded. This is because the levy is often small and subject to serious competition from multiple uses, nearly all of which can demonstrate more immediate benefits than can R&D. In fact, it could be argued that some important levied commodities receive less funding than what might have been provided through normal government budgets (coffee in Colombia, tea in Kenya, and cocoa in Côte d'Ivoire). Part of the problem is that most levies are volume based, and adjustment of the levy rate upward in line with rising prices has been slow. These findings on underfunding of R&D on commercial crops are especially critical, given that spill-ins of technology for these crops are likely to be much smaller than those for food crops, an area where CGIAR has a strong regional presence and an explicit mandate to foster spill-ins.

On the second objective of improving the demand orientation and accountability of research, the conclusions are universally positive. All of the cases have developed governance mechanisms to ensure that producers and processors have a strong say in research priorities. There is little evidence that some large and more politically powerful producers have distorted priorities in their favor. Of course, good governance goes with strong producer and industry organizations, and this is a weakness in many African countries.

Where the research is carried out by a research institute under the control of the producer association or industry board, the research institutes seem to be well managed and productive, relative to public research organizations. They generally have more flexibility in allocating funds between salaries and operating costs, and salaries are competitive—at least with the public sector. And while detailed impact evaluations are not generally available, all can point to significant successes in adoption of their research products.

Policy Options to Encourage Future Growth in Private Agricultural R&D

Several policy tools are available to African policymakers should they decide to encourage more private R&D (Table 7.4). The first set of factors involves the business climate for private firms. Although much has changed, as described above, a number of African countries could still stimulate growth by further liberalizing and privatizing agricultural input and processing industries. Kenya is one example. KARI still has a monopoly on foundation seed of public hybrids and varieties, and KSC remains a government corporation, which limits private firms' share of the hybrid maize seed market and suppresses seed prices. Six of the seven sugar mills in Kenya are owned by the government. The parastatal Central Artificial Insemination Station has a de facto monopoly on the cattle semen market (sustained by regulations limiting who can extract semen, and what foreign bulls are approved).

In some countries, governments need to use their industrial policy to ensure sufficient competition and incentives to conduct research. In South Africa, regulators have brought cases of price fixing against fertilizer producers. In the South African seed industry, antitrust authorities stopped the acquisition of Pannar by DuPont, but their ruling was overturned by the Supreme Court. During negotiations, DuPont agreed to invest in maize for smallholders and to make major new investments in R&D on maize for SSA (Kaskey 2013).

A second set of government policies is government investments and support for public research institutes and research universities. The growth of private seed-industry research in Southern Africa has been based on the research of the International Maize and Wheat Improvement Center (CIMMYT) and

TABLE 7.4 Important policies needed to support private agricultural R&D

Government policy and investment areas	Plantation/processing	Input industries	Levy-based research
1. Business climate for private firms	• Allow private investment by local and foreign firms, and reduce the size of parastatals • Introduce antimonopoly policies to ensure competition	• Allow private investment by local and foreign firms, and reduce the size of parastatals • Introduce antimonopoly policies to ensure competition	• Allow private investment by local and foreign firms, and reduce the size of parastatals • Introduce policies that facilitate collective research funding
2. Policies to increase the productivity and reduce the costs of private research	• Support NARS contract research and research facilities • Invest in PhD training and university-based research	• Support NARS provision of advance lines and germplasm to private seed firms • Invest in PhD training and university-based research • Subsidize venture capital funds and financing for R&D facilities	• Support NARS contract research and research facilities • Invest in PhD training and university-based research • Provide government funds to match commodity levies
3. Policies that influence market size	• Reduce agricultural import and export barriers and other measures that tax agriculture	• Reduce agricultural import and export barriers and other measures that tax agriculture • Reduce technical barriers on trade in agricultural inputs and harmonize regulations regionally	• Reduce agricultural import and export barriers and other measures that tax agriculture • Facilitate collective action on R&D at the subregional level
4. Intellectual property rights		• Improve the enforcement of patents and plant breeders' rights	
5. Technology regulations and quality control	• Establish government laboratories to ensure product quality	• Reduce efficacy testing on products like seeds • Improve control of fake and dangerous inputs	• Establish government laboratories to ensure product quality

Source: Authors.

Note: NARS = national agricultural research system; R&D = research and development..

national agricultural research systems. Seed companies are also supported by public–private partnerships, such as the Drought Tolerant Maize for Africa (DTMA) project, which partners CIMMYT and the International Institute for Tropical Agriculture with 13 national research programs and seed companies in the 13 countries. However, shortages of well-trained scientists are a major constraint to the growth of private R&D in all SSA countries—not only in terms of research, but also in terms of the technology regulatory system and science policies. Thus, continued expansion of higher education and PhD training is necessary (Chapters 8 and 9, this volume).

Other policies could increase potential market size for new technologies and stimulate research. Trade policies, exchange rates, and taxes that reduce the demand for agricultural products still need to be reformed in many SSA countries. Reductions of barriers to regional trade in fertilizer, seed, and other agricultural inputs would be particularly important for input research and technology transfer of inputs. The reduction of badly designed input subsidies that channel input trade through government tenders rather than markets could also provide more space for private-sector input markets. In addition, frequent short-term bans on the export of maize and other products cause sudden declines in prices to farmers and reduce their incentives to invest in modern inputs. Further, relaxation—or at minimum regional harmonization—of technical regulations on agricultural technologies could have a big impact on the pace at which cultivars are introduced and stimulate demand for R&D. Liberalizing tariff and nontariff barriers on technology should increase the quantity and efficiency of private R&D in Africa as a whole by allowing companies to expand their markets for new technology to the region.

Strengthening IPR could provide incentives for increased research in the input industries. For example, even though wheat, soybeans, and cotton are small crops in South Africa, the private sector conducts research on them, in part, because South Africa has strong IPR. In addition, global biotechnology companies are attracted to South Africa because the patent system is effective in protecting their proprietary biotechnology products.

Effective regulations to ensure farmers that the seed or pesticides they buy do in fact have the characteristics advertised on packages can increase the demand for modern inputs. Such regulations can also reduce exposure to dangerous pesticides and other chemicals by reducing or eliminating their use. In addition, regulations allowing the use of safe GM organisms could induce research by seed and biotech companies in some countries.

Finally, African governments and industry associations can do much to exploit the potential of industry levies, which need to be set high enough and, where possible, be dedicated to R&D. Matching government funding can provide a powerful incentive for an industry to impose a reasonable levy on itself. Matching funds can also be justified by spillovers associated with research that go beyond the specific industry. Further, legislating matching funds guarantees a place in national budgets. Despite these advantages, there is still no example of matching funds in Africa, although Côte d'Ivoire tried. Levy-funded research can deliver more efficient and demand-oriented results, whereby strong producer or industry associations ensure an important or even controlling interest in the governance of the funds collected. There are some short-term opportunities—such as cotton in francophone West Africa, where strong organizations already exist—but elsewhere the development of strong producer organizations is by nature a long-term process. The private sector must initiate the development of such organizations, but governments can help through supporting legislation and grants for capacity building. A second-best alternative would be a reformed parastatal with increasing producer influence, such as provided to cocoa research in Ghana.

Conclusion

SSA has immense potential in commercial agricultural production, but missed out significantly in the commodity boom of the early 21st century. Low productivity resulting from low investment in R&D is one of the main reasons for the poor competitiveness of African agriculture. This can be reversed through greater private investment. Signs indicate that this is starting to happen in the region. Privatization and liberalization combined with higher commodity prices have started the growth of R&D from a low base. Research in the seed industry is growing particularly rapidly, led by the maize seed industry and the entry of several multinational companies, as well as regional expansion of local companies. Most of these companies are emphasizing maize because they can protect their intellectual property through the use of hybrids, even in countries where plant breeders' rights do not exist or are not enforced.

Research by the plantation and agricultural processing industries is also increasing. African companies, such as Illovo for sugar, now operate in several countries of the region. Also, global companies based in Europe, the United States, India, and Brazil are investing in the African processing industry. These companies bring technology and, when needed, invest in research. Like the input industry, privatization and liberalization, along with increasing African and global demand for agricultural products, have led the way. In both the agricultural input and processing industries, promising trends could lead to rapid growth in R&D. Governments are working to create a better business environment and reduce barriers to trade and foreign investment. IPR and regulations are gradually improving. When India experienced similar trends, private R&D grew rapidly (Pray and Nagarajan 2012).

Collective action to levy commercially oriented industries, for both exports and domestic markets, can provide several times more funding for research than currently, but this outcome is not guaranteed. Strong industry associations, such as in Côte d'Ivoire, have shown the power of R&D funding provided and managed by the relevant industry.

Greater private investment offers much promise to enhance the funding of African R&D and the effectiveness of the R&D carried out with that funding. Combined with an improved policy and business environment, increased private R&D on commercial crops could allow SSA to regain competitiveness and tap a major opportunity for growth. Even so, increased private R&D needs to be combined with other options for promoting R&D on commercial crops, including public funding, enhanced regional collaboration, liberalized seed markets to encourage spill-ins, and technology transfer through foreign direct investment.

Finally, it is often assumed that the promotion of private research will emphasize commercial farming and neglect the technological needs of resource-poor farmers—but this need not be so. It is true that seed companies focus on developing hybrid seeds that are more expensive than open-pollinated seeds that can be saved from year to year. However, smallholder farmers have extensively adopted hybrid maize in Africa; practically all farmers in Zimbabwe use hybrid seed produced by private R&D. The research priorities of private input firms can also be refocused on the problems of poor people, such as drought focusing on DTMA through public–private research partnerships partly funded by donors or governments (for example, for Africa). Levy-based research organizations can be dominated by large farmers, but the review presented in this chapter found little evidence of this. Most important, by stimulating private investment in R&D in selected areas, such as hybrid maize, public resources can be freed to focus on the problems of resource-poor farmers, such as breeding for neglected crops in marginal areas.

References

Anderson, K., and E. Valenzuela. 2008. *Estimates of Global Distortions to Agricultural Incentives, 1955 to 2007.* Washington, DC: World Bank.

Bingen, R., and D. Brinkerhoff. 2000. *Agricultural Research in Africa and the Sustainable Financing Initiative: Review, Lessons and Proposed Next Steps.* Technical Paper 12. Washington, DC: United States Agency for International Development, Office of Sustainable Development, Bureau for Africa.

Boughton, D., and C. Poulton. 2011. "Cotton Research." In *Organization and Performance of Cotton Sectors in Africa: Learning from Reform Experience*, edited by D. Tschirley, C. Poulton, and P. Labaste. Washington, DC: World Bank.

Brennan, J., and J. Mullen, 2003. "Joint Funding of Agricultural Research by Producers and Government in Australia." In *Agricultural Research in an Era of Privatization,* edited by D. Byerlee and R. Echeverría. Wallingford, UK: CABI Publishing.

Byerlee, D. 2011. *Producer Funding of R&D in Africa: An Underutilized Opportunity to Boost Commercial Agriculture.* Conference Working Paper 4 from the ASTI/IFPRI–FARA Conference "Agricultural R&D: Investing in Africa's Future: Trends, Challenges, and Opportunities," Accra, Ghana, December 5–7. Accessed April 2014. www.asti.cgiar.org/pdf/conference/Theme1/Byerlee.pdf.

Erenstein, O., G. Kassie, and W. Mwangi. 2011. "Challenges and Opportunities for Maize Seed Sector Development in Eastern Africa." Paper presented at the conference "Increasing Agricultural Productivity and Enhancing Food Security in Africa: New Challenges and Opportunities," November 1–3, 2011, Africa Hall, United Nations Economic Commission for Africa, Addis Ababa, Ethiopia.

FAO (Food and Agriculture Organization of the United Nations). Various years. FAOSTAT database. Accessed May 2014. http://faostat.fao.org/site/452/default.aspx.

FIRCA (Interprofessional Fund for Agricultural Research and Advisory Services). Various years. *FIRCA Annual Reports.* Accessed March 2014. www.firca.ci/en/rapports.php.

Flaherty, K., F. Murithi, W. Mulinge, and E. Njuguna. 2010. *Kenya: Recent Developments in Public Agricultural Research.* ASTI Country Note. Washington, DC: International Food Policy Research Institute and Kenya Agricultural Research Institute.

Gouse, M. 2013. "Socioeconomic and Farm-Level Effects of Genetically Modified Crops: The Case of *Bt* Crops in South Africa." Chapter 1 in *Genetically Modified Crops in Africa: Economic and Policy Lessons from Countries South of the Sahara*, edited by J. Falck-Zepeda, G. Gruère, and I. Sithole-Niang. Washington, DC: International Food Policy Research Institute.

IFPRI–McGill–Rutgers (International Food Policy Research Institute, McGill University, and Rutgers University). 2010/11. Unpublished surveys on private agricultural R&D and innovation in South Africa and sub-Saharan Africa. Washington, DC: IFPRI; Montreal: McGill University; New Brunswick, NJ: Rutgers University.

Illovo Sugar. 2011. *Illovo Annual Report, 2011.* Accessed October 1, 2011. www.illovo.co.za/financial/annual_reports/annual_report2011.aspx.

Kangasniemi, J. 2003. "Financing Agricultural Research by Producers' Organizations in Africa." In *Agricultural Research in an Era of Privatization*, edited by D. Byerlee and R. Echeverría. Wallingford, UK: CABI Publishing.

Kaskey, J. 2013. "DuPont Gains Control of Africa's Pannar Seed after Delay." *Bloomberg News.* Accessed June 7, 2014. www.bloomberg.com/news/2013-07-31/dupont-gains-control-of-africa-s-pannar-seed-after-delay.html.

Kirsten, J., R. Stander, and C. Haankuku. 2011. *Measuring Private Research and Innovation in South Asia and Sub-Saharan Africa: A Country Report on South Africa.* Washington, DC: International Food Policy Research Institute.

Klerkx, L., and C. Leeuwis. 2008. "Institutionalizing End-User Demand Steering in Agricultural R&D in the Netherlands." *Research Policy* 37: 460–472.

Langyintuo, A., W. Mwangi, A. Diallo, J. MacRobert, J. Dixon, and M. Bänziger. 2010. "Challenges of the Maize Seed Industry in Eastern and Southern Africa: A Compelling Case for Private–Public Intervention to Promote Growth." *Food Policy* 35: 323–331.

Mapakame, E. 2013. "Seed Co Gives Nod to US$30M Deal." *The Sunday Mail* (November 3). Accessed January 2014. www.sundaymail.co.zw/index.php?option=com_content&view=article&id=39299:seed-co-gives-nod-to-us30m-deal&catid=41:business&Itemid=133#.Uyn1pk1OXIU.

Miller, C., S. Richter, P. McNellis, and N. Mhlanga. 2010. *Agricultural Investment Funds in Developing Countries.* Rome: Food and Agriculture Organization of the United Nations.

Mitchell, D. 2012. "Kenya Smallholder Coffee and Tea: Divergent Trends Following Liberalization." In *African Agricultural Reforms: The Role of Consensus and Institutions,* edited by M. Aksoy. Washington, DC: World Bank.

Monsanto. 2013. *Annual Report.* Accessed January 2014. www.monsanto.com/investors/pages/annual-report.aspx.

Mwala, M., and D. Gisselquist. 2012. *Private-Sector Agricultural Research and Innovation in Zambia: Overview, Impact, and Policy Options.* Background paper prepared for the ASTI/IFPRI–FARA Conference "Agricultural R&D: Investing in Africa's Future: Analyzing Trends, Challenges, and Opportunities," Accra, Ghana, December 5–7. Accessed January 2014. www.asti.cgiar.org/pdf/private-sector/Zambia-PS-Report.pdf.

Ndii, D., and D. Byerlee. 2004. "Realizing the Potential for Private-Sector Participation in Agricultural Research in Kenya." In *Transformation of Agricultural Research Systems in Africa: Lessons from Kenya,* edited by C. Ndiritu, J. Lynam, and A. Mbabu. East Lansing: Michigan State University Press.

Nieuwoudt, T., and W. Nieuwoudt. 2004. "Privatising Agricultural R&D: An Example from the South African Sugar Industry." *Agrekon* 43: 228–243.

Nzuma, J. 2011. *Producer Funding of Agricultural R&D: The Case of Kenya's Tea Industry.* Background paper prepared for the ASTI/IFPRI–FARA Conference "Agricultural R&D: Investing in Africa's Future: Analyzing Trends, Challenges, and Opportunities," Accra, Ghana, December 5–7. Accessed January 2014. www.asti.cgiar.org/pdf/conference/Theme1/CaseStudies/Nzuma.pdf.

Odame, H., E. Kangai, and D. Spielman. 2012. *A Kenya Country Report on Private Sector Innovation and Research.* Washington, DC: International Food Policy Research Institute.

Park, W. 2008. "International Patent Protection, 1960–2005." *Research Policy* 37 (4): 761–766.

Pray, C., D. Gisselquist, and L. Nagarajan. 2011. *Private Investment in Agricultural Research and Technology Transfer in Africa.* Conference Working Paper 13 from the ASTI/IFPRI-FARA Conference "Agricultural R&D: Investing in Africa's Future: Trends, Challenges, and Opportunities," Accra, Ghana, December 5–7. Accessed March 2013. www.asti.cgiar.org/pdf/conference/Theme4/Pray.pdf.

Pray, C., and L. Nagarajan. 2012. *Innovation and Research by Private Agribusiness in India.* IFPRI Discussion Paper 1181. Washington, DC: International Food Policy Research Institute.

Ruttan, V. 2001. *Technology, Growth and Development: An Induced Innovation Perspective.* New York: Oxford University Press.

Stads, G., and L. Sène. 2011. *Senegal: Private Agricultural Research and Innovation.* Washington, DC: International Food Policy Research Institute.

Walker, T. A. Alene, J. Ndjeunga, R. Labarta, Y. Yigezu, and A. Diagne. 2014. *Measuring the Effectiveness of Crop Improvement Research in Sub-Saharan Africa from the Perspective of Varietal Output, Adoption, and Change: 20 Crops, 30 Countries, and 1150 Cultivars in Farmers' Fields.* Rome: CGIAR Independent Science and Partnership Council, Standing Panel on Impact Assessment.

World Bank. 2007. World Development Report 2008: Agriculture for Development. Washington, DC. http://siteresources.worldbank.org/INTWDR2008/Resources/WDR_00_book.pdf.

———. 2011. "Country-Level Indicators Data." Accessed January 2014. http://data.worldbank.org/indicator/NV.AGR.TOTL.ZS.

———. 2013. *Growing Africa: Unlocking the Potential of Agribusiness.* Washington, DC.

PART 3

Human Resource Development and Tertiary Education

Chapter 8

ENSURING HUMAN RESOURCE CAPACITY TO SECURE AGRICULTURAL TRANSFORMATION

Nienke Beintema and Howard Elliott

Developing countries have shifted from simply being consumers of technologies developed by high-income countries (which are often no longer appropriate to their needs), to borrowing technologies from emerging economies and adapting them to local conditions, and even to developing their own innovations. Many low-income countries, however, have not been able to keep pace with the rapid developments in the various disciplines of science and technology because of a lack of appropriate human resource capacity. This is especially so in the field of agricultural science and in the countries of Africa south of the Sahara (SSA). Key causes are lack of training and other opportunities and high rates of staff attrition, both from developing to developed countries and from science to nonscience and technical sectors (IAC 2004; Beintema and Di Marcantonio 2010). Despite continuous growth in absolute numbers of agricultural researchers employed in SSA in recent decades, researcher numbers and qualification levels are among the more serious constraints facing national agricultural research systems (NARSs). Combined with insufficient and unstable funding and limitations in the scope and quality of their institutional infrastructure (Chapters 2 and 4, this volume), these constraints significantly compromise the region's ability to harness the potential of the agricultural sector to meet its economic development and social welfare goals.

This chapter provides an assessment of the current state of human resource capacity within African NARSs based on comprehensive datasets collected by Agricultural Science and Technology Indicators (ASTI), which is led by the International Food Policy Research Institute. The discussion also focuses on key human resource challenges, including high researcher attrition rates, the aging of senior researchers, and the recent influx of junior researchers who require significant training and experience in order to contribute effectively to the region's agricultural research and development (R&D) goals. The chapter concludes with an overview of some of the human resource strategies that have been adopted in African countries and could be replicated in others, and a summary of recent initiatives instituted at regional and international levels.

Existing Agricultural Research Capacity

Overview

As of 2011, SSA employed an estimated 14,300 agricultural researchers, measured in full-time equivalents (FTEs).[1] Beintema and Rahija (2011) estimated that these researchers were supported by 50,000 technicians (some of whom held university degrees), administrative staff, and other support staff (such as field workers, drivers, and guards). In 2011, three countries employed more than 1,000 researchers—Kenya (1,150 FTEs), Ethiopia (1,877 FTEs), and Nigeria (2,687 FTEs)—and accounted for 40 percent of the region's total number of FTEs (Figure 8.1). Ghana, South Africa, Sudan, and Tanzania employed 607, 746, 939, and 815 FTEs, respectively. A majority of SSA countries, however, have very small capacities in terms of agricultural research staff. For the 39 SSA countries for which data were available, 10 countries employed fewer than 100 FTE researchers. The smallest of these, in terms of FTE capacity, were Guinea-Bissau (9 FTEs), Cape Verde (21 FTEs), and Swaziland (27 FTEs).

For the most part, small countries have to deal with the same range of agricultural research issues as medium-sized and large countries, which raises questions about the effectiveness of their research (Chapter 2, this volume) and how their research strategies should be adapted. Fragmentation of limited resources is a key problem among small countries; for example, half of Guinea-Bissau's 9 FTE researchers focus on rice research, while the other half focus on a wide range of crop and livestock items (Table 8.1). Most small countries, therefore, focus on adapting existing technologies developed elsewhere, but this option can also be limited in SSA, given that neighboring countries often have similarly small capacities (Chapter 2, this volume).

Absolute numbers of researchers offer limited cross-country insights; comparative measures, such as the intensity of agricultural research, can be more revealing. Researcher numbers as a share of the agricultural labor force is one such indicator (Figure 8.2). Wide variation exists across countries, and the

---

1 ASTI bases its calculations of human resource capacity on FTEs, which take into account the proportion of time researchers spend on R&D activities. University staff members, for example, spend the bulk of their time on nonresearch-related activities, such as teaching, administration, and student supervision, which need to be excluded from research-related resource calculations. As a result, four faculty members estimated to spend 25 percent of their time on research would individually represent 0.25 FTEs and collectively be counted as 1 FTE.

FIGURE 8.1 National agricultural research systems by absolute number of researchers, 2011

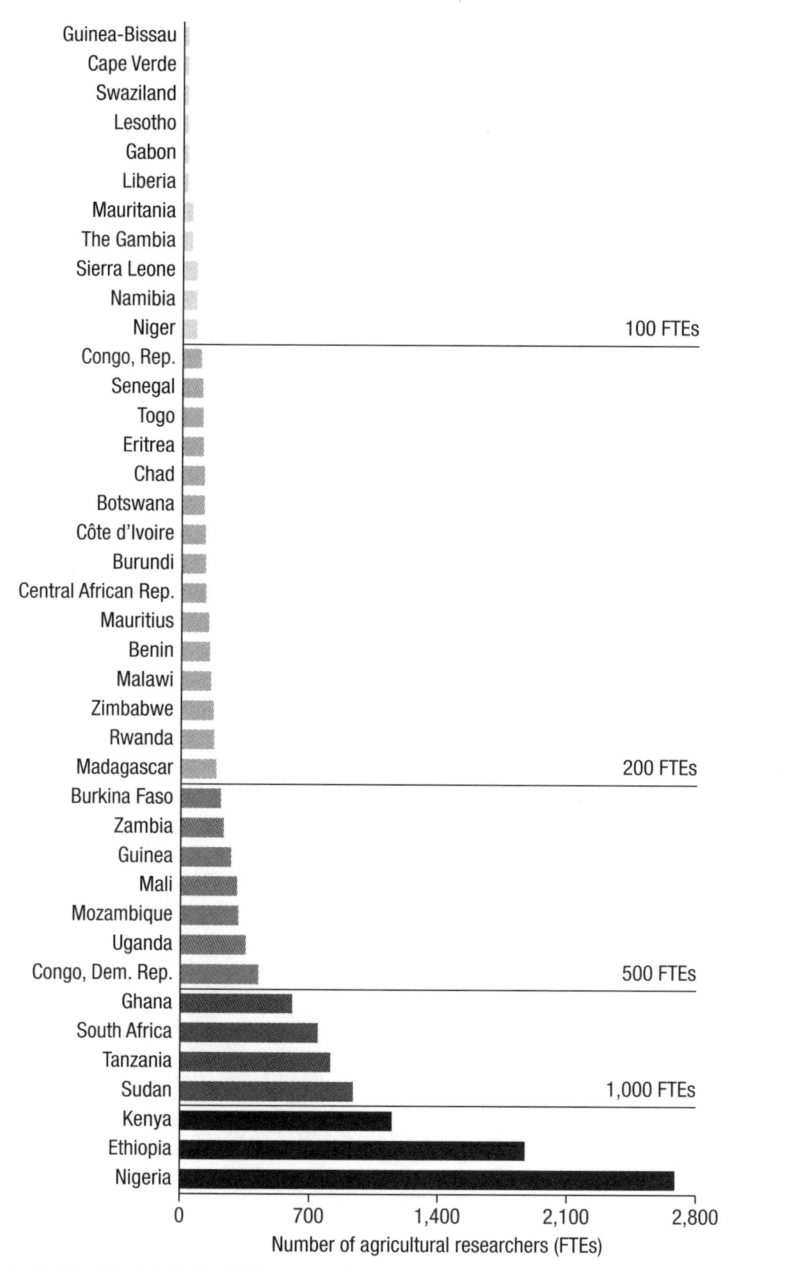

Source: Calculated by authors based on ASTI (2014).

Notes: Angola, Cameroon, Comoros, São Tomé and Príncipe, Seychelles, Somalia, and South Sudan are excluded because of lack of available data. FTEs = full-time equivalents.

TABLE 8.1 Fragmentation of research focus among selected countries with fewer than 100 full-time equivalent researchers, 2011

		Minimum number of commodities being researched and allocated shares of researchers							
		Targeted crop items/categories				Targeted livestock species/categories			
Country	Total number of researchers (FTEs)	Minimum number of items/groups	Share of FTE researchers (%)	Items with <5% of FTE researchers	Fragmentation index (%)	Minimum number of species/groups	Share of total FTE researchers (%)	Items with <5% of FTE researchers	Fragmentation index (%)
Cape Verde	21.0	6	50	1	17	2	20	—	—
Gabon	42.6	14	37	13	93	—	—	—	—
The Gambia	65.8	17	46	13	76	8	32	5	63
Guinea-Bissau	8.6	6	84	—	—	2	16	—	—
Lesotho	41.1	13	76	8	62	3	12	2	67
Liberia	45.1	19	72	11	58	4	9	3	75
Mauritania	62.9	17	25	16	94	6	17	5	83
Namibia	89.4	11	30	10	91	6	35	3	50
Sierra Leone	81.7	24	77	18	75	6	6	6	100

Source: Calculated by authors based on ASTI (2014).

Notes: Data were collected for 35 crops and crop categories (including other crops) and 8 livestock items or categories (including other livestock). The table excludes other research areas, such as forestry, fisheries, natural resources, and socioeconomics. The fragmentation index is calculated as the share of the number of crop or livestock items/categories constituting the focus of less than 5 percent of the country's total number of FTE researchers divided by the total number of crops or livestock items being researched.

FIGURE 8.2 Full-time equivalent agricultural researchers per 100,000 farmers, 2011

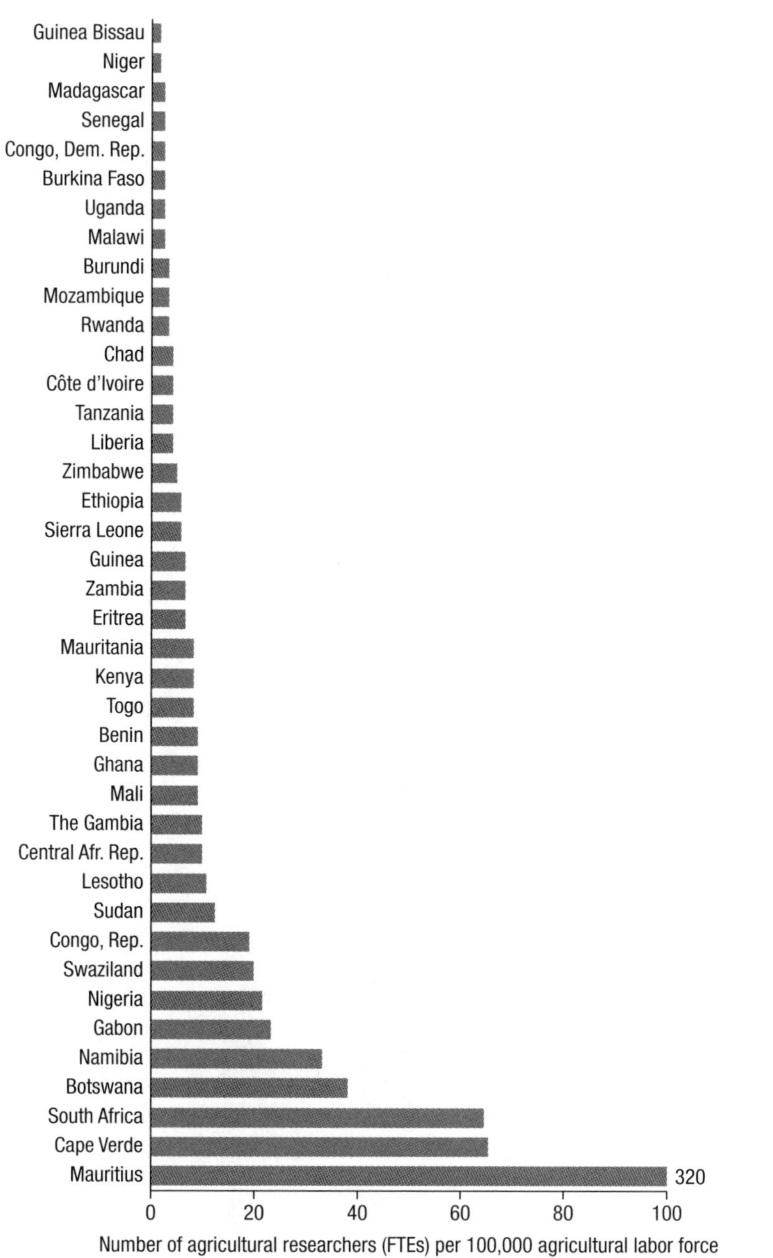

Source: Calculated by authors based on ASTI (2014) and FAO (2012).

Notes: Angola, Cameroon, Comoros, São Tomé and Príncipe, Seychelles, Somalia, and South Sudan are excluded because of lack of available data. FTEs = full-time equivalents.

need for a minimum number of researchers to reach a critical mass is reflected in the comparatively higher ratio of researchers to farmers in many of the smaller countries.

Growth in Researcher Numbers

Since attaining their independence, most SSA countries have made considerable progress in building their human resource capacity in agricultural R&D. In 1961, the region employed about 2,000 FTE researchers in agricultural sciences (Pardey, Roseboom, and Beintema 1995); this number increased to 9,000 FTEs in the mid-1990s and (as previously mentioned) to more than 14,000 FTEs in 2011. During the past four decades, most of the countries in the 40-country sample for which data were available modestly increased their total number of FTE researchers (Figure 8.3). In particular, the number of midsize to large systems increased (that is, those employing 100–499 and more than 500 FTE researchers, respectively). In 1971, 21 countries employed fewer than 25 FTE researchers, and 10 countries employed between 25 and 49 FTEs. In 2011, 8 of the 40 countries employed fewer than 50 FTE researchers (mostly the countries with smaller populations, such as the Seychelles, Cape Verde, and Guinea-Bissau).

FIGURE 8.3 Growth in national agricultural research capacity, 1971–2011

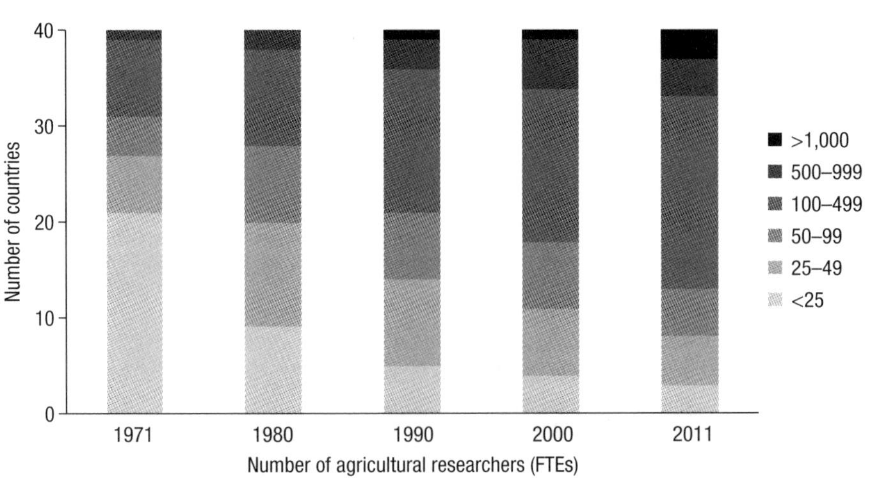

Source: Calculated by authors based on ASTI (2014).

Notes: Angola, Cameroon, Comoros, Eritrea (which did not gain independence until 1993), São Tomé and Príncipe, Somalia, and South Sudan are excluded because of lack of available data; were these countries included, the distribution would be different, but the long-term trend would not likely differ substantially. FTEs = full-time equivalents.

The number of countries with sizable teams of agricultural researchers (more than 500 FTEs) also increased substantially. South Africa, having one of the most established and well-funded research systems in SSA, was the only country to employ more than 500 FTEs in 1971. As of 2011, however, South Africa was outranked in terms of researcher numbers by Sudan (939 FTEs), Kenya (1,151 FTEs), Ethiopia (1,877 FTEs), and Nigeria (2,688 FTEs). Ghana and Tanzania also employed more than 500 FTE researchers that year.

Unsurprisingly, the countries with the largest absolute number of agricultural researchers are also the main drivers of the 50 percent growth in the regional number of researchers during 2000–2011 (Figure 8.4). Ethiopia and Nigeria together accounted for more than half of the region's increase of 4,352 FTEs during 2000–2011. Kenya, Sudan, and Tanzania also reported significant increases, and several small systems reported growth from a low base.

South Africa, however, recorded the largest absolute decline in its number of agricultural researchers during 2000–2011 (232 FTEs), which marked a period of change in orientation for the Agricultural Research Council (ARC). The Sahel also recorded decreases during this period, as did Mauritania, Niger, and Senegal; Burkina Faso and Mali recorded more recent decreases (during 2008–2011). These losses often resulted from the combination of recruitment restrictions and the retirement or departure of numerous highly qualified researchers.

Shifts in Qualification Levels

A minimum number of PhD-qualified scientists is generally considered fundamental to the conception, execution, and management of high-quality research; to effective communication with policymakers, donors, and other stakeholders, both locally and through regional and international forums; and for increasing an institute's chances of securing competitive funding. Most of the growth in agricultural researcher numbers during 2000–2011 stemmed from the recruitment of MSc- or BSc-qualified researchers (Figure 8.5). This contrasts with the trend in the decades preceding 2000, when the educational profiles of African agricultural researchers steadily improved, mostly as a result of substantial donor support for educational training, which was halted in the 1990s. More recently, overall growth in the number of PhD-qualified researchers has been slow. Of particular concern, 12 of 30 countries for which long-term data were available recorded a decline in the absolute number of PhD-qualified researchers employed during 2008–2011. This trend was particularly severe in Burkina Faso, Mali, Mauritania, Sudan, and Tanzania. For example, Burkina Faso, South Africa, and Tanzania lost about

FIGURE 8.4 Yearly growth in the number of researchers by country, 2000–2011

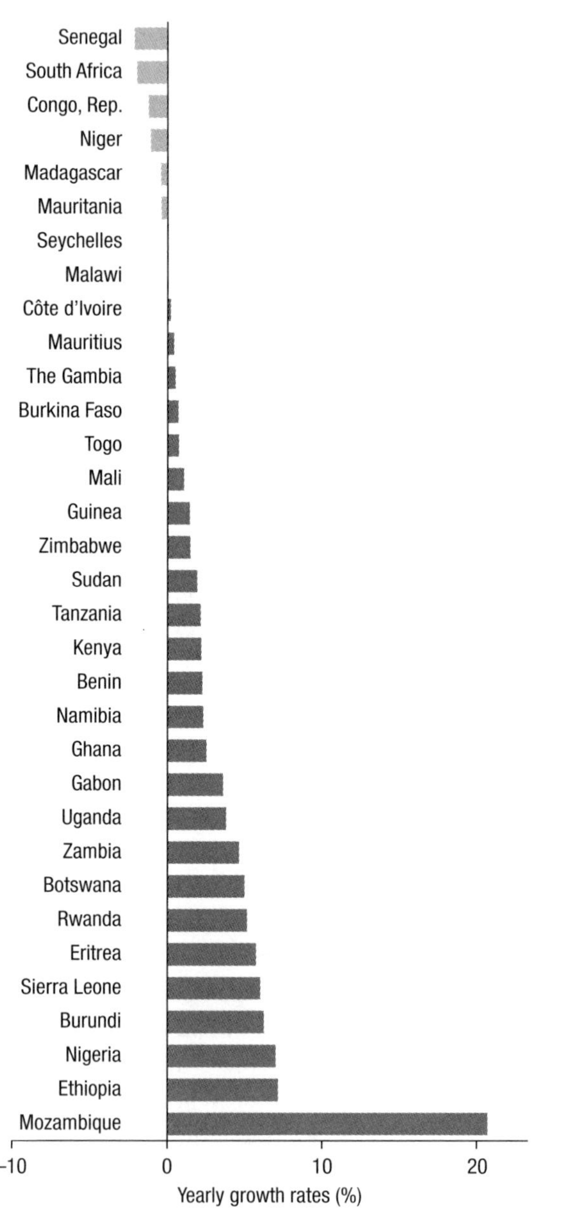

Source: Calculated by authors based on ASTI (2014).

FIGURE 8.5 Trends in researcher qualification levels, 2000–2011

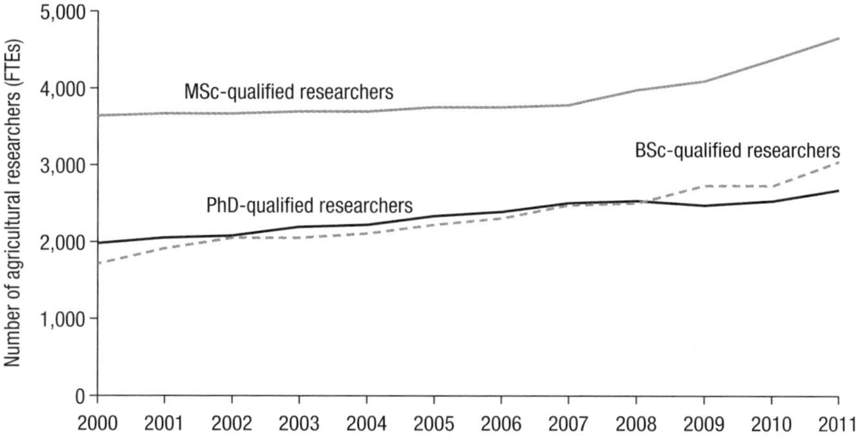

Number of agricultural researchers (FTEs)
5,000
4,000
3,000
2,000
1,000
0
MSc-qualified researchers
BSc-qualified researchers
PhD-qualified researchers
2000 2001 2002 2003 2004 2005 2006 2007 2008 2009 2010 2011

Source: Calculated by authors based on ASTI (2014).
Notes: Data underlying this figure are for a 39-country sample. FTEs = full-time equivalents.

20 PhD-qualified researchers during 2008–2011 (Beintema and Stads 2014). Of 37 countries for which a complete set of data on degree qualifications was available, five countries (Benin, Burkina Faso, Madagascar, Senegal, and Swaziland) reported shares of PhD-qualified researchers of 40 percent or higher, whereas another five countries (Ethiopia, The Gambia, Guinea-Bissau, Lesotho, and Mozambique) reported shares of 10 percent or lower.[2]

Many of the national agricultural research institutes (NARIs), especially those in smaller countries, lack the minimum number of PhD-qualified scientists needed to design and execute high-quality research programs. In fact, Eyzaguirre (1996) argued that smaller NARSs need comparatively higher numbers of MSc- and PhD-qualified researchers than do larger systems. Guinea-Bissau, for example, employed no agricultural researchers with PhD degrees in 2011, and the NARIs in Burundi and The Gambia employed just one and two PhD-qualified scientists, respectively. This highlights the need for regional initiatives focusing on issues relevant to small countries, resource sharing, and achieving economies of scale.

2 Universities generally employ a much higher share of PhD-qualified scientists compared with government agencies, but with high, and growing, student populations in the agricultural sciences, it is not surprising that they spend the vast majority of their time on their primary mandate—teaching—and not on research. Nonetheless, the growing core of PhD-qualified researchers within the higher-education sector is a valuable—and as yet largely untapped—future resource for African NARSs (Beintema and Stads 2014).

The number of junior researchers only qualified to the BSc-degree level has increased substantially in a number of key countries. For example, during 2000–2011, the share of researchers with only BSc degrees grew from 19 to 29 percent in Nigeria, from 33 to 52 percent in Botswana, and from 22 to 40 percent in Tanzania. These dramatic shifts are primarily the result of the end of long-term civil service recruitment freezes combined with uncompetitive remuneration packages, which make it difficult for institutes to attract highly qualified staff. In addition, PhD- (and to a lesser extend MSc-) qualified agricultural scientists are in short supply in many countries because of limited postgraduate training opportunities (Beintema and Stads 2014; Chapter 9, this volume). Despite the imbalance in the educational profile of agricultural researchers averaged across countries, shares of PhD-qualified researchers in a few countries, such as Benin and Burundi, increased between 2000 and 2011 (by 5 and 8 percent, respectively).

Capturing quantitative information on degree-qualified research support staff—who, given proper training and promotional opportunities, constitute a significant potential resource—is important in outlining a complete picture of the current status of agricultural research capacity in the countries of SSA. An increasing number of support staff (including technicians and research and laboratory assistants) have BSc, MSc, and occasionally PhD qualifications, but they are not officially classified as researchers. An MSc degree is the minimal requirement for a researcher in an increasing number of NARIs. Many research support staff at Uganda's National Agricultural Research Organization (NARO), for example, have MSc or BSc degrees, mostly attained without official NARO backing. In contrast, support staff members at other NARIs, such as Tanzania's Department of Research and Development (DRD), are promoted to researchers when they obtain their BSc degrees. The pool of degree-qualified support staff is sizable in a number of countries. As of 2011, 62 percent of all the degree-qualified research staff at the Senegalese Agricultural Research Institute (ISRA) were technical support staff with BSc and MSc degrees (Figure 8.6). For the NARIs in Benin, Burkina Faso, Mozambique, Sudan, and Uganda, this share ranged from 20 to 36 percent.

Low Female Participation

Women's roles and status in African agricultural research are often limited, which starkly contrasts with women's participation in the agricultural workforce in SSA. Female researchers are more likely to offer unique insights, perspectives, and skills that can help research institutions more effectively address the specific challenges of Africa's female farmers. Furthermore, attracting

FIGURE 8.6 Shares of researchers and research support staff with BSc, MSc, and PhD degrees, selected countries, 2011

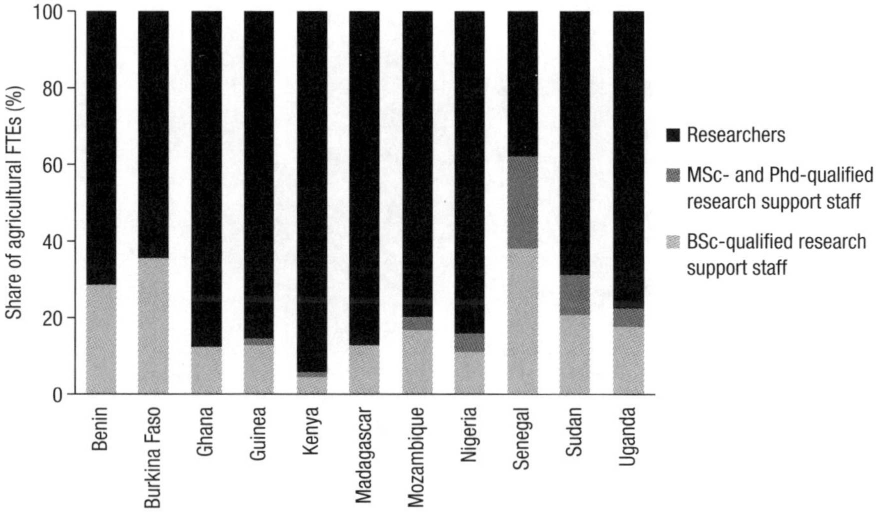

Source: Calculated by authors based on ASTI (2014).
Notes: Data only include each country's national agricultural research institute(s), or NARI(s); FTEs = full-time equivalents.

women into agricultural research would be a highly beneficial strategy for filling gaps in researcher capacity in many countries. Despite aggregate increases in the number of female agricultural researchers in SSA in recent decades—in both absolute and relative terms—shares of female researchers remain low in many countries. In 2011, 22 percent of the total number of researchers in a 37-country sample was female. Nevertheless, large national-level gender discrepancies remain. In general, Southern African countries employ relatively more female researchers than do other subregions. Women's representation is particularly low in West African countries, specifically in Togo (8 percent), Chad (7 percent), Guinea (4 percent), and Guinea-Bissau (zero). The Democratic Republic of the Congo (DR Congo), Eritrea, and Ethiopia also have shares of female researchers of 10 percent or less. Despite the low base, however, shares among West African countries appear to have grown in recent years.

On average, the shares of women with BSc and PhD degrees were almost equal, but were lower than the share of women with MSc degrees (Figure 8.7). Women are relatively more represented in the younger age brackets: a quarter of all researchers age 40 years or younger were female, whereas only 13 percent of researchers in their 50s and 60s were female. Although it is too early to tell,

FIGURE 8.7 Share of female researchers by degree, age bracket, and sector, 2011

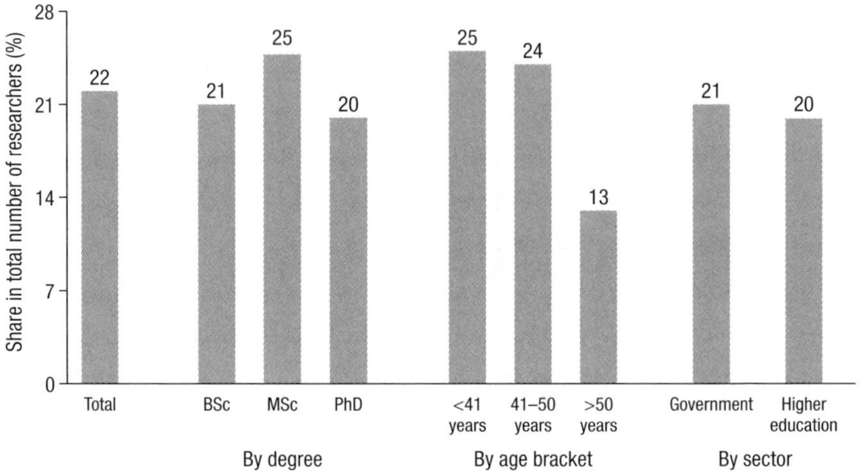

Source: Calculated by authors based on ASTI (2014).
Note: Data underlying this figure are for a 37-country sample.

and there is still a long way to go, this could be an encouraging indication that African agricultural R&D is becoming more gender balanced and that human resources are being used more effectively.

Despite women's greater involvement in agricultural R&D over time, high-level research and management positions are still mostly held by men. Consequently, women have less influence in policy- and decisionmaking processes. A study collecting data from key NARIs and agricultural higher-education agencies across SSA found that, as of 2008, the share of women in management positions was only 14 percent (Beintema and Di Marcantonio 2010). Furthermore, women face a number of specific challenges, such as unequal access to basic education; traditional beliefs that limit the roles of women in society; and balancing work and family responsibilities, especially while caring for infants and raising young children.

Key Vulnerabilities: Staff Remuneration, Turnover, and Aging

In many countries, NARIs are highly vulnerable when it comes to human resource capacity. To begin with, long-term civil service hiring freezes have left many institutes with aging pools of researchers, many of whom are nearing or have reached the official retirement age. Recruitment efforts in more recent years have led to an influx of junior staff in need of further training, mentoring, and supervision. As a result of these two factors, institutes lack appropriately trained and experienced staff to step into the roles left vacant by retiring (and departing) senior staff, while at the same time having too few senior staff to train and mentor their newly appointed junior colleagues. This issue is even more severe at institutes with numerous disciplines and areas of research focus, or where highly specialized training and experience are required.

The other key issue is high staff turnover to other agencies, sectors, and countries resulting from low salary levels, poor service conditions, limited career prospects, and inadequate facilities and equipment. This makes it difficult not only to attract new staff—especially highly qualified staff—but also to retain and motivate staff over time. Exacerbating this issue is the fact that many scientists sent abroad for training fail to return to their home countries and sponsoring institutes, are promoted into (often nonresearch-related) management positions, are seconded to other ministries and directorates, or are attracted by other sectors or into donor agencies and international organizations (Beintema and Stads 2014). It goes without saying that financial constraints are ultimately at the core of all of these issues, in terms of the ability to offer competitive salaries and conditions; to provide training and career opportunities; and to create the necessary incentives to attract, retain, and motivate staff over the long term.

Staff Remuneration and Turnover

Within NARIs, salaries and conditions of service are often linked to civil service scales that are insufficient to compete with universities, the private sector, and other organizations in recruiting and retaining well-qualified researchers. For example, many well-qualified scientists have left the National Institute of Agricultural Research of Benin in recent years as a result of the large differences in the institute's salary levels and benefit packages compared with those of universities and international organizations. Burkina Faso's Environment and Agricultural Research Institute lost 40 PhD-qualified researchers during 2006–2011, most of whom departed for better-paid positions at universities or other organizations. Salary packages are important and therefore need to remain competitive over time. For example, increases in the salary levels at South Africa's ARC did not kept pace with inflation or with salaries offered elsewhere. This, combined with relatively low adjustments in subsistence and travel allowances, may have been the cause of the high increase in the rate of staff departures during ARC's recent restructuring (Sène et al. 2011). With government support, several NARIs have been able to increase their salary levels. For example, the Senegalese government approved the salary levels of ISRA's researchers and improved their promotional opportunities. The government of Ghana instituted the "Single Spine Pay Policy," which introduced parity between the salaries of scientists employed at the Council for Scientific and Industrial Research and those of university-based scientists. Staff morale appears to have improved at both institutes, resulting in a decline in staff attrition. Both institutes have also become more desirable employers for university graduates (Stads and Beintema 2014).

Staff Aging

In a 37-country sample in 2011, more than half of the PhD-qualified researchers were more than 50 years old in 19 countries, and more than 70 percent of the PhD-qualified researchers were more than 50 years old in an additional 9 countries (Figure 8.8). The situation is particularly severe in West Africa and in Madagascar and a few other countries. For example, about three-quarters of all agricultural researchers in Guinea were more than 50 years old as of 2011. Despite a recent increase in the official retirement age from 60 to 65 years, the Guinean Agricultural Research Institute (IRAG) is expected to have to replace 90 percent of its PhD-qualified researchers by 2023 as a result of retirement. Filling these positions internally will not be possible because of the Institute's small pool of researchers overall (a result of past recruitment freezes) and the relatively small number of researchers with PhD degrees (a result of declining donor support for overseas training and lack of local postgraduate training in agricultural sciences). This means IRAG will need to recruit researchers externally; however, uncompetitive remuneration packages act as a constraint here as well. Mali is also severely challenged in having aging pools of agricultural researchers, particularly among those qualified to the PhD level (more than 80 percent of its PhD-qualified researchers are in their 50s or 60s). Madagascar is another country facing this challenge; as of 2011, more than 60 percent of all researchers and three-quarters of those employed at the National Center for Applied Research and Rural Development (FOFIFA) were more than 50 years old.

To address the aging challenge, a number of countries have increased the official retirement age from 60 to 65 years, and even from 65 to 70 years, although without large-scale recruitment this will only relieve the situation temporarily. The official retirement age is still quite low in many countries; only 15 of the 37 countries for which anecdotal information was available had

FIGURE 8.8 Share of agricultural researchers over 50 years old, 2011

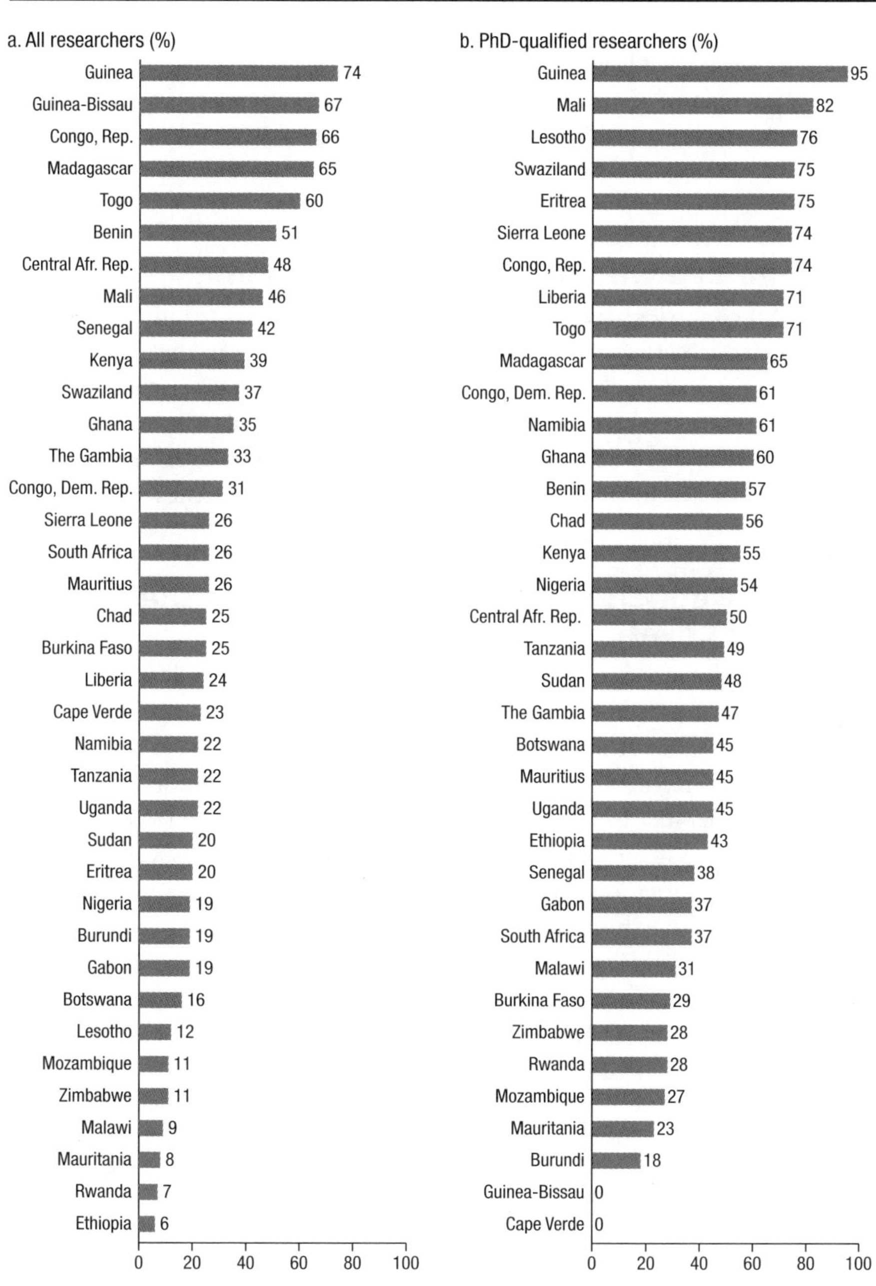

Source: Calculated by authors based on ASTI (2014).

official retirement ages of 65 years or higher. In Gabon and Zambia, researchers retire at 55 years old. In a few countries, the retirement age is higher at universities than at NARIs (and also sometimes higher for men than for women). The desire to extend their careers and earning opportunities acts as an additional incentive for researchers to leave NARIs in preference for university positions (Beintema and Stads 2014).

Lessons Learned at the Country Level

Several SSA countries have instituted strategies in response to human resource challenges, which offer valuable lessons for other countries in the region (Table 8.2). Such strategies include recruitment of (often junior) staff upon the cessation of recruitment freezes, improved training and mentoring opportunities, and improved benefits packages. Some of these strategies have been implemented at the institute level, while others have a broader reach.

Current Capacity-Building Initiatives at the Regional Level

Capacity building for agricultural research began soon after independence, and was mostly funded by donor organizations (Box 8.1). During the late 1980s and throughout the 1990s, donor interest in and support of agriculture and agricultural research waned. In recent years, however, a number of initiatives have been established, which could play an important role in rebuilding the region's cadre of agricultural researchers. While the prospect exists of increased collaboration between Africa and emerging national economies (that is, Brazil, China, and India), the major drivers of the newly rejuvenated capacity building in research for development remain the historical donor organizations, such as the World Bank, the European Commission, the United States Agency for International Development (USAID), and, more recently, the Bill & Melinda Gates Foundation.

Multidonor Trust Fund and Agricultural Productivity Programs

The Multi-Donor Trust Fund of the Comprehensive Africa Agricultural Development Programme (CAADP) was established in 2008 to increase and align donor support for CAADP activities. By 2014, the fund had grown to a cumulative total of US$53 million for strengthening the capacity of African agricultural research institutions to develop new technologies for Africa's farmers. A recent independent evaluation of the fund and CAADP

TABLE 8.2 Examples of human resource strategies adopted in specific countries or regionally

Problem	Strategy	Examples of adopting countries/programs
Inadequate numbers of agricultural researchers resulting from long-term hiring freezes and lack of funding	Recruiting large numbers of junior staff upon the cessation of hiring freezes (with attendant problems of further training, supervision, and mentoring)	Burkina Faso, Senegal, Sudan, and Togo
Need to upgrade researcher qualifications at NARIs	Developing local MSc and PhD programs in agricultural sciences, either at existing or new universities or by merging existing universities	Rwanda
Countries with large pools of junior agricultural researchers needing further training and mentoring	Promoting collaboration between the government and higher-education sectors (such as the NARI and main university) to create joint PhD programs, supported by the national government	Eritrea
	Developing regionally based collaborative training programs to achieve the necessary critical mass and economies of scale	The RUFORUM network operating in East, Central, and Southern Africa
	Retaining the expertise of retiring senior researchers by • Raising the retirement age from 60 to 65 years • Hiring retired researchers on contract	Madagascar and Sudan, Senegal and Tanzania
Lack of relevance of academic training to clients' needs	Promoting collaboration between government and higher-education sectors (such as NARIs and main universities) in reforming curriculums and conducting postgraduate field research	Uganda and the RUFORUM network operating in East, Central, and Southern Africa
High rate of researcher attrition in the government sector	Providing incentives by • Increasing salary levels to achieve parity with those offered in the higher-education sector	Ghana, Kenya, Nigeria, Senegal, and Sierra Leone
	• Offering scholarships and paid study leave for internal advancement	Kenya
	• Increasing freedom to engage in paid consultancies and sessional lecturing	Kenya
	• Increasing opportunities for merit-based career advancement (rather than through seniority only)	Kenya and Senegal
	• Encouraging the return of senior staff from neighboring countries by improving salaries to regionally competitive levels	Burundi
Lack of gender equity in research and senior management positions	Encouraging mentoring partnerships and development of women's scientific and leadership skills	AWARD program operating in various African countries

Source: Compiled by authors from ASTI country factsheets (various countries and years).

Note: AWARD = African Women in Agricultural Research and Development; NARI = national agricultural research institute; RUFORUM = Regional Universities Forum for Capacity Building in Agriculture.

BOX 8.1 A historical perspective on capacity building in Africa south of the Sahara

Capacity building for agriculture began in the early post-independence era. Training the first generation of African academics in France, the United Kingdom, and the United States was relatively inexpensive and was supported for foreign policy reasons. Most universities had yet to inflate foreign student fees as part of their business model, and hence welcomed the opportunity to receive project support for work abroad. Examples include the following:

1. Collaboration between US universities and African researchers began in the 1960s through exchanges that mostly focused on applied research and extension activities with farmers (Nesseth Tuttle, Wedding, and Applefield 2011). The experiences of the United States Agency for International Development in three Nigerian universities (Ahmadu Bello, Ife, and Nsukka) and attempts to establish a land-grant model following the US model are well documented and confirmed the ease of adopting the teaching component (Gamble et al. 1988). Ahmadu Bello was the only university that achieved some success by integrating research with extension.
2. The Rockefeller Foundation's Education for Development (originally the University Development Program) was a global program that attempted to create regional centers of excellence at three universities: the University of Ibadan, Nigeria; the University of East Africa; and the National University of Zaire. The 1960s and 1970s represented an era of experimentation, adaptation, absorption, and rejection. All three cases experienced institutional shocks at the national level that prevented their success as regional centers of excellence (Coleman and Court 1993). The Rockefeller Foundation and other private foundations withdrew from large-scale institutional development of universities when multilateral lenders and the major aid donors expanded support to universities.
3. France resisted demands for the creation of full national universities in francophone countries, but ultimately yielded; hence, national universities were established.
4. In the 1980s, the World Bank invested heavily in improving the policy environment for agriculture through (a) the design of organizational structures (such as planning units, extension and research services, and agricultural credit agencies); (b) the provision of short- and long-term technical assistance; and (c) training. The World Bank's Operations Evaluation Department (OED) noted that the "essence of capacity building is sustainability—the creation of institutions and practices that continue to perform after a project is completed." Unfortunately, special

incentives and support to make the programs effective were unsustainable after the conclusion of the project. The OED concluded that project-level interventions were an inappropriate approach to systemic problems. If low public service salaries were the problem, it did not help sustainability to shift to hiring contractual employees for the life of the project. Moreover, the large sums spent on technical assistance were often found to be not worth the cost. The OED concluded that "institutional weaknesses are no longer primarily the result of a shortage of trained nationals—more typically problems arise from failures to mobilize or effectively use nationals because of weaknesses in incentive systems. . . . There is a tradeoff between the use of technical assistance and the funding of local staff. Any such step must avoid the pitfalls of donor interventions distorting the public sector salary structure and the lack of sustainability of donor-supported institutions" (World Bank–OED 1999, 3–4).

5. In the 1980s, CGIAR centers still operated in gap-filling/training mode, providing short courses (such as in seed production and farming systems); they also collaborated in postgraduate training with nearby educational institutions (for example, the University of Ibadan and International Institute of Tropical Agriculture). Centers offered research facilities, co-supervision of thesis research, and regional networking. By the end of the 1990s, however, CGIAR was directed by its Technical Advisory Committee, to emphasize its comparative advantage in bridging basic and strategic research for applied research undertaken by national systems and to leave capacity building to universities and development to NARSs. Driven by reductions in core support, the centers were perceived to be competing downstream for funding with the NARSs they were supposed to support. Strong calls by research leaders for African centers supported by African universities led to the creation of a Sub-Saharan Africa Task Force, whose findings on better alignment were summarized in "The Tervuren Consensus" (CGIAR Secretariat 2005).

Source: Authors.

performance found that (1) they have increased the capacity of lead institutions to drive the CAADP processes forward, (2) their role in raising awareness at the national level "cannot be overstated," but (3) their role has "only modestly improved alignment in CAADP support." The evaluation recommended a re-launch of CAADP with countries and national-level stakeholders having a stronger role, and better mainstreaming of CAADP in the Regional Economic Communities (RECs) as a precondition for future effectiveness (ECDPM, ESRF, and LARES 2014, vii). Capacity building under the regional agricultural productivity programs for East, West, and Southern Africa (EAAPP, WAAPP, and APPSA), which are cofinanced by the World Bank, a multidonor trust fund, and national governments (Chapters 2 and 6, this volume), plays a major role in these programs, and many researchers and technicians are already benefiting from postgraduate and short-term training.

USAID's Feed the Future

The Feed the Future Food Security Innovation Center leads USAID's implementation of the Feed the Future (FTF) research strategy through seven interlinked research, policy, and capacity development programs designed to sustainably transform agricultural production systems (FTF no date). The Program on Human and Institutional Capacity Development invests in human and institutional capacity as the essential building blocks of growth and innovation in the agricultural sector. Programs address on request from country missions Innovation for Agricultural Training and Education (InnovATE); African Women in Agricultural Research and Development (AWARD); and research fellowships (Norman Borlaug Awards for Field Research). Modernizing Extension and Advisory Services brings a consortium of universities to help partners establish financially sustainable public–private rural extension and advisory service systems. Rigorous evaluation is practiced to learn what works. The Feed the Future Food Security Innovation Center, housed in the Bureau for Food Security, has interlinked research programs. As of September 2014, the center operated 24 innovation laboratories led by 15 universities and involving more than 60 US colleges and universities in 39 states.

European Commission's AGRINATURA

AGRINATURA is a group of European universities and research organizations with a common interest in supporting sustainable agricultural development. A member of the European Forum on Agricultural Research for Development, the group believes that research and higher education underpin the innovations needed to increase agricultural production, productivity, and sustainability. AGRINATURA works through a Platform for African–European Partnership in Agricultural Research for Development and in partnership with RUFORUM in the Educational Linkage (EDULINK) program that promotes regional integration in higher education. It seeks to nurture scientific excellence through joint research, educational, and training programs (Chancellor, Hauser, and Sarfatti 2014).

The Bill & Melinda Gates Foundation's Support to AGRA, AWARD, and RUFORUM

The Bill & Melinda Gates Foundation is one of the main donors of a large number of regional capacity-strengthening initiatives. For example, the Alliance for a Green Revolution in Africa (AGRA), founded in 2006, works across the continent to help smallholder farmers gain access to improved seeds, soils, and markets, with the aim of increasing farm productivity and incomes. AGRA's activities emphasize strengthening individual and institutional capacities. AWARD, as previously mentioned, provides fellowships to female agricultural researchers focusing on strengthening their research and leadership skills. AWARD was officially established in 2008 with funding from USAID and the Gates Foundation following a three-year pilot phase funded by the Rockefeller Foundation during 2005–2007. Since then, close to 400 women have received fellowships. RUFORUM, also as previously mentioned, is an association of 46 universities in East, Central, and Southern Africa with three new members from West Africa joining as of late 2014. Through RUFORUM's regional MSc and PhD programs, partner universities have made strides in building field-focused postgraduate training, ensuring the quality and relevance of curriculums, and improving the policy environment through national forums and intra-African mobility (Chapters 9 and 10, this volume).

CGIAR Consortium

One of the historical strengths of CGIAR has been the ability of small groups of innovative scientists to self-organize as "communities of practice," leading to major innovations that subsequently enter the mainstream. A community of practice on capacity development attempts to reframe and reposition "an orphaned CGIAR function" (ILRI 2013). The community has produced a "Capacity Development Framework: Working Draft" as part of the second round of CGIAR Research Programs (CRPs). It takes a broad, innovation-systems approach compatible with that of most CGIAR centers and comprising nine elements: (1) assessing needs, (2) designing and delivering innovative learning materials, (3) developing CRPs and centers' partnering capacities, (4) developing future leaders through fellowships, (5) adopting gender-sensitive approaches, (6) strengthening institutions, (7) monitoring and evaluating programs and practices, (8) developing organizations, and (9) conducting research on capacity development. This approach requires a long-term commitment and perspective. Donor concerns with "results" rather than "inputs" have focused attention on "impact pathways" and "intermediate development outcomes" achieved through multiactor programs with defined terms. Long-term capacity building is difficult to build into such programs.

The Forum for Agricultural Research in Africa–Led Science Agenda for Agriculture in Africa

The African heads of state approved the Science Agenda for Agriculture in Africa (S3A), led by the Forum for Agricultural Research in Africa (FARA), in Malabo, Equatorial Guinea in June 2014. S3A commits to investments in science for agriculture, inter-African solidarity, and inter-African mobility of scientists, with the CGIAR Consortium committing to better alignment. S3A strongly emphasizes the need for a basic level of human capacity at national levels; closer integration of science with education and extension; and the promotion of regional partnerships, including greater involvement of RECs and increased collaboration at the global level (FARA 2014).

The Way Forward

The preceding sections have assessed existing human resource capacity in SSA and key vulnerabilities needing urgent attention; the ability to build human resource capacity, however, depends on the longer-term financial and institutional capacities to do so, and on the supporting or limiting factors inherent in the policy environment. This section addresses the kinds of policy and institutional measures needed to support the development of human resource capacity in agricultural research across the region.

Increased strategic planning and coordination in human resource management could make a significant difference. NARSs, however, are limited in their choice of options to address the challenges they face in maintaining and developing their researcher capacity, not the least because of funding constraints (Chapter 4, this volume). Fundamental to building strong human resource capacity in agricultural research is the development of comprehensive recruitment, training, and succession plans to fill existing and anticipated medium- to long-term staffing gaps. Such plans should include assessments of gaps in specific skills and disciplines, the distribution of staffing by age and gender, and degree-level and short-term training needs. Skills assessments should also include fundamental skills (such as proposal writing) and targeted training requirements (such as program management and research design). An implementation plan is required for the management and provision of training and mentoring, especially given that many countries now have high shares of junior scientists, often only trained to the BSc degree level. In many countries, the expansion and strengthening of postgraduate training programs in agricultural and related sciences are needed to meet the demand for qualified agricultural scientists. Training opportunities should also be sought through bilateral cooperation with countries that already have strong NARSs and higher-education networks (Chapter 9, this volume).

Obviously, countries and institutions with uncompetitive salary and benefits packages need to take steps to redress these barriers. In a large number of countries, significant discrepancies exist in the salary and benefits packages offered to NARI researchers compared with their university-based colleagues. Furthermore, advocating an increase in the retirement age to 65 years for those agencies with lower mandatory retirement ages would ameliorate the impending loss of senior researchers to retirement in the short- to medium-term, and establish parity in retirement ages between the government and higher-education sectors in countries where it is lacking. Employing recently retired researchers as consultants on a contract basis is another valuable approach to training and mentoring junior scientists during a transition period and maintaining continuity in the stock of institutional knowledge.

Equally important is establishing a proper career path for researchers (whether those up and coming, returning upon completion of degree training, or even being attracted back home from overseas), as well as enforcing policies requiring that scientists return to their sponsoring organizations for a minimum term of employment upon completing their postgraduate training. This would not only require opportunities for merit-based promotion within the field of research (as opposed to purely management positions), but also improve service conditions, support, facilities, and equipment to enable researchers to carry out meaningful research.

Another challenge in this regard is to integrate NARIs into the broader agricultural research system and agricultural sector. For research to be effective, it must successfully address farmers' and other clients' needs, which demands a participatory approach. Similarly, universities and other training institutes need to be forward looking in their curriculums; adopt pedagogical methods; and collaborate with research institutes not only in training, but also in research. In addition to national collaboration between the government and higher-education sectors, regional collaboration is warranted to achieve critical mass and efficient resource use within training programs (Chapter 9, this volume).

The new, donor-funded regional capacity-building initiatives could play an important role in rebuilding the region's cadre of agricultural researchers, but they will need to be upscaled to compete with those of other countries and sectors. The development of S3A and its endorsement by the heads of state is an extremely promising development, although its success will require that national governments and donor organizations substantially increase their investments in agricultural research, education, and extension. Success will require the development of supportive policies to enable the region's agricultural R&D undertakings to become more effective in delivering the technologies needed to meet the objectives of CAADP and the regional development agenda (Chapter 1, this volume). Furthermore, aligning the numerous and disparate actors required to successfully build agricultural research capacity, in terms of both conducting research and training researchers, remains complicated. Donors are often working separately from each other in their own preferred countries, with their preferred partners, and through their preferred delivery mechanisms. At the regional level, better coordination is needed among national capacity-building programs, cross-country collaborative initiatives, and the various subregional programs, as well as between national-level research and higher-education entities.

References

ASTI (Agricultural Science and Technology Indicators). 2014. ASTI database. Washington, DC: International Food Policy Research Institute. Accessed September 2014. www.asti.cgiar.org.

Beintema, N., and F. Di Marcantonio. 2010. *Female Participation in African Agricultural Research and Higher Education: New Insights. Synthesis of the ASTI–Award Benchmarking Survey on Gender-Disaggregated Capacity Indicators.* IFPRI Discussion Paper 957. Washington, DC: International Food Policy Research Institute and African Women in Agricultural Research and Development.

Beintema, N., and M. Rahija. 2011. *Human Resource Allocations in African Agricultural Research: Revealing More of the Story behind the Regional Trends.* Conference Working Paper 12 prepared for the ASTI/IFPRI–FARA Conference "Agricultural R&D—Investing in Africa's Future: Analyzing Trends, Challenges, and Opportunities," Accra, Ghana, December 5–7.

Beintema, N., and G. Stads. 2014. *Taking Stock of National Agricultural R&D Capacity in Africa South of the Sahara: ASTI Synthesis Report.* Washington, DC: International Food Policy Research Institute.

CGIAR Secretariat. 2005. *Report of the CGIAR Sub-Saharan Africa Task Forces: The Tervuren Consensus.* Washington, DC.

Chancellor, T., M. Hauser, and P. Sarfatti. 2014. "European Partnerships for Demand-Led Agricultural Research and Capacity Development: The Case of Africa." Accessed September

2014. http://knowledge.cta.int/Dossiers/S-T-Policy/Research-collaboration-in-a-globalised-world/Feature-articles/European-partnerships-for-demand-led-agricultural-research-and-capacity-development-the-case-of-Africa.

Coleman, J., and D. Court. 1993. *University Development in the Third World: The Rockefeller Foundation Experience.* Oxford: Pergamon Press.

ECDPM, ESRF, and LARES (European Centre for Development Policy Management, Economic and Social Research Foundation, and Laboratoire d'Analyze Régionale et d'Expertise Sociale). 2014. *Independent Assessment of the CAADP Multi-Donor Trust Fund.* ECDPM Discussion Paper 158. Accessed December 2014. http://ecdpm.org/wp-content/uploads/DP-158-Independent-Assessment-CAADP-Multi-Donor-Trust-Fund-2014.pdf.

Eyzaguirre, P. 1996. *Agriculture and Environmental Research in Small Countries: Innovative Approaches to Strategic Planning.* Chichester, UK: John Wiley & Sons.

FAO (Food and Agriculture Organization of the United Nations). 2012. FAOSTAT database. Accessed September 2012. www.fao.org/.

FARA (Forum for Agricultural Research in Africa). 2014. *Science Agenda for Agriculture in Africa (S3A): "Connecting Science" to Transform Agriculture in Africa.* Accra, Ghana.

FTF (Feed the Future). No date. "Research." Accessed September 2014. www.feedthefuture.gov/research.

Gamble, W., R. Blumberg, V. Johnson, and N. Raun. 1988. *Three Nigerian Universities and Their Role in Agricultural Development.* AID Project Impact Evaluation 66. Washington, DC: US Agency for International Development. Accessed September 2014. http://pdf.usaid.gov/pdf_docs/PNAAX200.pdf.

IAC (InterAcademy Council). 2004. *Inventing a Better Future: A Strategy for Building Worldwide Capacities in Science and Technology.* Amsterdam.

ILRI (International Livestock Research Institute). 2013. "Capacity Development Is 'Back': Reframing and Repositioning an 'Orphaned' CGIAR Function for an Expanded Future." *ILRI News,* October 22. Accessed September 2014. www.ilri.org/ilrinews/index.php/archives/12216.

Nesseth Tuttle, J., K. Wedding, and A. Applefield. 2011. *Strategic Partnerships to Build African Scientific Capacity for Agriculture.* A Report of the CSIS Global Food Security Project. Washington, DC: Center for Strategic and International Studies.

Pardey, P., J. Roseboom, and N. Beintema. 1995. *Agricultural Research in Africa: Three Decades of Development.* ISNAR Briefing Paper 19. The Hague: International Service for National Agricultural Research.

Sène, L., F. Liebenberg, M. Mwala, F. Murithi, S. Kaboré, and N. Beintema. 2011. *Staff Aging and Turnover in African Agricultural R&D: Lessons from Five National Agricultural Research*

Institutes. Conference Working Paper 17, prepared for the ASTI/IFPRI–FARA Conference "Agricultural R&D—Investing in Africa's Future: Analyzing Trends, Challenges, and Opportunities," Accra, Ghana, December 5–7.

Stads, G., and N. Beintema. 2014. "An Assessment of the Critical Human, Financial, and Institutional Capacity Issues Affecting West African Agricultural R&D: Synthesis and Policy Considerations." Washington, DC: International Food Policy Research Institute; Dakar, Senegal: West and Central African Council for Agricultural Research and Development.

World Bank–OED (Operations Evaluation Department). 1999. "Capacity Building in the Agricultural Sector in Africa." *Précis* 180 (Spring). Washington, DC. Accessed September 2014. http://ieg.worldbank.org/Data/reports/180precis.pdf.

Chapter 9

AFRICAN FACULTIES OF AGRICULTURE WITHIN AN EXPANDING UNIVERSITY SECTOR

Moses Osiru, Paul Nampala, and Adipala Ekwamu

Lack of human resources, particularly senior academic staff and university graduates, still places a major constraint on the quality and volume of agricultural research and development (R&D) output in Africa (Chapter 8, this volume) and hence on the supply of technological innovation in African smallholder agriculture. Despite many attempts to expand (MSc- and PhD-level) training programs in agricultural sciences across Africa, such programs remain relatively underdeveloped. Most postgraduate training programs in agricultural sciences in Africa are still very small in terms of student enrollment, and they themselves suffer from staff limitations, exacerbated by losses of qualified and experienced staff to retirement or more attractive job opportunities outside academia in Africa or elsewhere. Moreover, programs often have difficulty establishing a credible research culture because of lack of research facilities, institutional incentives, and funding.

Where PhD programs are offered, they are usually limited to research only and do not require coursework. Students from diverse and often weak MSc and undergraduate programs have limited or no opportunity to strengthen their disciplinary competencies. Development and delivery of PhD course learning materials require a pool of knowledge emanating from locally relevant research for which senior faculty in a range of disciplines is required. Investment in faculties of agriculture, and in particular in postgraduate programs in agricultural sciences, is critical to enhancing agricultural research and innovation and hence agricultural development across Africa.

This chapter identifies key challenges for universities and the higher-education sector in Africa south of the Sahara (SSA) and, more specifically, explores trends in African faculties of agriculture and their impact on postgraduate training programs. Selected mechanisms by faculties of agriculture and educational networks to respond to the identified challenges are presented and implications of the current trends within the selected faculties are discussed. Finally, the chapter provides lessons and recommendations for improving the quality of postgraduate programs at African faculties.

Faculties of Agriculture in a Changing World

Faculties of agriculture across Africa primarily have a mandate for teaching, research, and outreach, and for producing a core workforce to staff and run agricultural research, extension, and training institutions that support development. Nevertheless, lack of consensus on the role of universities, on the part of both faculty members and their institutions, continues to impede the pursuit of the mission of many universities (Hawkins and Osiru 2012). For many years, African universities, including faculties of agriculture, were labeled "ivory towers" and perceived as making an insignificant contribution to societal needs, including those of rural farmers (Chakaredza et al. 2008). The context underpinning the higher-education sector itself has changed significantly as a result of changes in funding availability, the revolution in information and communications technologies (ICTs), and policy changes that have led to a rapid expansion of universities across the region. These issues prompt a review of the current status of African universities—more specifically of faculties of agriculture—with a consideration for the need to "retool" them to address the demands of a changing world.[1]

Following independence, many African nations assumed that universities would primarily train candidates for the civil service and public professions, filling the gap left by departing colonial administrators and professionals (Cloete and Maasen 2015). In 1972, the Association of African Universities organized a conference that led to the Accra Declaration, calling all universities to become "development universities." However, an economic downturn in the late 1970s and into the 1980s kept policymakers from reforming higher-education institutions and instead prompted them to reduce their expenditures and adopt a policy of diversifying the funding base for universities beyond just government funding (Hayward 2010). For universities, this reform of the education sector opened the door for the establishment of private universities and programs as universities sought for alternative sources of income. Parallel paid programs (discussed later in this chapter) and other commercial activities at public universities were used to generate much-needed income. Emboldened by the rapid increase in the demand for a university education, universities quickly increased in number, many colleges evolved into universities, and teaching institutions became colleges or universities.[2]

This trend occurred in many African countries. In Kenya, for example, the number rose from 6 public and 18 private universities in 2006 (Ngugi 2007),

1 Note that, for the purpose of this chapter, "higher education" refers to university education or training contributing to the award of an undergraduate or postgraduate degree.

2 At independence, and for many years afterward, most African countries had only one public university, largely modeled on the British and French higher-education systems.

to 22 public and 24 private universities in 2013. In Uganda, the number rose from 5 public and 12 private universities in 2003, to 9 public and at least 30 private universities as of 2013. The danger with the proliferation of universities is that, with so many higher-education institutions, none is able to obtain a critical level of funding and staffing to deliver its mandate (Eicher 2004). Most of the private universities focus on humanities and social sciences, and only a few focus on natural and applied sciences. Nevertheless, some private universities have recently established faculties of agriculture, including postgraduate programs in agricultural sciences.

The massification of higher education—defined by Jansen (2003, 292) as "absolute growth in student enrolments, as well as a more egalitarian distribution of students"—was further brought about by the recognition that (higher) education was necessary for national economic growth and seen as a fundamental human right that should be available to all who aspire for such an education. The focus emanated from previous policy shifts to make available to all primary[3] and then secondary education, with two major issues emerging: (1) demand from a growing pool of secondary school graduates required that universities were able to absorb and provide opportunities for eligible students, and (2) the shift in resources toward primary and, later, secondary education stretched the overall envelope of resources, including staff and infrastructure available for university education (Hayward 2010). So while in Africa the number of students in higher education more than tripled between 1991 and 2006 (from 2.7 to 9.3 million), funding for higher education has only doubled, unlike in the rest of the world, where higher-education funding has kept pace with student enrollment (World Bank 2010). In a study of 33 SSA countries, the World Bank noted that public expenditures per student at the higher-education level had fallen drastically, from US$6,800 in 1980 to $981 in 2007–2008 (World Bank 2009).

Based on prevailing trends in SSA, although actual primary and secondary enrollments have been much higher, tertiary enrollments have the highest rates of yearly growth (Table 9.1). The World Bank (2010) estimated that Africa would have 18–20 million higher-education students by 2015, which would require a doubling of the teaching capacity between 2006 and 2015. Although the majority of students favor humanities and social sciences, it is assumed that faculties of agriculture will also attract a growing student population. Nevertheless, agriculture's share of overall student enrollments is

3 This was also prompted in part by Millennium Development Goal 2a, which called for ensuring that male and female children everywhere would be able to complete a full course of primary schooling by 2015.

TABLE 9.1 Enrollments in Africa south of the Africa by level of education, 1999–2012

Level of education	1999	2005 (millions)	2012	Average yearly growth, 1999–2012 (%)
Primary	82.2	111.9	144.1	5.8
Secondary	21.6	31.5	48.6	9.6
Tertiary	2.3	3.9	6.3	13.9
Total	106.1	147.4	199.0	6.7
Share of tertiary in total (%)	2.1	2.7	3.2	

Source: Calculated by authors based on UNESCO (no date).

generally low across SSA (4.4 percent on average), ranging from 0.5 percent in Cape Verde to 7.1 percent in Rwanda (Table 9.2). The fact that science faculties, including faculties of agriculture, are more expensive (because they require larger investments to establish) may act as a constraint. Haramaya University in Ethiopia offers an example of the evolution of higher-education institutions in Africa over time (Box 9.1).

TABLE 9.2 Number of tertiary students enrolled in agricultural programs, 2010–2012 average

Country	Number of students	Agriculture's share of enrollments (%)
Benin	1,589	1.4
Burkina Faso	527	0.9
Cape Verde	51	0.5
Central African Republic	343	3.1
Congo, Democratic Republic	33,879	6.8
Ethiopia	33,007	5.7
The Gambia	58	1.3
Ghana	9,304	3.3
Liberia	2,085	5.4
Madagascar	1,980	2.4
Mali	1,003	1.2
Mauritius	359	0.9
Mozambique	5,213	4.6
Rwanda	5,102	7.1
Tanzania	1,636	1.0
Zimbabwe	2,296	2.4
16-country average	6,152	4.4

Source: Calculated by authors based on UNESCO (no date).
Note: Averages were calculated based on one to three data points.

BOX 9.1 Haramaya University, Ethiopia

The first university in Ethiopia, Addis Ababa University, was established in 1950, and with its constituent colleges it remained Ethiopia's only university for close to 50 years. In the past 10 years, however, several new universities have been formed, and many of Addis Ababa University's colleges have expanded into fully fledged universities in their own right. Haramaya University College of Agriculture (originally the Alemaya College of Agriculture) made this transition in 1985, became a multidisciplinary university in 1996, and was renamed Haramaya University in 2006. From 1952 to 1968, the institution received substantial support from the the United States Agency for International Development (USAID). Since 1997, the university has run a BSc program for students who work in the public sector and hold either agriculture or forestry diplomas from accredited colleges. The program's objectives are to upgrade the skills of frontline, midcareer extension workers. The university was initially Ethiopia's only higher-education institution offering MSc training in major agricultural specialties, and in 2002 it launched PhD programs. Haramaya University originally had the national mandate for both agricultural research and extension, but currently its research and extension activities operate under the umbrella of the Ethiopian Institute of Agricultural Research.

Haramaya University has ambitious plans for new BSc-, MSc-, and PhD-level degrees. Additionally, in collaboration with the International Food Policy Research Institute, a Center for Agricultural Research Management and Policy Learning for Eastern Africa (CARMPoLEA) was established. This regional center served as home to a capacity-building initiative to improve the management, organization, and leadership of agricultural research and policymaking and, ultimately, to support national and regional agricultural innovation systems. The center organized a series of workshops intended to respond to regional needs for knowledge and skills in the areas of agricultural research management and policy, including in-person and virtual courses, as well as targeted follow-up services, for researchers, policymakers, and other stakeholders in East Africa.

As of December 2014, Ethiopia had 31 public universities, many with infrastructure development still underway. Enrollment in the 2009/10 academic year comprised 203,455 students, plus intakes of 78,822 students in 2010/11 and 94,000 in 2011/12. National reforms have included curriculum revision; the development of new programs; and dedicated support of institutions focused on enhancing quality, such as the Higher Education Relevance and Quality Agency and the Higher Education Strategy Centre, in line with the 2009 Higher Education Proclamation. The Ethiopian government took other steps to enhance quality, although concerns remain that their implementation has been politicized, and questions have been raised as to whether the implementation process conforms closely enough with the Proclamation.

Sources: Blackie, Mutema, and Ward (2009); Areaya (2010); Haramaya University (2013).

Staffing, Infrastructure, and Funding Constraints

Staffing and Infrastructure

As faculties of agriculture respond to local and national demand for agricultural specialists as well as the need for broader knowledge to intensify agricultural production, MSc- and PhD-level programs have diversified, expanded, and increased in number. Nevertheless, the bulk of the student population (and hence the teaching load) at most faculties of agriculture in Africa is still concentrated in undergraduate programs (Table 9.3).

Staffing at faculties of agriculture varies greatly, indicating the relative strengths of the various faculties. A study (RUFORUM 2009) assessed staffing levels and postgraduate programs at six faculties of agriculture (Table 9.3). Staffing levels varied from 188 academic staff at Makerere University to 33 at the National University of Rwanda, which at the time of the study ran only one MSc program with an enrollment of 23 students but no PhD program. As of 2011, Egerton University in Kenya employed 121 academic staff and had 47 MSc students enrolled; Makerere University's Faculty of Agriculture[4] employed 63 PhD-qualified staff and had 108 MSc students enrolled. As of 2012, Lilongwe University of Agriculture and Natural Resources employed 143 academic staff, and all are expected to be PhD qualified by 2018 (Blackie, Mutema, and Ward 2009). Cloete, Bunting, and Maasen (2015) showed that the low proportions of PhD-qualified and senior academic staff were linked to low knowledge production at universities. In general, many universities have shifted away from offering the general programs common in the past toward more specialized programs. Several programs, however, are challenged by the low number of students—such as at Egerton University and the National University of Rwanda, which had 1 and 0 PhD students enrolled, respectively. Programs require a minimum number of students to be cost-effective. Academic staff members spend most of their time on the large number of BSc students enrolled, leaving little time for research-oriented activities.

Of particular importance, the pool of academic staff at faculties of agriculture is aging. The share of PhD-qualified higher-education researchers older than 50 years ranged from 14 percent in Burundi to 98 percent in Guinea (Figure 9.1). In close to 80 percent of the countries (21 of 27 countries sampled), the share of PhD-qualified staff over 50 years old was more than

4 Makerere University reorganized its faculties into colleges and departments in 2010; the Faculty of Agriculture is now under the College of Agricultural and Environmental Sciences.

TABLE 9.3 Academic staff and postgraduate programs at selected faculties of agriculture, 2011/12

University	Qualification			Total	No. of MSc. programs	No. of PhD programs	No. of BSc students	No. of MSc students	No. of PhD students
	PhD	MSc	BSc						
Egerton University, Kenya	56	46	19	121	14	9	302	47	1
Makerere University, Kenya	63	103	22	188	9	8	1,175	108	39
National University of Rwanda	18	13	2	33	1	0	822	23	0
University of Zambia	18	17	0	35	4	4	450	24	8
University of Zimbabwe	23	29	10	62	4	3	452	42	na
Lilongwe University of Agriculture and Natural Resources, Malawi	58	70	15	143	8	2	900	na	na
Université d'Abomey-Calavi, Benin	55	15	2	72	na	na	na	53	116
University of Ghana	88	30	10	128	na	na	na	25	99
Université de Lomé, Togo	37	6	1	44	na	na	na	na	na

Sources: Calculated by authors based on ASTI (2014); RUFORUM (2009); and Blackie, Mutema, and Ward (2009).

Note: na = data were not available.

FIGURE 9.1 Share of PhD-qualified researchers in the higher-education sector over 50 years old, 2011

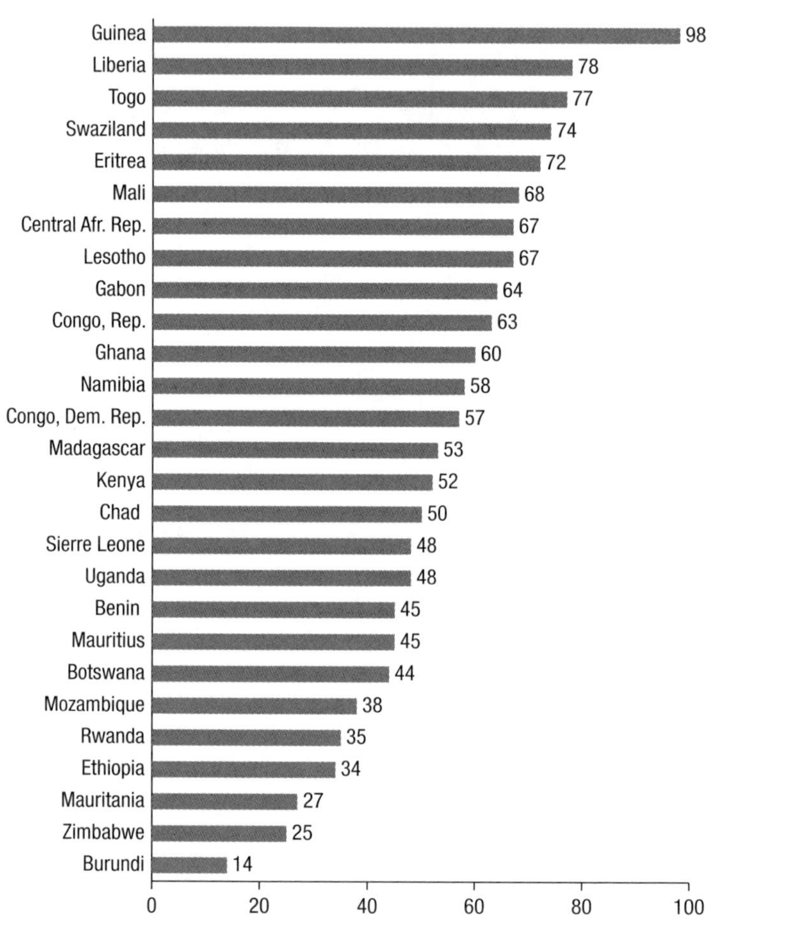

Source: Calculated by authors based on ASTI (2014).

40 percent of all staff. Most of the current staff were recruited during the establishment of the faculties of agriculture in the 1970s and 1980s and trained abroad, often up to the PhD level[5] this cohort of staff is nearing or at retirement age. Later recruitments have generally been fewer and have benefitted less from overseas training opportunities because of limited funding for higher education.

5 During the 1990s, USAID supported 9,128 developing-country students, 310 of whom were in agriculture; by 2000, this number had dropped to fewer than 1,700 (BIFAD 2003).

In most faculties of agriculture, BSc classes are overcrowded as a result of the increasing enrollments without the requisite increase in the number of staff or facilities. Classroom, library, and laboratory space per student is low, averaging 1.3 square meters in African public universities compared with 4–10 square meters at universities in Organisation for Economic Co-operation and Development countries (World Bank 2010). Basic classes are usually taken by all first-year students prior to the selection of electives in later years, which only exacerbates this situation. At Malawi's Lilongwe University of Agriculture and Natural Resources (formerly Bunda College)—with a significantly increased first-year intake of more than 2,500 students in 2009—most classes comprised at least 250 students (Blackie, Mutema, and Ward 2009). At Makerere University, undergraduate classes often have more than 300—and even up to 500—students, especially for common courses (housed within structures planned for significantly lower numbers). Excessive class size, which is seemingly becoming more common, challenges faculty staff in terms of promoting discussion and group activities (Hawkins and Osiru 2012). Most postgraduate programs in agricultural sciences in Africa, however, do not suffer from overcrowded classrooms; they often have so few students as to lack a critical mass and be cost-inefficient.

Rather than each faculty attempting to offer a wide range of programs—and being unable to excel at any of them—it would seem to make more sense to cluster specialized postgraduate programs at national and regional levels. The Regional Universities Forum for Capacity Building in Agriculture (RUFORUM) has established centers of leadership that are now offering high-quality MSc and PhD programs jointly implemented by two or more of its 55 member universities. The programs are designed to include the full participation of faculty and stakeholders in the region and are a response to the articulation of demand by stakeholders in higher education (RUFORUM 2014).

Although most faculties of agriculture have ample PhD-qualified staff members (Table 9.3), undergraduate teaching is mainly being carried out by MSc-level staff and, in some cases, by postgraduate students. Staff pointed to the low salaries, limited opportunities for professional development (sabbaticals, postdoctoral fellowships, PhD training opportunities), and limited research funding as key constraints to their professional advancement (Blackie, Mutema, and Ward 2009). Moreover, some postgraduate programs surveyed had too few students enrolled for the programs to be cost-effective (RUFORUM 2009). Staff member-to-student ratios are low but highly variable. Nevertheless, the challenge remains to upgrade the qualification and experience levels of the more junior staff. The challenges have encouraged the development of regional efforts to strengthen postgraduate training in the region, such as through the RUFORUM network.[6]

Access to basic infrastructure and reading materials remains a constraint. Overcrowding and lack of investment at faculties of agriculture and universities, in general, has resulted in poor access to library facilities (Kanyengo 2009). The University of Zambia, for example, opened in 1966 with 312 students and a planned doubling of its student population within five years. Nevertheless, the student body grew to more than 1,000 by 1970 and to 10,008 by 2014. The university's library was forced to develop new strategies in response to the expanded student numbers, including employing students to work in the library; enhancing the use of electronic resources and databases, such as The Essential Electronic Agriculture Library (TEEAL) and Access to Global Online Research in Agriculture; moving resources from open shelves to a reserve collection to enhance student access to materials; restricting access to serial collections to postgraduate students; and increasing library opening hours. Other challenges included a lack of shelving, reading tables, reliable Internet connectivity, and insufficient funding. Similarly, libraries in Ghana and Uganda reported difficulty in sustaining the cost of subscribing to journals (Marty 2002; Were 2002), a key resource in conducting high-quality research.

Funding

Funding, or lack thereof, has been a key driver of reform at African universities. Following the policy changes in the late 1980s and early 1990s, which resulted in increased demand and enrollments at universities in general and faculties of agriculture in particular, African universities were forced to innovate by diversifying their funding streams. However, most faculties of agriculture struggled to balance the increasing demand for higher education with the need to improve quality, and many have grown beyond their financial capacity. Expenditure per student in Africa has declined by 30 percent over the past 15 years (World Bank 2009), raising concerns about the capacity of African universities to enhance the quality and relevance of training.

6 RUFORUM, based at Makerere University, Uganda, is a consortium of 55 universities in 22 countries. Recognizing the important (and largely unfulfilled) role universities play in contributing to the well-being of small-scale farmers and economic development throughout SSA, RUFORUM's mandate is to oversee graduate training and networks of specialization in the universities and countries in which it works (Chapter 10, this volume).

At the same time, only a quarter of all international aid to the education sector is spent on higher education. Of this support, less than a quarter goes directly to the African institutions; the rest is spent on scholarships and associated costs abroad (World Bank 2010).

Donor funding has evolved significantly over the years since independence. Initially, funding was often targeted toward support for infrastructure development, including the construction of buildings, laboratories, and libraries in the 1960s and early 1970s (DDRN 2010). This then evolved in the 1970s and 1980s into a combination of technical assistance to local African universities and long-term training at universities in donor countries. The focus shifted to strengthening the requisite human resources to support local research and build training institutions. An example was the Manpower for Agricultural Development program in Uganda, supported by USAID. As lessons from these programs emerged, programs were tweaked in response to challenges. One recurrent challenge was difficulty in getting students who received training to return to their sponsoring institutions, and equally to resettle them into their jobs. A second challenge was the focus of the research and results of programs provided by external universities were not usually directly applicable to the local contexts of the students being trained. Graduates often did not return to Africa and, when they did, resettlement remained a challenge. "Sandwich" programs, in which students do their PhD coursework and defense of their dissertation abroad, but their research at home, were designed to make the research experience and its outputs more relevant to the trainee and his or her sponsor, maintain the bond between the two, and at the same time contribute solutions to local development challenges in integrated agricultural research for development. Currently, the training and research support available is delivered in various ways, but is largely less institutionalized and mostly managed by the target beneficiaries as individual scholarships.

At the end of the 1980s, funding agencies began to rethink their support following a World Bank report that estimated the social rates of return to higher education to be lower than those of basic education (Psacharopoulos and Patrinos 2004). This was in line with World Bank suggestions that higher education was largely a "luxury" for Africa (Mamdani 1993). Subsequently, World Bank funding for basic education increased dramatically. Other donors followed this strategy and national governments were (implicitly or explicitly) advised to prioritize basic education over higher education. Donor funding to higher education increased slightly from the 1990s, but not to the levels needed to fill the gap left by national governments or to meet the increased demand for higher education. Donor support to higher education in SSA stood at approximately US$600 million per year in 2010 (World Bank 2010). With highly fragmented support, it was difficult to undertake the type of longer-term initiatives needed to strengthen faculties of agriculture.

African universities have responded to the above challenges by adopting new management techniques, developing more demand-oriented curriculums, creating new courses, and taking steps to strengthen public–private partnerships (Mihyo 2008). Other innovations have been through various mechanisms to strengthen fundraising, including privatization and outsourcing of services. Innovations have often exposed further challenges, however, and quality concerns have remained. Recognizing that government funding is stretched (World Bank 2009), today higher education is paid for through a variety of mechanisms, including tuition, examination, and other related fees that generate an average of close to 30 percent of university income—ranging from 5 percent in Madagascar and Zimbabwe, to 56 percent in Uganda, and 75 percent in Guinea-Bissau (World Bank 2010). However, although a large portion of financing for faculties of agriculture is from overseas development assistance, overall donor support to agricultural higher education and training in Africa has remained low (World Bank 2009). For example, the World Bank allocated only 20 percent ($170 million) of its agricultural budget in Africa (Eicher 2006). Despite this, African institutions have largely recognized the need to train students locally because (1) students would often not return from training abroad, and when they did they had difficulty fitting back into local conditions; (2) research undertaken during training at foreign universities was often not relevant to local problems; and (3) if the objective was staff development, staff would remain in the workforce for a significant part of the training period.

Many faculties of agriculture (like universities generally) have diversified their training programs. In addition, they have started to offer parallel (that is, duplicate) programs in the evening and on weekends in order to accommodate students with day jobs—an important and growing market (Box 9.2). Distance education and adult-learning programs have also been initiated in recent years. Unfortunately, teaching facilities and resources—including lecture rooms, library facilities, and teaching staff—did not increase proportionately with the number of students enrolled. In addition, faculty members often accept additional part-time employment at new faculties of agriculture.

BOX 9.2 A comment on parallel programs

A study commissioned by the Association for Strengthening Agricultural Research in Eastern and Central Africa and RUFORUM reported that several faculties of agriculture in the region are running parallel programs for paying students, a practice that boosts university finances. Students in these programs have special privileges in that they can choose their first preference for their degree programs, whereas regular students can be assigned to programs of little interest to them if their first choices are oversubscribed or they do not meet the minimum entry requirements. Students in parallel programs do not have to wait for government sponsorship processes and can enroll a year earlier than regular-entry students. An important outcome is that student enrollment in parallel programs takes priority to maximize the revenues generated, but the quality of the education suffers; it is common, for example, for a university administration to give directives on student numbers, ignoring available resources and capacity.

Source: Blackie, Mutema, and Ward (2009).

Research and Postgraduate Degree Programs

Research undertaken by higher-education institutions is a significant and growing component of national agricultural research in Africa (Table 9.4). In 2011, on average, higher-education establishments constituted 25 percent of SSA's national agricultural research capacity, up from 20 percent a decade earlier. Over this decade, agricultural research capacity at higher-education institutions grew much faster (6 percent per year, on average) than that at government and nonprofit agencies (2.6 percent per year on average). While much of the growth in full-time equivalent (FTE) researchers at universities is the result of an expansion of faculty staff, some of it may also be attributed to the fact that staff have begun to spend more time on research. In a study of eight flagship universities in SSA, Cloete, Bunting, and Maasen (2015) reported an increase in the number of MSc graduates from 2,268 in 2001 to 7,156 in 2011 and in the number of PhD graduates from 154 in 2001 to 367 in 2011.[7] RUFORUM supported the training of more than 1,300 MSc students during 1992–2013 and 212 PhD students during 2008–2013 (RUFORUM 2014).

---

7 The eight flagship universities involved were the University of Botswana, University of Cape Town, University of Dar es Salaam, Eduardo Mondlane University, University of Ghana, Mauritius University, Makerere University, and University of Nairobi.

TABLE 9.4 Agricultural researchers in selected African countries, 2001–2011

Indicator	2001	2008	2011
Total number of agricultural researchers employed at government and nonprofit agencies (FTEs)	7,260	8,281	9,412
Total number of agricultural researchers employed at higher-education institutions (FTEs)	1,769	2,565	3,179
Total number of agricultural researchers employed at higher-education institutions as a share of publicly employed researchers (%)	20	24	25
Share of PhD-qualified agricultural researchers employed at government and nonprofit agencies (%)	25	26	23
Share of PhD-qualified agricultural researchers employed at higher-education institutions (%)	48	53	52
Share of female agricultural researchers employed at government and nonprofit agencies (%)	na	24	25
Share of female agricultural researchers employed at higher-education institutions (%)	na	22	21

Source: Calculated by authors based on ASTI (2014).

Notes: Overall researcher data are based on 34 countries of SSA; sample shares for researchers by degree and gender are for 24 and 26 countries, respectively. Note that public agricultural researchers are defined here to include those employed in the government, nonprofit, and higher-education sectors. FTEs = full-time equivalents; na = data were not available.

Time allocated to research varies widely by faculty and staff members within faculties; on average, however, faculty members spend around 25 percent of their time on research (Beintema and Stads 2011). Notably, agricultural researchers within higher-education institutions generally have a far better educational profile compared with those employed within government and nonprofit agencies. In 2011, 52 percent of the researchers in the higher-education sector were PhD qualified, compared with an average of only 23 percent in the government and nonprofit sectors (Table 9.4).

Faculty members from relevant disciplines offer a valuable untapped resource in the conduct of multidisciplinary research, as do female scientists, whose participation has been low across the board (suggesting considerable gender disparity; Chapter 8, this volume). Thus, faculties of agriculture could play an even greater role in performing research and solving rural farmers' problems, while at the same time ensuring that agriculture-related programs produce graduates with the required levels of skills and experience relevant to development needs.

Africa's relative share of the global output of scientific research, measured in terms of international publications, was declining prior to 2002 (Tijssen 2007). For the period 2002–2008, this share increased from 1.6 to 2.2 percent, largely through international collaborations with non-African partners (Tijssen 2015). At the same time, the share of researchers marginally declined, from 2.2 to 2.1 percent (Zeleza 2014). Low government support for university-based research—with available funding being allocated to recurrent expenditures (mainly staff salaries and utilities)—has not helped the situation. Funding for research has to be mobilized mainly from external sources. To date, few African governments have established competitive funds to support research at universities. Hence, most external funding for research is derived from foreign donors, which makes it more difficult for countries to set their own research agendas. As a result, many faculties of agriculture have established grant management offices to facilitate resource mobilization.

In particular, universities need to focus on developing and strengthening high-quality PhD and MSc programs in order to strengthen research and training. Huge challenges remain, including limited and unsustainable funding, low numbers of PhD-qualified staff to run training programs, and inadequate policy regimes to support a culture of research. Funding for publicly supported students in universities is usually paid as tuition at a rate that is often below the market rate for universities. Setting student tuition fees has become a political issue at many universities; those that receive government funding are often not allowed to set their own fees, which is seen as a political tool that few governments would want to lose control of.

Most universities still focus their primary efforts on teaching, followed by research (which is reinforced by eligibility criteria for career advancement); there is little incentive for academic staff to focus on the third facet of the university mandate, which is outreach. In addition, funding is a major limitation to the conduct of research in most African faculties of agriculture. This contrasts with the situation in Brazil, for example, where graduate programs are assessed for funding based on their research output (UNESCO 2009). Cloete and Maasen (2015) recommend that African governments must consider differentiating higher-education institutions because mandates are often contradictory; they cite Mohamedbhai's (2012) review, showing that only five African universities—none of which is in SSA—were in the two main rankings of the top 500 universities worldwide.[8] Hence, it may be more practical for universities to have multiple mandated areas.[9]

8 The five universities were the University of Cape Town, Stellenbosch University, the University of the Witwatersrand, Alexandria University, and the University of KwaZulu-Natal.

9 South Africa's higher-education system has three categories of universities (Bunting et al. 2015): (1) traditional universities focusing on general academic and professional programs, (2) universities of technology focusing on vocational programs, and (3) comprehensive universities offering a mix of (1) and (2).

Postgraduate programs are the center for research at most universities, and research is coordinated through a school of graduate and postgraduate studies. Research at faculties of agriculture is usually based on discipline-related mandates, student training, and donor objectives. This is partly changing, as several universities have established various platforms, such as RUFORUM's national forums, to ensure greater alignment and responsiveness to national research priorities.[10] Although universities have significant staffing capacity compared with public agricultural research institutions, they are often unable to access sufficient public research funding. Ongoing efforts to address this problem include the restructuring of national agricultural research systems and the use of competitive funding mechanisms to separate research funding from its delivery. This allows other research actors, such as faculties of agriculture, to tap into national research funding. If universities are to actively pursue research, they must provide their staff with research facilities that meet international standards, such as laboratories, research fields, and access to up-to-date literature (UNESCO 2009); train them in research methods and in writing research proposals; and provide them with appropriate incentives.

Most PhD programs at African faculties of agriculture follow the old European model of delivering a PhD dissertation based on the candidate's own research, with no formal coursework required. This works when professors have only one or two PhD students at a time (and training is primarily one on one), but it becomes unmanageable with larger groups of students. Also, most European universities currently require formal coursework as part of their PhD programs, but the research-only model is still the most commonly used at many many African universities. The research-only model is increasingly coming under scrutiny because the level of students entering universities is uneven. The low proportion of staff undertaking research (or qualified to conduct research) hinders the capacity of the faculties to support high-quality postgraduate training and has limited the scope of the programs, such that less contextually specific material is being used. In addition, many degrees still take longer than their prescribed duration; MSc programs usually take three to four years, rather than two, and a PhD program can take as long as six years.

---

10 Thirteen national forums are currently operating in Burundi, Democratic Republic of the Congo, Kenya, Malawi, Mozambique, Rwanda, Swaziland, Tanzania, Uganda, Zambia, and Zimbabwe. They are designed to provide a platform for stakeholders to articulate demand for university services, advocate for change, and provide feedback on the utility of RUFORUM's activities. The forums typically comprise member universities; policymakers; and representatives from farmers' organizations, the private sector, and the national agricultural research and extension system (more information is available at www.ruforum.org).

TABLE 9.5 RUFORUM's regional PhD programs, 2014

Program	Commencement	Host university, country	No. of students
Dryland Resource Management	2008	University of Nairobi, Kenya	46
Food Science and Nutrition	2013	Jomo Kenyatta University of Agriculture and Technology, Kenya	7
Agricultural and Rural Innovation Studies	2012 2012 2013	Makerere University, Uganda Egerton University, Kenya Sokoine University of Agriculture, Tanzania	46
Plant Breeding and Biotechnology	2009	Makerere University, Uganda	47
Soil and Water Management	2010	Sokoine University of Agriculture, Tanzania	31
Agricultural and Resource Economics	2009	Lilongwe University of Agriculture and Natural Resources, Malawi	21
Aquaculture and Fisheries	2009	Lilongwe University of Agriculture and Natural Resources, Malawi	14
Total			212

Source: Compiled by authors based on RUFORUM (2014).

Recognizing the need to enhance the quality of research and teaching at African universities, and in light of the low proportion of PhD-qualified staff at African faculties of agriculture, RUFORUM and its 46 member universities have initiated a model based on the US system, maximizing the regional comparative advantages of the different universities initiated between 2008 and 2013. The RUFORUM networking model has provided some important instruction for strengthening universities under the current weak public financing. Working in partnership with universities in Europe, the United States, and the United Kingdom, RUFORUM has worked with six universities to build seven regional PhD programs and has provided support and facilitated training for 212 PhD students (Table 9.5) and 211 MSc students since 2008. The programs have demonstrated the strength of networking capacity, highlighting the potential for matching weak and stronger universities, building cost-effective graduate programs, strengthening the internationalization of programs, improving staff engagement and curriculum review, creating partnerships with research institutions for graduate research, and supporting capacity building for newer universities.

RUFORUM has also shown that African faculties of agriculture can produce quality graduates within a specified and reasonable timeframe, which is a major concern in the region. Makerere University's MSc plant-breeding program is a case in point. Since its inception in 2009, the program has produced graduate students—within the specified timeframe of 24 months—with strong technical and practical skills, acceptance in peer-reviewed publications, and success in finding employment. This success has been possible because the program was designed to respond to market needs, is well linked to national plant-breeding programs and the emerging seed industry in the region, and has appropriate levels of well-qualified, full-time academic staff allocated to and invested in the program. The regional nature of the program allows it to draw students and staff from across Africa and to attract international scholars. The lessons from this MSc program, as well as from Makerere's PhD plant-breeding and biotechnology program, have catalyzed the emergence of joint graduate training programs in SSA. These include, for example, the PhD program in agricultural and rural innovations, jointly run by Egerton University, Makerere University, and Sokoine University, which now also offers joint courses with three European universities (Montpellier SupAgro, the University of Copenhagen, and Wageningen University and Research Centre). Each year an average of 40 PhD students from Africa and Europe take joint field classes concurrently with online courses.

The RUFORUM model has led to other benefits, such as the infusion of new ideas into university research by attracting teaching staff from foreign universities. Several universities have been challenged by "inbreeding" in the sense that a majority of faculty are former students joining the teaching staff of their universities upon graduation. RUFORUM's programs have also provided much-needed and affordable training for the smaller, often newer and private, universities that usually do not run postgraduate programs and have numerous staff in need of further education. New universities have enhanced the competitiveness of universities in terms of both teaching and research. The downside has been the exodus of staff from established universities to the newer ones as part-time teachers. Retired academics have also been reabsorbed into the newer private universities.

Programs that enhance the opportunity to share academic knowledge, talent, and experience will be critical in the future. As African faculties become more interconnected through increased access to ICTs, it is critical that they develop research cooperation that responds directly to the needs of African faculty. The longstanding cooperation that has supported and strengthened a variety of programs, such as the Norwegian and Swedish support to Sokoine University and the University of Malawi, and the Rockefeller Foundation's support of Makerere University, will continue to be relevant. To date, research cooperation has been limited to a few African universities (Shabani 2008), but RUFORUM is addressing this by facilitating regional learning through cross-country research among its member universities.

The advent of the African Union, the New Partnership for Africa's Development, and the Forum for Agricultural Research in Africa has strengthened research cooperation among national agricultural research institutes in Africa, and universities need to be linked to these efforts (Chapters 1 and 9 this volume). Research is a powerful tool for strengthening the quality of teaching. In a study of three African universities in Kenya, Malawi, and Uganda, Hawkins (2010) found that most lecturers in faculties of agriculture had little or no formal training in learning theory or teaching, and hence their understanding of learning concepts, experiential learning, action research, and outreach concepts was highly variable. The study found that, for many staff, experiential learning was a process of practicing what had been learned or of experimenting. Outreach and action research were often viewed as processes for disseminating technology generated through conventional research, or as deeper engagement with other development stakeholders. Field attachments were highlighted as the key method for enhancing experiential learning. This being the case, universities and faculties maintain a variety of teaching and delivery methods, dependent on the individual staff involved. Younger staff members were more likely to attempt to incorporate group methods, including classroom discussions and student research on course material. Faculty members need to shift their focus from teaching approaches based on pedagogy to those based on andragogy (Table 9.6).

TABLE 9.6 The difference between pedagogy and andragogy

Pedagogy	Andragogy
The learner is dependent, and the teacher determines what is learned, when it is learned, and how learning is evaluated.	The learner is more independent, and the teacher encourages this independence and guides the learner.
The experience of the learner is not considered to be significant; teaching methods are didactic.	Experience is valued as a rich resource for learning and forms the basis of discussions and problem solving.
People learn what society expects them to learn; the curriculum is standardized.	People learn what they need to; the curriculum is organized around their needs.
Learning themes are organized around abstract disciplines.	Learning themes are organized around experiences, problems, or expected competencies.

Source: Hawkins (2010).

Tomorrow's Faculties of Agriculture

Africa needs a far larger pool of talent in the agricultural sciences to support knowledge generation and its application to intensify and raise agricultural production and productivity. Faculties of agriculture and related sciences within the broader university context have a greater role to play both in training innovators and problem solvers and in creating knowledge to enhance the competitiveness of African agriculture. Integration of innovation into the research process will require universities to develop their collaboration with other partners—such as national and international research institutions, extension and rural development agencies, and the private sector—to ensure demonstrable outcomes (Chapter 13, this volume).

Higher-education institutions in Africa are rapidly increasing and expanding to meet the growing demands for higher education in the region. The prevailing general trends and developments apply equally to faculties of agriculture in Africa:

- Both the number and the size of higher-education institutions are expanding rapidly in Africa to meet the growing demand for higher education. Moreover, because such a low share of the African population has a higher degree (less than 5 percent compared with more than 40 percent in the more advanced developed countries), demand is expected to remain high for some time to come (that is, substantially above population growth). Meeting this demand will also require a significant increase in teaching capacity.
- Most higher-education institutions in Africa began with undergraduate education only, but many are now gradually beginning to offer postgraduate degrees. As a prerequisite for this upgrade, these institutions must substantially strengthen their research profile. In the years to come, growth in postgraduate education will most likely exceed growth in undergraduate education. The push toward the "development university" model has meant that some countries (such as Ethiopia and Kenya) are supporting the emergence of research-based universities.
- For the past few decades, public funding for higher-education institutions in Africa has grown far more slowly than the number of students enrolled. Higher-education institutions have tried to close this funding gap by mobilizing external funding (including student fees) and enhancing efficiency; however, according to many observers, the quality of the education programs has suffered.

- Overwhelmed by the numbers of students and lack of funding and facilities, many higher-education institutions in Africa have been unable to properly address both their teaching and their nonteaching mandates.

Growth in funding of higher-education institutions (including faculties of agriculture) should be increased considerably, at least to match growth in student enrollments. Funding for nonteaching activities, such as research and outreach, should be decoupled from student enrollments and given dedicated funding in the form of government grants and contracts or competitive funding schemes. Faculties of agriculture need to be more competitive, for example, by marketing their success stories and developing the necessary capacity and databases to ensure that information on their status is more widely accessible. Modernization of teaching methods and curriculums should continue.

A closer look at the postgraduate programs of selected faculties of agriculture in East, Central, and Southern Africa reveals that many of these programs are too small to be cost-efficient or assemble sufficient critical mass to excel. This problem seems to be common in other parts of Africa as well. To some extent, these are typical growth issues that may disappear over time. Nevertheless, collaboration between national and regional faculties of agriculture is warranted in order to streamline, through clustering and specialization, the range of postgraduate programs being offered. This means that students completing their undergraduate degree may need to transfer to a different faculty of agriculture to pursue the postgraduate degree of their choice. In some cases this may mean that students will have to pursue postgraduate training in a neighboring country. At the same time, recipient programs will need to make the necessary arrangements for foreign students, including accommodations and visas. Such cross-institutional and cross-border collaboration has the potential to facilitate emergence of centers of leadership, as this will greatly increase the overall efficiency and quality of postgraduate training in agricultural sciences in the participating countries. Efforts should be made, however, to ensure that these regional "centers of leadership" are underpinned by networks to avoid a repeat of past incidences whereby higher-education institutions were ravaged by civil conflict and never recovered. Specialized postgraduate programs in agricultural sciences will also have an impact on the research profile of the participating faculties of agriculture. Rather than offering an overview of all research topics, faculties of agriculture may become national or regional specialists in certain thematic areas.

References

Areaya, S. 2010. "Tension between Massification and Intensification Reforms and Implications for Teaching and Learning in Ethiopian Public Universities." *Journal of Higher Education in Africa* 8: 93–115.

ASTI (Agricultural Science and Technology Indicators). 2014. ASTI database. Accessed May 2014. http://asti.cgiar.org/data/.

Beintema, N., and G. Stads. 2011. *African Agricultural R&D in the New Millennium: Progress for Some, Challenges for Many.* IFPRI Food Policy Report. Washington, DC: International Food Policy Research Institute.

BIFAD (Board for International Food and Agricultural Development). 2003. *Renewing USAID Investment in Global Long-Term Training and Capacity Building in Agriculture and Rural Development.* Washington, DC.

Blackie, M., M. Mutema, and A. Ward. 2009. *A Study of Agricultural Graduates in Eastern, Central and Southern Africa: Demand, Quality and Job Performance Issues.* A report to ASARECA and RUFORUM. Kampala, Uganda: Regional Universities Forum for Capacity Building in Agriculture.

Bunting, I., N. Cloete, H. Li Kam Wah, and F. Nakayiwa-Mayega, 2015. "Assessing the Performance of African Flagship Universities." In *Knowledge Production and Contradictory Functions in African Higher Education*, edited by N. Cloete, P. Maasen, and T. Bailey. Cape Town: African Minds.

Chakeredza, S., A. Temu, J. Saka, D. Munthali, K. Muir-Leresche, F. Akinnifesi, O. Ajayi, and G. Sileshi. 2008. "Tailoring Tertiary Agricultural Education for Sustainable Development in Sub-Saharan Africa: Opportunities and Challenges." *Scientific Research and Essays* 3 (8): 326–332.

Cloete, N., I. Bunting, and P. Maasen. 2015. "Research Universities in Africa: An Empirical Overview of Eight Flagship Universities." In *Knowledge Production and Contradictory Functions in African Higher Education*, edited by N. Cloete, P. Maasen, and T. Bailey. Cape Town: African Minds.

Cloete, N., and P. Maasen. 2015. "Roles of Universities and the African University Context." In *Knowledge Production and Contradictory Functions in African Higher Education*, edited by N. Cloete, P. Maasen, and T. Bailey. Cape Town: African Minds.

DDRN (Danish Development Research Network). 2010. *Mapping the World of Higher Education and Research Funders: Actors, Models, Mechanisms and Programs.* A Report to the Danish Development Research Network and Universities. Copenhagen.

Eicher, C. 2004. *Rebuilding Africa's Scientific Capacity in Food and Agriculture.* Background Paper 4 commissioned by the InterAcademy Council Study Panel on Science and Technology

Strategies for Improving Agricultural Productivity and Food Security in Africa. Amsterdam: InterAcademy Council.

———. 2006. *The Evolution of Agricultural Education and Training: Global Insights of Relevance for Africa*. Staff Paper 2006-26. East Lansing: Michigan State University, Department of Agricultural Economics.

Haramaya University. 2013. "Haramaya University Facts and Figures 2012/13." Accessed April 2016. http://www.haramaya.edu.et/about/facts-and-figures/.

Hawkins, R. 2010. *Experiential Learning, Action Research and Outreach: A Comparative and Gap Analysis*. A Consultancy Report submitted to the Regional Universities Forum for Capacity Building in Agriculture. Kampala, Uganda.

Hawkins, R., and M. Osiru. 2012. "Experiential Learning: Theory and Practice at Three RUFORUM Universities." In *Proceedings of the Third RUFORUM Biennial Conference*, edited by G. Tusiime, P. Nampala, and E. Adipala. Kampala, Uganda: Regional Universities Forum for Capacity Building in Agriculture.

Hayward, F. 2010. "Graduate Education in Sub-Saharan Africa." In *Higher Education and Globalisation: Challenges, Threats and Opportunities for Africa*, edited by D. Tefera and H. Greijn. Maasstricht: Maastricht University; Boston: International Network for Higher Education in Africa.

Jansen, J. 2003. "The State of Higher Education in South Africa: From Massification to Mergers." Chapter 13 in *State of the Nation: South Africa 2003–2004*, edited by J. Daniel, J., A. Habib, and R. Southall. Cape Town: Human Sciences Research Council Press.

Kanyengo, C. 2009. "A Library Response to the Massification of Higher Education: The Case of the University of Zambia Library." *Higher Education Policy* 22: 373–387.

Mamdani, M. 1993. "University Crisis and Reform: A Reflection of the African Experience." *Review of the African Political Economy* 58: 7–19.

Marty, A. 2002. "The Programme for Enhancement of Research Information (PERI) Expected Impact in Africa," *INASP Newsletter* 19 (February): 3.

Mihyo, P. 2008. "Staff Retention in African Universities and Links with Diaspora Study." Paper presented at the Association for the Development of Education in Africa (ADEA) Biennial Conference on Beyond Primary Education, Maputo, Mozambique, May 5.

Ngugi, C., ed. 2007. *ICTs and Higher Education in Africa*. Cape Town: University of Cape Town, Centre for Innovation in Learning and Teaching. Accessed April 2013. www.cet.uct.ac.za/projects#PHEA.

Psacharopoulos, G., and H. Patrinos. 2004. "Returns to Investment in Education: A Further Update." *Education Economics* 12: 111–134.

RUFORUM (Regional Universities Forum for Capacity Building in Agriculture). 2009. *An Evaluation of Higher Agricultural Education Institutions in Eastern and Southern Africa.* Final Report submitted to the Technical Centre for Agricultural and Rural Cooperation (CTA). Kampala, Uganda.

———. 2014. "Celebrating RUFORUM@10." *The Monthly Brief of the Regional Universities Forum for Capacity Building in Agriculture* 8 (6). Accessed July 28 2015. April 2016. http://repository.ruforum.org/sites/default/files/RUFORUM%4010-June%2C%202014%20Newsletter%20Issue.pdf.

Shabani, J. 2008. "The Role of Key Actors and Programs." In *Higher Education in Africa: The International Dimension,* edited by D. Teferra and J. Knight. Accra, Ghana: Quali Type Limited.

Tijssen, R. 2007. "Africa's Contribution to the Worldwide Research Literature: New Analytical Perspectives, Trends and Performance Indicators." *Scientometrics* 71: 303–327.

———. 2015. "Research Output and International Research Cooperation in African Flagship Universities." In *Knowledge Production and Contradictory Functions in African Higher Education,* edited by N. Cloete, P. Maasen, and T. Bailey. Cape Town: African Minds.

UNESCO (United Nations Educational, Scientific and Cultural Organization/Institute of Statistics). No date. UNESCO statistical database. Accessed October 2014. http://data.uis.unesco.org/.

———. 2009. *Trends in Global Higher Education: Tracking an Academic Revolution.* Report prepared for UNESCO's 2009 "World Conference on Higher Education," Paris, July 5–8.

Were, J. 2002. "The Programme for Enhancement of Research Information (PERI) Expected Impact in Africa." *INASP Newsletter* 19 (February): 2.

World Bank. 2009. "Accelerating Catch Up: Tertiary Education for Growth in Sub-Saharan Africa." Accessed April 2016. https://openknowledge.worldbank.org/handle/10986/2589.

———. 2010. *Financing Higher Education in Africa.* Washington, DC.

Zeleza, P. 2014. "The Development of STEM in Africa: Mobilizing the Potential of the Diaspora." Paper presented at Michigan State University at the third annual conference "Effective US Strategies for African STEM Collaborations, Capacity Building and Diaspora Engagement," East Lansing, April 2–4.

Chapter 10

NETWORK INNOVATIONS: BUILDING THE NEXT GENERATION OF AGRICULTURAL SCIENTISTS IN AFRICA

Joyce Lewinger Moock

The past two decades have ushered in one of the most colossal revolutions of knowledge and information in human history. Digital information and communications technologies (ICTs) have transformed the way knowledge and technical know-how move around the world. Genetics and biotechnology are bringing about a new epoch of innovation in the sciences. And the emergence of new finance and investment models, such as social enterprise and venture capital, has helped turn knowledge into both great wealth creation and a widening wealth divide.

In the agricultural sector, recent advances in biotechnology—such as breeding of higher-yielding and better-adapted crop varieties, along with market-friendly policies and improved national research institutions—are helping to create a new platform for progress in Africa south of the Sahara (SSA). Strengthened commodity value chains that boost productivity, coupled with new forms of collective action and seismic change in farmer accessibility to low-cost information technologies, offer exciting opportunities to use agriculture to promote development.

In the face of this proliferation of new knowledge and scientific breakthroughs, the volume has been turned up on calls from African governments, the international funding community, and African scientists alike

This chapter was originally prepared for the ASTI/IFPRI–FARA Conference "Agricultural R&D: Investing in Africa's Future: Analyzing Trends, Challenges, and Opportunities," held in Accra, Ghana, December 5–11, 2011. The chapter greatly benefitted from the input and guidance of Adipala Ekwamu (RUFORUM); William Lyakurwa (AERC); Innocent Matshe (CMAAE); Jean Claude Kayisinga (SPREAD); John Lynam (Trustee, World Agroforestry Centre and former program officer of the Rockefeller Foundation); Alex Ezeh (CARTA); Timothy Schilling (PEARL/SPREAD/Borlaug Institute, Texas A&M University); Howard Elliott (former Deputy Director General of ISNAR); Moses Osiru (RUFORUM); Nienke Beintema (IFPRI); Jeffrey C. Fine (independent consultant); Carl Eicher (Michigan State University); Peter Moock (former lead education economist, World Bank); and Daniel Clay (Michigan State University). The chapter was also informed by the Biosciences eastern and central Africa–International Livestock Research Institute (BecA–ILRI, Kenya) Business Plan development under the guidance of Gabrielle Persley of the Doyle Foundation and then senior advisor to the Director General of ILRI.

for a response to the challenges facing resource-poor institutions in building research and development (R&D) capacity. An abundance of essays and reports—perhaps best typified by the catalytic 2008 *World Development Report*[1] and by Calestous Juma's book, *The New Harvest: Agricultural Innovation in Africa* (Juma 2011)—argue that most needed is a transformation that will connect the mission and vision of advanced learning institutions with new local and global contexts. University-derived research is now commonly touted as essential to agricultural performance, from rapid appraisal of delivery services, marketing, and policy, to strategic research aimed at the creation and testing of new products appropriate for the African environment.

Despite the past two or three decades of crises in higher education, there has been major improvement. Many universities and research institutes are abandoning outmoded ways of conducting business and devising new structures, behaviors, and incentives. Especially important are initiatives that advance the process of knowledge production and application, and encourage fresh thinking about building agricultural systems that adjust to change. Yet these gains are often inadequate to produce a new generation of agricultural scientists and leaders with the knowledge and skills to replace the large numbers in the agricultural sector now close to retirement, and spur the agricultural growth needed to reduce poverty (Beintema and Stads 2011; Chapter 8, this volume). At the MSc and PhD training levels especially, where staffing and other resource constraints are most severely felt, individual universities are hard-pressed to generate a critical mass of graduates with the requisite qualifications to catalyze social and economic progress (Chapter 9, this volume).

One increasingly popular way of building a strong human capital development infrastructure and harnessing gains from innovation in the research process is investment in networks. For the purposes of this discussion, networks refers to postgraduate training *and* collaborations that strengthen institutions, unimpeded by geography—such as a collection of agricultural scientists capitalizing on greatly improved mobility and telecommunications to transcend institutional and national boundaries. But while several such agricultural networks now exist in Africa, most have a scale or scope of operation too small and too poorly resourced to realize their potential for creativity and innovation (Fine 2007a).

---

1 By providing evidence that increasing agricultural productivity is three times more effective at reducing poverty in poor countries than growth in nonagricultural productivity, the 2008 *World Development Report* (World Bank 2007a) helped to make agriculture, once again, a high priority for African governments and the international development community.

This chapter identifies five models of strategic networks making progress toward the stated goals of bolstering university-based training and research, and enhancing the productivity of the agricultural sector. These models, while different in their composition, offer key principles and approaches of networks that are scalable and have the potential to be sustained.[2] Each model has a base secretariat or management group within a host institution that provides coordination and technical assistance, and promotes the use of low-cost (and in some cases, more advanced) information technologies. Each network is primarily based on one or more disciplinary fields, but offers an array of subject matter that encourages systems thinking; provides professional career structures necessary to develop a stable cadre of African research leaders; and creates network services that build economies of scale. These networks are fortified by linkages to local stakeholders, such as the private sector, nongovernmental organizations (NGOs), and government bodies; to continental alliances, such as the African Union (AU), Forum for Agricultural Research in Africa (FARA), and Comprehensive Africa Agriculture Development Programme (CAADP) under the auspices of the New Partnership for Africa's Development (NEPAD); and to global agricultural entities, such as the CGIAR Consortium, world-class universities outside the region, and international markets.

Background Issues

The network concept offers great appeal as a vehicle for fostering advanced knowledge and knowledge applications, and for extending limited resources. It creates enduring institutional relationships based on a common mission and standard of effectiveness and relevance that can attract the attention of African governments, the private sector, and external funders.[3]

---

2 This chapter draws on many of the insights offered by Jeffrey C. Fine's study commissioned by the Partnership for Higher Education in Africa (an alliance of seven US foundations) on regional networks engaged in research and postgraduate education on the continent (Fine 2007a). Peter Szyszlo developed a database of 120 networks for this work, which can be found at the Partnership website: www.foundation-partnership.org.

3 Several agricultural networks in various parts of the world, such as the Asian Rice Biology Network, were created to reinforce already strong institutional research or service delivery structures, and extend their impact. However, the majority of networks in Africa have evolved as compensatory mechanisms for fragile, neglected institutions and structural defects in national systems of agricultural research and higher education. They are designed to ensure depth of analysis and critical mass within strategic research fields that would otherwise be extremely difficult and costly to achieve on a country-by-country basis (Moock 2005).

The focus on training and research networks springs from broad shifts in world forces that affect higher education everywhere, including

1. the unfolding of the knowledge economy, which places a premium on intellectual capital, as reflected in boundary-crossing disciplines that few universities can properly cover;
2. the drive by funders of advanced learning—governments, donors, and students and their families—to unite knowledge with practical skill employment;
3. less expensive, more obtainable bandwidth that can exploit new modes of communicating information in various electronic formats;
4. burgeoning private investment in higher education, resulting in a free range of education providers and growing public concern about quality-control issues;
5. world trade in education services, including the flow of faculty and advanced graduate students across national borders that, if not resulting in permanent brain drain, can still cause periodic gaps in quality staffing; and
6. increasing "knowledge prospecting" (identifying new technologies and using them to create new businesses) across academia, government, and the private sector that offer universities an opportunity to step up their role in shaping Africa's future (Juma 2011).

Within Africa, reasons to invest in cross-institutional networks are especially compelling, including

1. generating economies of scale among research universities that are small and unable to attain the necessary expertise, equipment, and financial resources to cover core and specialized courses in most postgraduate agricultural fields;
2. building credibility and legitimacy for African governments and donors in demonstrating solid academic programs that engage with other stakeholders in the agricultural sector and produce employable graduates;
3. exploiting both the lessened rigidity of faculties under more democratized and decentralized university management, and the complementarities and synergies in innovation;

4. promoting quality assurance through interaction, information sharing, and peer review;
5. strengthening links between academic research centers and the (re-) emerging private sector;
6. building a critical mass of female scientists in the face of the narrow pipeline of female students surfacing from undergraduate studies at individual universities[4]; and
7. harnessing movements toward regional integration that present opportunities for reducing the costs of research and training,[5] avoiding duplication, and simultaneously providing safety nets in the event of political strife in any one geographic site.

Similar reasons have led to recent calls for large-scale investment in "centers of excellence," which are also intended to build economies of scale in producing qualified staffing and facilities. Such initiatives can be attractive to funders, as they hold the promise of sidestepping the high transaction costs of bringing together different actors and institutions with diverse capacities; however, there is a major downside to the creation of these *insulated* regional entities. As Jeffrey Fine points out, "Past experience . . . dictates that the lack of a genuine buy-in by national institutions, in particular leading universities, will prove fatal. Once external funding disappears, local support also evaporates. Unless these collaborative efforts complement rather than substitute for investment in national systems of higher education and research, they will also fail" (Fine 2007b: 3).

In contrast, well-designed institutional collaborations can have a longer shelf life.[6] If the primary need in producing the next generation of agricultural scientists is a rapid increase in numbers, then networks and insulated centers of excellence can be equally powerful, with the advantage perhaps going to the more

4 According to Beintema and Di Marcantonio (2010), in 2007 an average of one-third of the students enrolled in and graduating from 28 higher-education faculties or colleges in a sample of 12 SSA countries were female.

5 In 2006, estimated total costs of a two-year MSc degree in agricultural economics at a US university with a fellowship from the United States Agency for International Development was $60,000, while a US sandwich course—with coursework in the US and thesis research in Africa—was $30,000, and a degree program offered by the Collaborative MSc in Agriculture and Applied Economics (CMAAE) was $20,000 (Eicher 2006).

6 Eicher (2009, 252) argues that "Regional models of agricultural training and research were productive during the colonial period and the early years of Africa's independence. But development specialists have few answers to the difficult problem of financing regional organizations and regional centers of excellence. The wave of the future should be to encourage regional knowledge networks and regional training programs and increase the use of ICT."

easily managed, unencumbered centers. However, solid network approaches—especially those backed by world-class overseas universities or high-quality local institutions serving as regional postgraduate program hubs—may have the edge in the long run in attracting funding from African governments on the basis of unlocking innovation customized to the dynamics of the national environment.[7] In this regard, perhaps the greatest attraction of networks is their ability to serve as leverage points for restructuring domains of training and research to relate more significantly to complex social and economic dynamics.

The most promising networks for agricultural development are based on a notion of capacity building that is undergoing enormous change. This involves consideration of a much broader range of influences and consequences than were included in traditional definitions. In the context of competitive and knowledge-intensive agricultural economies, capacity building must refer to more than technical training and transfer of skills. While these are necessary, they are not sufficient for fostering capacity that can be well used, retained, and replenished. A more systemic definition of capacity building would include, *in addition to technical skills transfer:* institution strengthening, the improvement of inter- or intraorganizational structures, and the imparting of entrepreneurial competencies and business acumen necessary to develop vision and strategies (Figure 10.1). Thus, the emphasis must be on doing and accomplishing, not just on training and learning. This extended definition enables a program to be assessed based on whether its design is adequate to produce the desired outcome.

Sustained capacity building in Africa today requires flexible, low-cost approaches that (1) spark not only conventional skills, but also improvisational, experimental, management, and leadership talents; (2) strengthen universities and provide transition mechanisms, such as mentoring and apprenticeships, for graduates to access opportunities for meaningful work; (3) offer effective use of skills through alignment of the various components of the agricultural system and chances for joint action; and (4) promote

7 Such hubs may be hosted by universities or research institutes with strength in narrowly specialized or newly emerging areas (for example, the Dryland Resource Management regional PhD program, University of Nairobi; the Aquaculture and Fisheries Science program, Bunda College, University of Malawi; the MSc Research Methods course at Jomo Kenyatta University of Agriculture and Technology, Kenya; the Soil and Water Management regional PhD program, Sokoine University, Tanzania; or BecA-ILRI, Kenya). These hubs differ from insulated centers of excellence. Although resources are concentrated on these subregional catchment centers, all university members benefit from institution-strengthening grants, scholarships, curriculum development, and participation in research supervision and teaching. Under the auspices of the AU, the Pan-African University is now in the midst of establishing five thematic centers of excellence on the continent. If each of these centers is eventually linked to 10 existing African institutions, as planned, the resulting regional networks may achieve sustained political backing, reliable financial resources, and—most important—credible grassroots support.

FIGURE 10.1 Three dimensions of entrepreneurial capacity

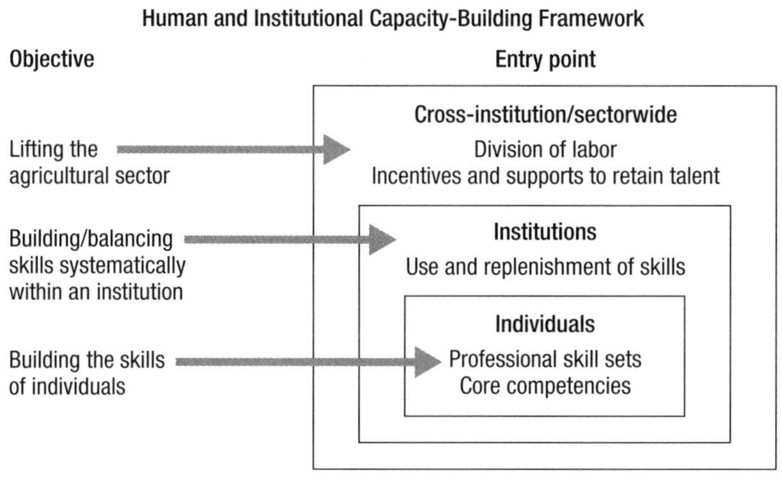

Source: Author.

retention by professional community development, network-based knowledge dissemination, incentives, and output rewards.

While professionalism is critical, skilled individuals cannot produce public goods in a vacuum. Attention needs to be given to quality training, the development of institutions, intelligent policymaking, and well-functioning national agricultural systems. Africa's next generation of agricultural scientists will need to be *scientist entrepreneurs*—technologically sophisticated people capable of bold thinking with a primary question in mind: how can *high*-impact innovations be adapted to the growth of agriculture with a view to poverty alleviation and environmental sustainability? The next generation will need to join the ranks of sharp, savvy entrepreneurs who are emerging across the span of African enterprise. They are the catalysts of change, conceiving new products and services, and the means to produce, market, and appraise them. Another way of looking at the role of postgraduate education systems and networks within the essential elements of a national agricultural innovation system can be depicted through the linkages among the various components, and the agencies and policies that make up the enabling environment in which they function (Figure 10.2).[8]

8 "In essence, an agricultural innovation system is a blending of institutional capacities, coordination mechanisms, communications networks, and policy incentives that fosters innovation-led gains in agricultural productivity" (World Bank 2007b, 6).

FIGURE 10.2 Capacity building for scientists as a critical part of an agricultural system

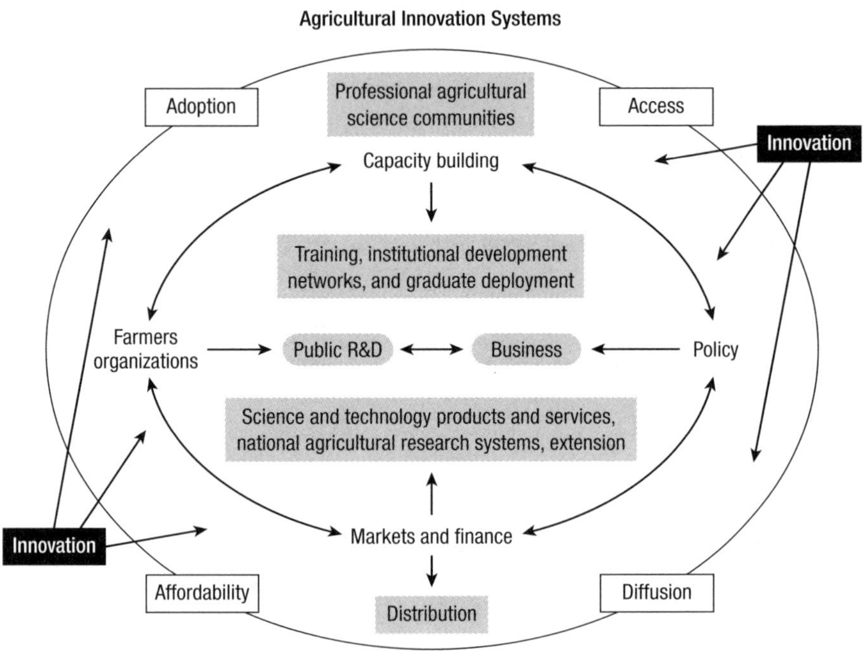

Agricultural Innovation Systems

Adoption

Professional agricultural science communities

Access

Innovation

Capacity building

Training, institutional development networks, and graduate deployment

Farmers organizations

Public R&D

Business

Policy

Science and technology products and services, national agricultural research systems, extension

Markets and finance

Innovation

Affordability

Distribution

Diffusion

Source: Adapted from Morel et al. (2005).
Notes: NARSs = national agricultural research systems; R&D = research and development; S&T = science and technology.

This complex system of diverse actors and their interactions has enormous implications for higher-education reform, especially in unleashing talent and innovation, and integrating educators and researchers into professional networks with other agricultural system agents (Spielman et al. 2008; Lynam 2012). Africa needs to increase both the supply of and demand for quality graduates through a supportive environment for agricultural enterprise at all levels (Blackie et al. 2010).

Faculties of agriculture certainly cannot be held accountable for all of these components, but they can set up the essential learning platforms to accommodate continued learning and high performance following graduation. This is the nexus between research and practice or policy that some of the more dynamic networks are reaching for. To achieve these ends requires thinking differently about institutional arrangements and reconsidering not only the creation of economies of scale, but also how advanced learning centers can serve as pivotal supports in local knowledge and innovation systems.

The following section explores the key characteristics of five leading agricultural capacity-building research networks in Africa.[9] There are several other networks, but these stand out in terms of their scale, scope, and potential for replication and sustainability:

1. Regional Universities Forum for Capacity Building in Agriculture (RUFORUM)
   Status: NGO
   Secretariat location: Makerere University Campus, Uganda
   Coverage: 55 universities in 22 East, Central, Southern, and West African countries
   Internet address: www.ruforum.org

2. Collaborative MSc Program in Agriculture and Applied Economics (CMAAE)
   Status: Program of the African Economic Research Consortium (AERC), an NGO
   Secretariat location: AERC, Kenya
   Coverage: 17 universities in 13 East and Southern African countries
   Internet address: www. aercafrica.org

3. Education for African Crop Improvement (EACI)
   Status: Program of the Alliance for a Green Revolution in Africa (AGRA)
   Central management location: AGRA, Kenya
   Coverage: 10 MSc universities and 2 PhD training centers at the University of Ghana (West African Center for Crop Improvement) and the University of Kwa-Zulu Natal (African Center for Crop Improvement) serving 16 countries
   Internet address: www.agra-alliance.org

4. Biosciences eastern and central Africa (BecA)
   Status: NEPAD-endorsed initiative hosted and managed by the International Livestock Research Institute (ILRI)
   Central management location: ILRI Campus, Kenya
   Coverage: One central hub and six institutional nodes serving 18 African countries
   Internet address: hub.africabiosciences.org

---

9 Information on each network is derived from extensive documentation on its history, objectives, structure, and activities. Additional information came from exchanges with leadership and management staff, and with funding organizations and external advisers and evaluators.

5. Partnership to Enhance Agriculture in Rwanda through Linkages (PEARL), 2000–2006/Sustaining Partnerships to Enhance Rural Enterprise and Agribusiness Development (SPREAD), 2006–2011[10]
   Status: Rwanda institutional partnership
   Secretariat location: National University of Rwanda
   Coverage: National University of Rwanda, Kigali Institute of Science and Technology, National Institute of Agriculture Research, NGOs that target agricultural cooperatives with more than 15,000 member farmers in Rwanda
   Former Internet address: www.spreadproject.org

Network Characteristics, Underlying Principles, and Challenges

The formation of networks in Africa has been a relatively autonomous process, often with considerable spontaneity and good fortune involved in their emergence. The result has been important differences in their format and use, both across and within sectors. Clearly, not every postgraduate training and research network in Africa requires a similar design. However, there are a number of prerequisites for building capacity under fragile institutional circumstances that boost quality and relevance and lay the foundation for sustained expansion of the pool of qualified researchers. Such fundamentals generally fall into three categories: (1) quality, access, and relevance; (2) systems orientation; and (3) scalability and sustainability.

Network Characteristics

Figure 10.3 illustrates these categories as they relate to well-functioning networks engaged in postgraduate training; to research and institution strengthening in the agricultural sector; and, by extension, to cross-border networks in other fields.

The five agricultural training and research collaborations selected for closer examination offer the advantage of lifting *all* nodes in the network, significantly increasing the talent pool beyond the postgraduate fellowships provided and putting in place the conditions that lead to ongoing regeneration of human capital. Table 10.1 reviews the components listed above as they relate to each of the five networks.

10 Note that PEARL ended in 2006, and SPREAD closed in 2012.

FIGURE 10.3 Components of viable network programs under tenuous institutional conditions

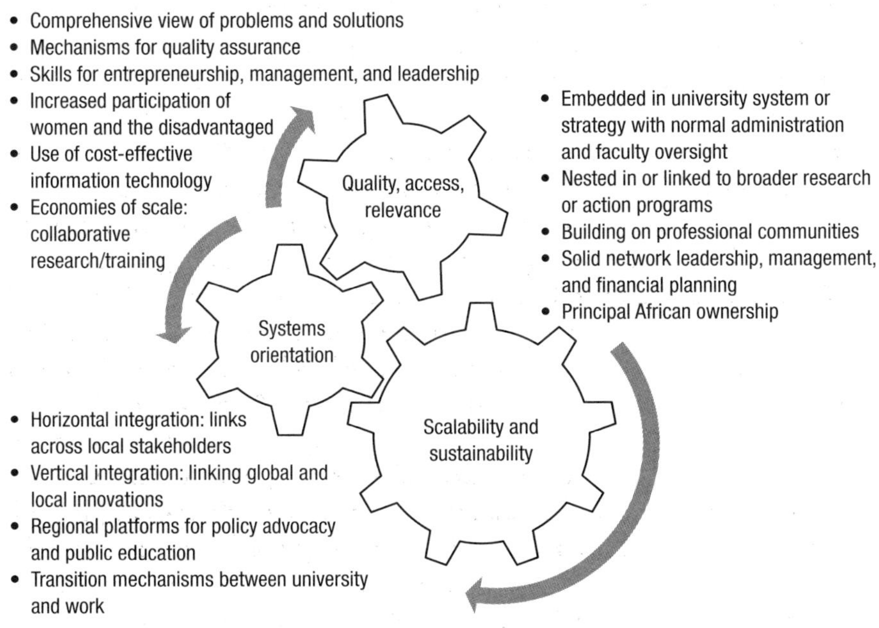

Source: Author.

These networks are demonstrations of key mechanisms for fast-tracking training and for building research capacity through collaborative arrangements among better-endowed institutions and those less well-off in Africa. As noted by Juma (2011: 63), the flow of knowledge among institutions of advanced learning and between them and enterprises through networking facilitates the formation of "dynamic self-teaching systems" that speed up innovation.

Together, these collaborations reflect a remarkable change in learning strategies by cash-strapped African universities. The sample networks are not alone. Other *current* capacity-building networks of note within agriculture include (1) the African Network for Agriculture, Agroforestry and Natural Resources Education (ANAFE), which assists university faculties, particularly in West Africa, to undertake curriculum reviews, facilitate staff exchanges, and develop teaching materials; (2) the Building African Scientific and Institutional Capacity (BASIC) network, initiated by FARA to improve teaching methods and course content; (3) the AGRA soils network, offering PhD courses at two training hubs, as well as MSc studies at individual universities; (4) African Women in Agricultural Research and Development

TABLE 10.1 Properties of promising agricultural research and development capacity-building programs in Africa

Design properties	Program: Regional Universities Forum for Capacity Building in Agriculture (RUFORUM)	African Economic Research Consortium (AERC), Collaborative MSc in Agricultural and Applied Economics (CMAAE)	Alliance for a Green Revolution in Africa (AGRA), Education for African Crop Improvement (EACI)	Biosciences eastern and central Africa (BecA)	Partnership to Enhance Agriculture in Rwanda through Linkages (PEARL)/Sustaining Partnerships to Enhance Rural Enterprise and Agribusiness Development (SPREAD)
Description					
	University network offering mentoring and research opportunities through competitive MSc research grants (290), collaborative MSc training (1,200), regional PhDs (212), a community-action research program, and institutional grants across 55 universities in 22 East, Central, Southern, and West African nations	Network of departments across 17 universities in 13 countries offering collaborative MSc (800), shared electives facility, African and external subject specialists, PhD (11) support, department-building grants, senior policy workshops, faculty exchange and retooling, and professional peer review	Network of 11 universities and 2 regional hubs feeding 16 countries, redeploying 192 PhD candidates and 207 MSc crop scientists and agronomist candidates to national programs, and providing overseas university backup	Hub and nodes model of top scientific expertise and facilities accessible by visiting scientists (92) and MSc/PhD students (more than 110) from 18 African countries; 6 institutional nodes; alumni lead research programs at home universities and return to BecA as visiting scientists supervising own graduate students	Alliance of Rwandan universities, government, Rwanda/foreign industry, local NGOs targeting cooperatives in a niche agricultural initiative; MSc training (19) at Texas A&M and Michigan State universities; graduates train producers in new production techniques; based on US land-grant model
Quality, access, relevance					
Comprehensive view of problems and solutions	• Seven regional PhD and five MSc programs • Value-chain orientation • Joint research methods courses	• Emerging topics of environment and natural resource management, and agribusiness	• Disciplinary concentration but value-chain orientation	• Wide variety of bioscience areas and related disciples	• Food sciences courses; new outreach center • Value-chain orientation • Associated health program

Mechanisms for quality assurance	• Competitive research awards • Peer review by a technical committee • Faculty exchange program • Strong universities teamed with weaker universities	• Rigorous study/ supervision • Instructors' workshops • External examiners, remedial courses • Sandwich programs (University of Pretoria and Cornell University)	• Rigorous study/ supervision • External technical assistance (Cornell University, AGRA, and flagship hub staff)	• Top-line labs • "Hosted programs" by senior African or international scientists with own postdoctoral and graduate students	• Initial MSc courses in the United States for local staff development
Skills for entrepreneurship, management, leadership	• "Soft skills" relevant to the needs of farmers and business • Hands-on problem solving • Participatory research	• Attachments to organizations that provide management and leadership proficiency	• 1- to 4-month hands-on training and attachments to private sector or international organizations with management expertise	• Emerging science leaders as BecA affiliates who then lead/manage own research teams	• Work with Rwanda Coffee Board adds management and marketing skills to help grower cooperatives to build export businesses
Initiatives to increase female participation	• 42 percent female participation • Multiple mechanisms for recruitment, career launch, and skills use/retention • MOU with the AWARD program	• 46 percent female participation • Multiple mechanisms for recruitment, career launch, and skills use/retention	• 24 percent of graduates with 40 percent goal; recruitment assisted by the AWARD program	• Currently 40 percent of graduate fellows are female	• 31 percent of graduate students are female
Use of cost-effective information technologies	• Improved information technologies applications and knowledge access systems • 26 OER courses created and licensed • Major regional repository in SSA	• Improved information technology applications and knowledge access systems • Work with OER Africa • Blended learning courses	• Digital networking • Electronic libraries • Video-recorded lecture series	• Latest bioinformatics tools and services • BecANet online resources	• Internet café • Connectivity via fiber optic cable • Geographic information systems and remote sensing
Economies of scale: joint research and training facilities	• Competitive MSc research schema • 12 joint PhD and MSc programs each at lead university	• Standardized curriculum • Joint electives facility • Joint faculty grants	• Standardized curriculum • Two "flagships" of excellence for PhD training	• Linked with six institutional nodes • Cosupervision of theses • Traveling seminars	• Collaboration across universities, Agricultural Research Institute of Rwanda, and farmer cooperatives

TABLE 10.1 Continued

Design properties	Program				
	Regional Universities Forum for Capacity Building in Agriculture (RUFORUM)	African Economic Research Consortium (AERC), Collaborative MSc in Agricultural and Applied Economics (CMAAE)	Alliance for a Green Revolution in Africa (AGRA), Education for African Crop Improvement	Biosciences eastern and central Africa (BecA)	Partnership to Enhance Agriculture in Rwanda through Linkages (PEARL)/Sustaining Partnerships to Enhance Rural Enterprise and Agribusiness Development (SPREAD)
Systems orientation					
Horizontal integration (across local stakeholders)	• National forums engage farmers organizations to keep university research on track	• Links with national governments, NARSs, the local private sector, CGIAR centers, and NGOs	• Training with CGIAR centers • Investments in NARSs and seed business services	• Collaboration with NEPAD, biosciences networks, local universities in NARSs, industry	• In-country partner institutions • Demonstration sites and market outlets for agricultural products
Vertical integration (across global and local innovation)	• Collaboration with World Bank on African Centers of Excellence program • Backup by Costa Rica's Earth University and others for fieldwork innovations	• World-class universities, World Bank Institute, United Nations bodies, and industry ensure international standards	• Global agricultural initiatives and funding streams • Backup by world-class research institutes	• Industry partnerships, African diaspora, and global science communities for product incubation/innovation	• Global coffee and other crop markets • Backup by world-class universities
Regional platform for policy advocacy, public education	• Alignment with new African agricultural frameworks (the CAADP process, AU, FARA's Science Agenda, and many others)	• High-level public policy analysis seminars • Faculty research and thesis local dissemination	• Use of AGRA's strong communications channels • Links with all major regional policy bodies	• Publications with wide dissemination in Africa and overseas • Broad policy links	• Demonstrations, outreach center, radio, and extension • Internet access to agribusiness community
Transition mechanisms between university and work	• Three-month field attachments to CGIAR centers, NARSs, or firms • Earth University to help prepare graduates for research or start-up enterprises	• Consultation with public- and private-sector employers • Provision for internships and job placement	• CGIAR and NARS programs mentor graduate students • Absorption of graduates in AGRA-supported programs	• Exposure to international expertise offers career development opportunities	• Graduates placed in universities and outreach positions providing technical assistance to 95,000 smallholder growers

Scalability and sustainability

Nested in university system/strategy; normal administration and faculty oversight	• Deliberate mechanisms to diffuse new ideas and practices across the university system • Content organically grown to fit faculty strategic plans	• Offers spillovers from focal centers to similar disciplinary departments in other universities • Content organically grown to fit strategic plans of faculty	• Focal centers stand out from others • No obvious mechanisms for influencing broader training standards at host universities	• Decentralized capacity in individual BecA nodes in specified fields	• Engaging students in community work influences broader teaching styles at universities • No obvious mechanism for systematic spillovers
Building professional communities	• Alumni involved in groups attached to each research theme, electronic networks, biennial conferences, yearly networking workshops, and so on	• Created African Association of Agricultural Economists that sponsors global conferences for young and senior scientists to share knowledge	• Graduates generally become part of the AGRA family	• Synergies with university nodes extend to other biosciences networks • Predicted 15 percent student growth per year over next five years	• Professional links across academia, government, and industry • Through outreach center, graduates eventually handle all marketing and exporting of crops
Solid network leadership, management, and financial planning	• Strong director • Streamlined secretariat • Knowledge/evaluation management • Careful finance/auditing	• Strong leadership • Streamlined secretariat under AERC • Knowledge/evaluation management • Careful finance/auditing	• Strong AGRA staff provides overall leadership • Strong program operations management at participating universities	• Strong director • Leadership unit manages institutional arrangements under ILRI • New business plan	• Recent change in leadership; melded into Rwanda national agricultural program in 2012 • Secretariat at NUR linked with Texas A&M University

TABLE 10.1 Continued

Design properties	Program				
	Regional Universities Forum for Capacity Building in Agriculture (RUFORUM)	African Economic Research Consortium (AERC), Collaborative MSc in Agricultural and Applied Economics (CMAAE)	Alliance for a Green Revolution in Africa (AGRA), Education for African Crop Improvement	Biosciences eastern and central Africa (BecA)	Partnership to Enhance Agriculture in Rwanda through Linkages (PEARL)/Sustaining Partnerships to Enhance Rural Enterprise and Agribusiness Development (SPREAD)
	• New business plan	• New business plan • New Forum of Governors of Central Banks with funding contributions and representation on the AERC board			• The coffee value chain has become self-sustaining
Principal African ownership	• African secretariat • Vice Chancellor Board • Technical and deans committees • Thematic working groups • 18,000 network members • Association of 1,282 alumni	• African secretariat • Board composed of funders and independent directors • Content oversight by African program committee • Association of 575 alumni	• Sponsored by AGRA • Absence of a local participatory academic board	• African leadership • ILRI board provides fiscal oversight, policy, and strategic guidance • Intends to build interactions among ILRI, AU/NEPAD, African NARSs, and universities	• Sponsored by USAID with external university technical assistance • Program has been Rwanda led and managed
Principal funding source	• Rockefeller Foundation (initially) • Currently the Bill & Melinda Gates Foundation with others • University members contribute $5,000 in annual fees • Uganda government contribution • $70 million mobilized in direct support of network universities	• 13 donor agencies and three African governments support AERC • Earmarks for CMAAE mainly from the Bill & Melinda Gates Foundation, AGRA, ACBF • A new AERC Forum of African Governors of Central Banks joined the consortium with the provision of core support	• The Bill & Melinda Gates and Rockefeller foundations	• CIDA, AusAID, Syngenta Foundation for Sustainable Agriculture, Bill & Melinda Gates Foundation, Swedish goverment • Aims to generate 50 percent of future support from research projects	• USAID

Major agricultural sector results	• More than 300 technologies developed in cooperation with NARSs, farmers' associations, and the private sector reaching more than 1 million farmers • Examples: cowpea project in Uganda; soybean project in Zimbabwe; cereal banking in Kenya	• Executive policymaker guidance • Policymaker career pipeline • Management hub of World Food Programme's Purchase for Progress project to address small-scale farmers' market access	• PhD research led to release of 110 new crop varieties with national/international research institutes for farmer cultivation • Related projects include start-up capital for more than 90 seed enterprises and training 10,730 agro-dealers for input provision • More than 15 million farmers using improved seed varieties	• Many bioscience projects on crops, livestock, and food safety • Patented discoveries and numerous cutting-edge publications that feed the work of CGIAR, NARSs, other research institutes in Africa • Offers a "top 10" list of BecA innovations that benefit African farmers	• Around 400,000 coffee farm families making more than three times what they earned prior to these projects • More than 160 US, European, Asian, and Australian companies, such as Starbucks and Costco, buying almost 5,000 tons annually, directly from the growers • International demand for Rwandan's world-renowned specialty coffee exceeds supply.

Source: Author.

Note: ACBF = African Capacity Building Foundation; AU = African Union; AWARD = African Women in Agricultural Research and Development; CAADP = Comprehensive Africa Agriculture Development Program; CIDA = Canadian International Development Agency; FARA = Forum for Agricultural Research in Africa; ILRI = International Livestock Research Institute; MOU = memorandum of understanding; NARS = national agricultural research system; NEPAD = New Partnership for Africa's Development; NGOs = nongovernmental organizations; NUR = National University of Rwanda; OER = open educational resources; SSA = Africa south of the Sahara; USAID = United States Agency for International Development.

(AWARD), a two-year fellowship for fast-tracking the careers of female agricultural scientists hosted by the World Agroforestry Centre (ICRAF); and (5) the Association of African Business Schools, which offers quality control and an added focus on the smaller-scale and informal private sector in delivering essential public goods, and on the not-for-profit sector providing public health and agricultural business services.

Models of network approaches in other sectors also offer the potential for adaptation in agriculture. They provide direct links from the individual through the institution to the larger sectorial space (as in Figure 10.1). Three of these models seem particularly germane for this discussion:

- The AERC PhD economics program offers subregional catchment zones involving host and associated universities from which students and the bulk of teaching faculty are drawn. The design, involving professional peer review, enables capacity-building spillover to a large number of universities. Program oversight comes from a PhD academic board comprising the heads of member departments and senior African scholars who contribute to the maintenance of international standards (www.aerc.org).
- The INDEPTH Network is a learning platform of multisite demographic surveillance collaborations offering on-site training and internships, standardized research methods, and mechanisms for translating research on public health priorities into policy outcomes. By sharing data and results, these collaborations allow researchers to form the "big picture" from multiple experiments and venues. An associated MSc degree is accredited by the University of Witwatersrand in South Africa. INDEPTH has scaled up as many as 52 sites in 20 countries, with 23 sites in Africa (www.indepth-network.org).
- The Consortium for Advanced Research Training in Africa (CARTA), a program of the African Population and Health Research Center (APHRC) and the University of the Witwatersrand in South Africa, fosters the development of viable training and training hubs at nine universities across Africa. Its major features are a first-rate, joint advanced seminar package, opportunities for mentored research at any one of the network universities, and program backup through expertise from four leading African research institutes and seven northern institutions (www.aphrc.org).

It should be noted that a major impetus for strengthening agricultural training and research networks comes from improvements in national higher-education policy and from individual universities that attempt to align university studies with national development priorities, especially with regard to agribusiness.[11]

Underlying Principles and Challenges

Despite an array of strong agricultural postgraduate and research networks, the networking concept is still evolving. All too often, for a variety of reasons, emerging networks fall short of meeting their promise to advance higher learning and, ultimately, agricultural performance. First, the number of qualified universities for advanced training and participation in research networks is still small, with many aspirants unable to meet fundamental standards for teaching and research and, hence, for accreditation or world recognition of degrees.[12] Second, rushed planning under heavy pressure from potential funders can result in poor design and impeded implementation. Third, attempts to build alliances among universities and between them and the larger agricultural innovation system can lead to frustration if they fail to create added value for all members. Fourth, many networks never reach the takeoff point because they do not use their assets strategically to produce significant public goods. Fifth, collaborative arrangements may easily break down if partners do not reach early agreement on common interests, expectations, and contributions. Such prior negotiations offer high organizational payoff, especially in the event of tight fiscal conditions.

The shortcomings of many networks provide the backdrop for a set of general principles underlying the construction and improvement of postgraduate training and research collaborations in agriculture. In general, these networks need to concentrate on problems requiring collective action, and need to pool their talents to reach critical mass and synergy, and realize creative solutions. Specific actions include the following.

---

11 University innovations aimed at better links with agribusiness and markets include (1) agribusiness incubators (for example, Jomo Kenyatta University of Agriculture and Technology, Kenya; Makerere University, Uganda; Institut Polytechnique Rurale, Mali; and University of Zimbabwe); (2) development of student agribusiness plans (for example, United States International University, Kenya; University of Swaziland, University of Malawi, and University of Ghana); (3) science parks (for example, Egerton University, Kenya, and Institute of Food Technology, University of Pretoria); (4) memorandums of understanding with district agricultural offices (for example, Makerere University, Uganda); and (5) agricultural partnerships with cooperatives (for example, National University of Rwanda) and with companies (for example, University of Agriculture Abeokuta, Nigeria).

12 Professional networking and institutional linkages appear to be better among anglophone countries, given that they possess nearly four times as many agricultural researchers as do francophone countries (World Bank 2007b).

ALIGNING VISION AND MANDATE WITH NATIONAL ASPIRATIONS

A network is defined by its interaction with the professional field in which it operates and by the benefits that it affords its members. For agricultural networks, a key goal is to establish productive relationships with other actors in a country's innovation system through an ongoing consultative process (Spielman et al. 2008). Of the sample networks, one is designed specifically to build an export business in several crops to revitalize the agricultural sector. Thus, for PEARL/SPREAD, turning higher education toward understanding the dynamics of Rwanda's government and commercial sector has been paramount (Kayisinga 2010; Kitzantides 2010; Schilling 2008). BecA and EACI are seeking capacity strengthening through research and product incubation or varietal releases. Both are gearing themselves to complement parallel reforms occurring in CGIAR and national research systems in support of smallholder commercialization and public- and private-sector investments. RUFORUM holds the conviction that the research results of well-trained scientists are more likely to be applied when based on a demand-driven research agenda. Thus, it has created national forums now operational in seven countries that serve as stakeholder discussion platforms and policy advocacy units. For CMAAE, the task is to remedy mixed quality standards in a well-established field and ensure that sound research draws the attention of policymakers and helps to structure the policy debate. In each of these cases, the success of the network is a function of how closely its vision fits within the political and organizational context. Of the five networks, one (RUFORUM) was conceptually pretested in a pilot effort; each of the others emerged following a serious reconnaissance of the landscape in which it would function.

DETERMINING CORE COMPETENCIES AND COMPARATIVE ADVANTAGES

While fitting into the national agricultural system landscape is critical, a network also needs to establish a clear strategy for staff and stakeholders to follow to avoid inefficient opportunism and missed opportunities for impact. Building on core competencies may involve growth by adding new services to current members, or by a balanced or sequenced growth strategy adding new activities or regions while making careful trade-offs among activities to avoid dilution of effort, strain on management, and loss of brand value.[13]

13 Both BecA and RUFORUM have carefully laid out various pathways for growth in their new business plans; CMAAE/AERC has commissioned a study on ways to reformat its collaborative research activities and is developing a business plan; and EACI are proposing new lines of work in their next phases. See Elliott 2012 for a description of the BecA and RUFORUM business plans stressing balanced growth.

ENSURING THAT NEW APPROACHES IN ACADEMIA CAN BE MAINSTREAMED WITHIN THE UNIVERSITY SYSTEM

Networks featuring highly innovative characteristics that attract members and keep them intellectually stimulated may find that they are not well aligned with member university processes and normal faculty strategic planning. Without the engagement of a local academic board or similar body, network-induced reforms can provoke resistance from administrators and hinder spillover effects to other departments and universities.[14] There are deeper structural challenges to spillover from networks into institutional strengthening of the *larger* university system. In particular, four of the networks, with the exception of RUFORUM, are grounded in disciplinary professions, with principal links outside universities to clientele using those disciplines that have an interest in the quality of the graduate and the research on which much of the value and relevance is based.

Capacity to produce spillovers into the wider university space runs along a continuum, with the highly disciplinary-focused EACI on one end, and RUFORUM with its multiple disciplines and cross-disciplines on the other.[15] RUFORUM is the only network of the five deliberately designed to connect investments in individuals and faculties to improvements in the wider university body. It does so mainly in three ways: (1) focusing on commonalities at the margins of agricultural disciplines and overlapping methodologies (for example, its highly popular networkwide research methods courses); (2) working with a wide-ranging committee of university deans; and (3) instituting a board composed of vice chancellors of member universities who pay annual membership fees and cover their own travel expenses to meetings. It may be argued that with such layering, RUFORUM operates at too broad a level and that viable networks are best grounded in single professional disciplines with reach to external constituencies that provide essential feedback loops. In the end, however, lasting gains in strengthening institutions and raising

---

14 CMAAE, for example, receives oversight from an academic board, a body consisting of the heads of departments participating in the program and other senior African scholars actively involved in graduate teaching and research. This body (1) contributes to the establishment and maintenance of international standards by making recommendations on operating policy (such as the criteria and procedures for accrediting collaborating departments to offer the program) and (2) conducts various evaluations to ensure continued quality and relevance.

15 Under the PEARL/SPREAD programs, changing the curriculum of the agricultural faculty under a participatory, step-by-step approach to link with local enterprises has, according to SPREAD's director, more broadly affected the way teaching takes place at the National University of Rwanda. One example is the recent launch of an integrated health component within the agribusiness program (Kayisinga, personal communication 2011). The programs, however, have no explicit mechanisms for generating these effects.

professional standards may best be realized if networks put a premium on diffusing new ideas and practices throughout individual universities and across them to a variety of agricultural system stakeholders.

INCREASING THE PARTICIPATION AND VOICE OF WOMEN

There are multiple mechanisms for drawing women into postgraduate programs, helping them with career development, and ensuring their use and retention of skills. These mechanisms include creating a database of active female researchers, inviting women to participate in various network committees and activities, providing faculty deans with incentives for recruiting women, using role models and mentoring, bolstering women in entrepreneurial initiatives, and creating backup supports for female scientists who work with female farmers.

There is no doubt about the desire of women to enter a professional career track in agricultural science. The AWARD program reports that since its inception in 2008, it has received applications from more than 3,500 women for 390 available fellowships. On average, only the top 9 percent of applicants is selected each year. To date, African women from 11 countries have benefited directly as AWARD fellows. In the future, AWARD aims to place more emphasis on working with research and academic institutions to help fellows build their capacity for gender-responsive research.

INVESTING IN APPROPRIATE INFORMATION TECHNOLOGIES

Wise investment in low-cost technologies provides unprecedented opportunities for building network capacity to support effective, decentralized learning and knowledge sharing. All of the networks featured here are harnessing powerful new ICTs to improve the performance of management of the universities and other entities they serve. The uses include technology-mediated learning, teaching, and research; employment of open educational resources; dissemination of agricultural research information[16]; and network information management systems. In addition, cell phones, handheld computer devices, video, and radio provide relatively cost-effective distribution of scarce specialist teaching resources to reach many students conducting field research, community organizations, and other network stakeholders.

---

16 Lower-cost online and offline journals, such as Access to Global Online Research in Agriculture and the Essential Electronic Agriculture Library, wiki-type platforms, blogs, and other knowledge-sharing technologies offer potential for an enormous increase in collaborative learning. Recently, the Google Foundation has undertaken to use its technical expertise free of charge to help African networks set up information technology platforms for digital libraries and online forums.

IDENTIFYING THE FULL RANGE OF CLIENTELE

In its new business plan, RUFORUM recognizes the different demands of three types of clients: (1) member universities; (2) users of the outputs of RUFORUM programs, such as new graduates; and (3) global and regional partners and funding agencies. The distinction is important, as it differentiates among (1) RUFORUM members who derive special benefits from membership and may be willing to pay higher membership fees for "club goods"; (2) employers of graduates who pay market rates to the individual graduate that may include a quality premium; and (3) demand for public-good knowledge about universities and networks as bridging organizations among academia, national agricultural research systems (NARSs), policymakers, and the private sector. The finance dilemma is getting the customers to cover the full cost of having and maintaining RUFORUM (RUFORUM 2015). The network now meets the demand for products and services by adding value to the contribution of each of its customers in different ways. A decision to change the balance of its services in favor of new customers will have important implications for funding, as well as for the nature of its core functions.

DESIGNING STRATEGIES FOR COST RECOVERY AND GROWTH AT A MANAGEABLE SCALE

Without core funding, networks cannot function on a sufficiently strong footing to negotiate agreements among partners, establish priorities, invest in serious planning, and build organizational integrity to stay on course. However, many donors tend to prefer short-term project support, which can redirect priorities, overextend management, and leave the organization without the necessary funds to cover direct and indirect costs.[17] While changing local and global contexts drive the need for networks to evolve, growth will require full cost recovery for staff and operations, so that the core is progressively strengthened. Networks face three classical problems: (1) public goods are always underfunded because everyone can have access to them without paying ("free riding"), (2) the users of graduates from network programs do not have to finance the fellowships of students because they can hire the products on the market, and (3) member institutions seldom have

---

17 For example, RUFORUM has found itself pulled in many directions by its supporters and, in some cases, without the necessary project funding to cover the full costs of its operations. Its attractive concept has also resulted in rapid scaling (from 10 members in 2004, to 25 in 2009, to 55 in 2015) by universities, several quite weak, wishing to benefit from spillover knowledge from stronger institutions.

independent resources to fund a network (RUFORUM 2015).[18] These organizations, then, must design a differentiated resource mobilization strategy for each market segment, while recognizing that a majority share of support will need to come from donors or through governments by way of multilateral organization loans for some time to come.[19]

CONTRIBUTING TO ENHANCEMENT OF THE POLICY ENVIRONMENT

Networks can play a leading role in building knowledge and skills for improving policymaking in a shifting policy and institutional environment. They can serve as a convening force, bringing researchers and other agricultural stakeholders in closer contact with policymakers, channeling cross-country experience into national policy debates, and making those debates more evidence-based. Still, there is always the danger of naive assumptions on the part of researchers that strong scientific findings are by virtue of their "dispassionate" observations and analyses routinely used in the policy formulation. The process of implementing networks in the context of policymaking is far more complex, especially the impact of broader contextual factors, such as the political and institutional environment. Understanding of realistic policy options is facilitated by interaction with those charged with making policy decisions. Knowledgeable grasp of the nuances within which policymaking takes place can assist networks in having much greater impact on policy formulation and implementation (Bailey 2010).

BUILDING STRONG MANAGEMENT AND GOVERNANCE

Responding to the divergent demands and capacities of the various stakeholders and raising funds are only two of the major pressures on network managers. These alliances require efficient and transparent governance and advisory structures, often involving representatives from membership countries and

---

18 Private-sector funding of research in African universities is very limited. Expansion would require a strategic framework in universities to encourage university–industry linkages, and government policy support. To explore this potential, the Association of African Universities has formed a partnership with the Association of Universities and Colleges of Canada (Mohamedbhai 2011).

19 BecA's new business plan aims to reach a breakeven point and ensure its financial sustainability based on three core business areas: (1) capacity strengthening predominantly funded by donors, (2) research and research services through hosted programs funded by various clients' research grants or hosted institutions, and (3) product incubation and innovation funded by clients with product development programs (BecA 2013). Fundraising plans by other networks include developing university cluster proposals using the network platform; bringing research proposals in line with large-scale, country-level agricultural initiatives; assisting member universities to establish memoranda of understanding with district agricultural offices to upgrade staff under a fee-based service arrangement; providing indirect grants via partner networks (possibly as subcontractors); creating an innovation fund; establishing an endowment from member and alumni contributions; and placing heads of ministries, directors of central banks, or private-sector chief executive officers on boards.

institutions. Network managers are responsible for setting priorities, investing in financial and reporting systems, convening meetings, communicating with members and funders, developing multiyear business plans that sustain the organization, and administering network activities.

Clearly, networks have high transaction costs associated with assembling people from multiple institutions and geographies. These intricate organizations require a secretariat or host institution steeped in talent, especially at the leadership level, and with appropriate facilities. Yet quality management, which funders demand, entails administrative overheads, which they find objectionable. While there is no simple solution to this problem, overheads should be treated as a legitimate cost that is reflected within an approved business plan and budget.

FOSTERING SUSTAINABILITY THROUGH BETTER EVALUATION AND RISK MANAGEMENT

As most networks are donor dependent, their longevity and potential scaling are linked with changing funder preferences. While this is difficult to alter under current African circumstances and probably into the foreseeable future, at least four constructive steps can be taken to lessen funding shocks: (1) gathering momentum and attracting funding by building a common "brand" of excellence and reliability that gains legitimacy and financial support; (2) developing an evaluation strategy codesigned by management and funders that, while not necessarily settling the sustainability issue, can reduce what may appear as random decisions by funders based on inadequate information (Prewitt 1997); (3) having in place a practical business plan to identify customer segments, a viable growth model, legitimate costs, potential funding streams, and risk-mitigation strategies; and (4) recognizing that scaling up, with reference to breadth of operations and financing, may present risks for individual funders, especially in the context of long-term commitment. Spreading the burden among a broad group of supporters can provide a solution (Fine 2007b), as well as hold the line on core funding. However, the funding base should be diversified as early possible to avoid the impression of network "ownership" by a single donor agency.

Future Considerations

The purpose of this discussion is to highlight some key features and guiding principles of assistance to those engaged in forming, fortifying, and supporting professional capacity-building networks in the agricultural sector. The type of networks featured here are critical mechanisms for building the next generation of innovation-minded agricultural scientists in Africa. They are major vehicles for launching and maintaining scientific careers. Their uniqueness as organizational forms comes from features embedded within profession-enhancing strategies.

In the future, such strategies will need to accommodate global market forces, given that scientists are more likely in their professional lifetimes to move from place to place or work for multiple employers simultaneously. Many networks are already helping their members to initiate reforms, especially in terms of institutional flexibility and innovation, that will position them to face new competitive challenges. This may include transferability of qualifications and course harmonization across universities, organization of research universities within ever more differentiated systems, joint faculty appointments, "split-site" doctoral training within and outside Africa, shared facilities under a common research and training platform, and simplified administrative mechanisms.[20]

Evolving ICTs may enable faculty to be somewhat independent of their universities. The best faculty with multiple chairs in Africa and overseas may be able to video-in their lectures while sitting at a base other than their home university. In addition, future faculty—unfettered by traditional university procedures—may be primarily based in nonuniversity settings, such as government ministries, NGOs, NARSs, private businesses, think tanks, and so on, and may work on contract for universities for a portion of their time. Alternatively, universities with advanced technologies and equipment can outsource services to commercial providers or public-sector facilities, as a means of both raising cash and exposing students and staff to new learning environments.

The future restructuring of agricultural higher education in Africa may rest on new levers for transformation, including (1) populist movements toward tackling long-standing problems of inequities and exclusion; (2) the reorganization of knowledge systems to accommodate emerging complex fields, such as climate change, that demand overcoming disciplinary barriers to problem formulation and problem solving and require renewed appreciation of indigenous bodies of knowledge; (3) the growing importance of the private sector and value chains compelling the incorporation of a business school *optique* into research and training; and (4) the effects of globalization as the reduction of time and space influences relationships among institutions,

20 See Aina (2010) for a general discussion of the politics of higher-education transformation in Africa.

knowledge production, and other agents of the agricultural innovation system.[21]

In the long term, a successful professional network will be characterized by its ability to keep researchers in Africa, keep them scientifically active, and focus them on making measurable contributions to the broader system of innovation in the agricultural sector. Yet, even with evidence that networks are critical elements of the institutional landscape of professional capacity building in Africa, their role is reinforcing. They cannot take full responsibility for the rejuvenation of universities and research institutes. Networks support and complement, but do not replace, these essential institutions. The crucial role of networks over the next decade is to ensure that the bond between higher education and practical, problem-solving science and technology capacity in Africa is sturdy and backed by expanded access to technical resources, peers, reliable finances, and genuine local buy-in for sustained political support.

Funding agencies and others have an opportunity to play a more active role in strengthening how education and research contribute to enhancing innovative capacity in the agricultural sector. Over the past two to three decades, international development agencies have tended to focus more on building professional skills than on building institutional capability. They have stressed technical and analytical tools over problem solving and policy relevance; they have placed greater emphasis on pipeline production of professionals, rather than on their career tracks and skill use; and they have promoted the strengthening of individual institutions over the coordination among multiple, differentiated institutions that can advance and sustain entire professional fields (Moock 2005).

The examples offered here of current collaborative initiatives in agricultural R&D capacity building testify to creative thinking about the serious challenges at hand. These networks have in their DNA the recognition that success depends on translating knowledge into innovation and application. They are responding to a new realism voiced by Africa's political, business, and science leaders who recognize the need to devise fresh, bold, even radical approaches to fields of learning and research appropriate to the times, and to

21 The dynamics of globalization inherently compel durable, mutually supporting partnerships with advanced learning institutions outside Africa. These may include staff and student attachments in both directions and shared research. A major advantage of strong cross-institution, Africa-based networks is the portal they offer world-class external institutions for joint learning and intellectual exchange. The problem is how to seize this benefit without allowing powerful external bodies to have undue impact on the network's core agenda and comparative advantages.

invest in credible yardsticks for appraising these investments. It is a safe bet that the number of such networks will continue to grow.

References

Aina, T. 2010. "Beyond Reforms: The Politics of Higher Education Transformation in Africa." *African Studies Review* 53 (1): 21–40.

Bailey, T. 2010. *The Research–Policy Nexus: Mapping the Terrain of the Literature. Higher Education Research and Advocacy Network in Africa (HERANA).* Wynberg, South Africa: Centre for Higher Education Transformation.

BecA (Biosciences eastern and central Africa). 2013. *BecA–ILRI Hub Business Plan 2013–2018.* Nairobi, Kenya: International Livestock Research Institute.

Beintema, N., and F. Di Marcantonio. 2010. *Female Participation in African Agricultural Research and Higher Education: New Insights. Synthesis of the ASTI-Award Benchmarking Survey on Gender-Disaggregated Capacity Indicators.* IFPRI Discussion Paper 957. Washington, DC: International Food Policy Research Institute.

Beintema, N., and G. Stads. 2011. *African Agricultural R&D in the New Millennium: Progress for Some, Challenges for Many.* IFPRI Food Policy Report. Washington, DC: International Food Policy Research Institute.

Blackie, M., R. Blackie, U. Lele, and N. Beintema. 2010. "Capacity Development and Investment in Agricultural R&D in Africa." Lead Background Paper for the Ministerial Conference on Higher Education in Africa, Kampala, Uganda, November 15–19.

Eicher, C. 2006. *The Evolution of Agricultural Education and Training: Global Insights of Relevance for Africa.* Staff Paper 2006-26. East Lansing: Michigan State University, Department of Agricultural Economics.

———. 2009. "Building African Scientific Capacity in Food and Agriculture." *Review of Business Economics* LIV (3): 238–257.

Elliott, H. 2012. "Regional Research in an Agricultural Innovation System Framework: Bringing Order to Complexity." In *Agricultural Innovation Systems: An Investment Sourcebook.* Washington, DC: World Bank.

Fine, J. 2007a. "Competitive Challenges for Regional Research Networks Serving African Universities." Paper commissioned by the Rockefeller Foundation, New York.

———. 2007b. "Investing in STI in Sub-Saharan Africa: Lessons from Collaborative Initiatives in Research and Higher Education." Paper presented at the "Global Forum: Building Science, Technology and Innovation Capacity for Sustainable Development and Poverty Reduction," World Bank, Washington, DC, February 13–15.

Juma, C. 2011. *The New Harvest: Agricultural Innovations in Africa*. New York: Oxford University Press.

Kayisinga, J. 2010. "Improving Livelihood through Strategic Partnerships among U.S., Rwanda High Education Institutions and Industry." Powerpoint presentation on the SPREAD program, The Norman Borlaug Institute, Texas A&M University, November 26.

Kitzantides, I. 2010. "USAID Sustaining Partnerships to Enhance Rural Enterprise and Agribusiness Development (SPREAD) Project: Integrated Community Health Program, Mid-Term Program Evaluation." Paper commissioned by the United States Agency for International Development.

Lynam, J. 2012. "Agricultural Research within an Agricultural Innovation System." *Agricultural Innovation Systems: An Investment Sourcebook*. Washington, DC: World Bank.

Mohamedbhai, G. 2011. "Strengthening the Research Component of the Space of Higher Education in Africa." AAU Conference of Rectors, Vice-Chancellors and Presidents (COREVIP), Stellenbosch, South Africa, May 30–June 3.

Moock, J. 2005. "Rockefeller Foundation: How We Invest in Capacity Building." Rockefeller Foundation, New York (unpublished paper).

Morel, C., T. Acharya, D. Broun, A. Dangi, C. Elias, N. Ganguly, C. Gardner, et al. 2005. "Health Innovation Networks to Help Developing Countries Address Neglected Diseases." *Science* 309 (5733): 401-404.

Prewitt, K. 1997. *Networks in International Capacity Building: Cases from Sub-Saharan Africa*. New York: The Social Science Research Council.

RUFORUM (Regional Universities Forum for Capacity Building in Agriculture). 2015. *Coming of Age: RUFORUM Strategic Business Plan 2015–2020*. Kampala, Uganda: RUFORUM.

Schilling, T. 2008. "Through Strategic Partnerships among U.S. and Rwandan Higher Education Institutions and Industry." Presentation at the Africa Regional Higher Education Summit, Kigali, Rwanda, October 21–24.

Spielman, D., J. Ekboir, K. Davis, and C. Ochieng. 2008. "An Innovation Systems Perspective on Strengthening Agricultural Education and Training in Sub-Saharan Africa." *Agricultural Systems* 8 (1): 1–9.

World Bank. 2007a. *World Development Report 2008*. Washington, DC.

———. 2007b. *Cultivating Knowledge and Skills to Grow African Agriculture: A Synthesis of an Institutional, Regional, and International Review*. Washington, DC.

PART 4

Measuring and Improving Effectiveness

Chapter 11

THE ROLE OF IMPACT ASSESSMENT IN EVALUATING AGRICULTURAL R&D

George Norton and Jeffrey Alwang

Assessments of the impacts of agricultural research after the fact (that is, ex post) are conducted for many reasons. For example, results can be used to determine the effectiveness of previous investments, provide accountability, or justify future research. Several impact studies, many focusing on Africa, have assessed the impacts of national agricultural research system (NARS) programs and projects over time. Donors and governments want to measure the contribution of agricultural research to their own objectives and to compare that contribution with alternative investments.

To facilitate these comparisons, research evaluators need a clear understanding of donor objectives, including income gains to producers and consumers—termed "efficiency objective"—and other goals, such as reducing poverty, enhancing food security, or improving nutrition and health outcomes—all of which are termed "nonefficiency objectives." To facilitate decisionmaking when allocating funds to agricultural research programs, impacts of agricultural research can be projected prior to research investments being made (that is, ex ante) and these alternative investment choices can be prioritized through structured impact assessment. Effective impact analysis linked to a robust monitoring and evaluation (M&E) system can instruct research managers about how and why certain investments have larger impacts than others; this information can then be used to improve the design of research programs (Chapter 12, this volume).

Evaluations of previous or projected research benefits entail costs, require skills in impact assessment, and necessitate attention to data collection and analysis. Confounding factors influence research benefits, and the outcomes and impacts of many types of research—especially those related to research-induced institutional change—are difficult to measure. The impact pathway leading from agricultural research investments to nonefficiency outcomes,

The authors thank Mary Jane Banks, Nienke Beintema, John Lynam, and anonymous reviewers for comments on this chapter, but the authors, of course, are responsible for any remaining errors.

such as poverty reduction or nutritional improvement, is long and winding. Analysts need a clear map of these pathways if they are to successfully account for confounding factors and identify causal linkages. Many models of research evaluation exist, including in-house and independent assessments, each of which has its own advantages and disadvantages. Research evaluation often focuses on quantifying impacts, usually across multiple dimensions; however, decisionmakers generally want to know why research succeeds or fails. Conventional agricultural research evaluation often falls short of providing this information. Despite these challenges, progress has been made in assessing the impacts of agricultural research in Africa.

The purpose of this chapter is to review experience with impact assessment of agricultural research in Africa south of the Sahara (SSA). The analysis emphasizes ex post assessment, but lessons for ex ante analysis and priority setting are also presented. Methods used to evaluate agricultural research are briefly described and critiqued. Thereafter, empirical evidence on the benefits of agricultural research in Africa is summarized and categorized by type of research. Finally, lessons are drawn for the role of impact assessment for agricultural research in Africa.

Measuring Impacts of Agricultural Research

Impact assessments of agricultural research must identify the appropriate counterfactual (what would have happened without the research), measure the effect(s) of the research intervention, and add up those effects over the target population. The counterfactual can be a moving target because multiple simultaneous and sequenced interventions occur over time. Some research-based technology or institutional interventions depend on relatively few complementary factors, but others will fail to be adopted without them. Observed outcomes may be caused primarily by nonresearch factors, and accounting for these factors is essential to establishing impact.

Various methods have been used to identify the effects of research, but how carefully those methods are applied can influence the credibility of ex post research evaluation studies. The time it takes to complete agricultural research and to diffuse its results differs by technology, commodity, and regulatory process, and the adoption time path must be carefully estimated to generate credible impact assessments (Box 11.1). Spillovers of technologies across countries further complicate the analysis (Chapter 14, this volume).

Assessments of agricultural research impacts are under increased scrutiny because policymakers demand increasingly convincing evidence of impact and

BOX 11.1 Three basic approaches to quantifying ex post agricultural research impacts

Three basic approaches have been used to quantify ex post impacts of agricultural research. One is to use secondary (usually time-series) data at the national level and assess the aggregate productivity effects of research using a production function, cost function, or profit function approach. Results may then be included in a benefit-cost analysis that assesses benefits to producers and consumers over time. The second method is to gather producer-level data and estimate rates of adoption and farm-level impacts to determine the micro-level effects of research interventions. These effects are then combined with market-level data and models to produce estimates of aggregated impacts on producers and consumers, which again are included in a benefit-cost analysis. The third approach is to use data from experimental trials to estimate impacts with and without the effects of the research. Such data are then used to construct budgets or are combined with adoption estimates and market-level data and models to calculate the aggregate effects on producers and consumers and rates of return to research investments in a cost–benefit analysis. The first two methods use econometric methods, whereas the third simply uses calculations.

Source: Authors.

accountability. From a methodological perspective, this pressure builds from two related threads. First, statistical methods for assessing treatment effects (that is, the first-level effects of a research intervention) using observational data have improved dramatically over time (Imbens and Wooldridge 2009), and general conditions for the identification of causal effects are now widely understood. Second, the revolution in behavioral economics and the now-widespread application of randomized controlled trials (RCTs) have led to credible estimates of causal effects of many development investments (Datta and Mullainathan 2012).

Advances in causal estimation mean that measures of agricultural research impacts need to compete with highly credible estimates from other realms. For example, the impacts of bed nets, deworming in schools, and other interventions are widely accepted in the health and education sectors. Agricultural research impacts need to be similarly credible. Impact assessment outside of agricultural research is moving beyond solely measuring economic impacts to identifying linkages along an impact pathway embedded in a "theory of change" (see, for example, the International Initiative for Impact Evaluation [3ie] website at www.3ieimpact.org/en and Chapter 12, this volume). As linkages along the chain of changes often depend on behavioral responses, behavioral economics has a large role to play in these assessments. The theory of change approach changes the focus from pure impacts to how and why impacts occur.

A subtle shift has also occurred in the interest of policymakers beyond rates of return and estimated net present values of producer and consumer income change. Increased evidence of nonefficiency impacts (such as impacts on poverty and on the value of nutritional, environmental, and health benefits) is now needed. In response, the agricultural research evaluation community is developing improved methods for estimating nonefficiency impacts, but the challenge of establishing clear causality complicates their application.

Assessing the Counterfactual

When secondary, national-level data are used to estimate production, productivity, cost, or profit levels, establishing clear causal links between research investments and these outcomes is difficult. Aggregate estimates subsume the entire research impact pathway and provide no information about factors affecting the variability of returns to research. They can also fail to control for how unobservable variables affect both research expenditure and outcome variables, which likely leads to an upward bias in measuring the returns to research. This bias is compounded by rate of return computations, as noted below. In the African context, the quality of secondary data is also a constraining factor.

When producer-level data are used to calculate rates of adoption, levels of impact, or budgets, observed outcomes may be affected by nonrandomly assigned confounding factors that can bias the results. When observational survey data are used to measure the impacts of research-generated technologies, establishing the counterfactual is difficult because it is impossible to observe the same farmers as both adopters and nonadopters at the same time. Measurement of the causal effect of adoption on the outcome must include a credible counterfactual.

One means of including a credible counterfactual is through matching—identifying nonadopter(s) in the sample who have characteristics sufficiently similar to each adopter, and comparing the difference in outcomes between the adopter and the matched group of nonadopters. An alternative approach to eliminate bias is through a two-stage analysis using instrumental variables. In the first stage, the determinants of adoption are estimated, and in the second stage, the impacts of adoption on the outcome are estimated. These and other alternatives require different assumptions, but are used to purge the impact estimates of nonrandom selection bias. The external validity of measured effects depends on the credibility of the counterfactual and the strategy to identify the effect of interest.

With many nonresearch interventions, RCTs have been run to hold confounding factors constant and reduce the potential for selection bias. RCTs are less practical for measuring ex post impacts of agricultural research for many reasons, including the need to conduct the research in the fields of farmers who are willing to cooperate over multiple years. Agricultural research treatments are often complex, with multiple interventions made simultaneously or sequenced over several years. The impacts of technology adoption, such as changes in market prices, are often only evident over many years, and the long time lag between a variety's release and manifestation of its full impacts makes RCTs less suitable (Norton and Alwang 2012). Hence, approaches using observational data are more frequently used for assessing the impacts of agricultural research.

One of the few studies that has used RCTs to obtain estimated impacts of technology adoption was conducted by Duflo, Kremer, and Robinson (2008), who employed an RCT to examine the effects of fertilizer adoption in Kenya. The study examined a simple fertilizer technology, and the evaluation was very expensive. RCTs are potentially useful in agricultural research evaluation for assessing microlevel impacts of simple interventions that have been developed but not yet disseminated. However, they are less useful for other research evaluations because of

1. spillovers from the treated to the untreated group;
2. the difficulty in convincing subjects to participate and, if they do, keeping them in full compliance or in the trial at all;
3. ethical considerations, such as those associated with keeping a potentially valuable intervention away from part of the population during the trial;
4. their high cost once the diversity of the smallholder population of farm households is considered in combination with the complexity of some of the interventions;
5. the difficulty of running RCTs when multiple interventions are sequenced into the population during the assessment period; and
6. the need to set up the RCT before any participant households are selected, which can be a long time before micro-impacts can be assessed (Norton and Alwang 2012).

Some of these issues can be addressed by using pilot programs or phasing in an intervention, randomizing villages rather than individuals to reduce spillovers, or randomly assigning subjects who receive an announcement or incentive to encourage participation (Duflo, Kremer, and Robinson 2008). However, the severity of the problems differs by type of intervention. The reality is that, given the complexity of and length of time required to develop agricultural research interventions, RCTs are of limited use for agricultural research evaluation. In most cases, use of observational data and non-RCT approaches is the only option for ex post research impact evaluation.

Another means of assessing the counterfactual is to combine estimates of technology adoption or dissemination with simple per-unit budgets using data generated from randomized experimental plots (as opposed to randomized farms or villages). This method uses various techniques to measure the technology spread, and assigns each adopting land unit a treatment effect that corresponds to the difference between the unit cost of production of the new technology and the unit cost of production of the control (usually representing standard farmer practices). The advantage of this approach is that it uses both experimental data from randomized trials and observational data on adoption to produce an upper-bound estimate of economic impacts of adoption. The main disadvantage is that the phenomenon of yield gaps between experimental trials and actual outcomes in farmers' fields is nearly universally recognized. The sizes of these gaps can be large, leading to a substantial upward bias.

Agricultural Market and General Equilibrium Effects

Whether RCTs or alternative approaches are used to control for the first-level counterfactual, results are often combined with market models, such as economic surplus models, to assess the aggregate level and distribution of economic benefits of research to producers and consumers, as output prices as well as production may change. The lower commodity price resulting from additional technology-induced production is a major reason why consumers are often the primary beneficiaries of agricultural research. These price effects are especially large for basic staples, which have inelastic demands. Agricultural productivity growth can also lead to general equilibrium effects in the rest of the economy, as the nonagricultural sector is stimulated by the lower food prices and labor markets are affected (Hareau et al. 2005). Relatively few impact studies on agricultural research in Africa have assessed the effects of specific research programs and technologies on labor markets and nonfarm growth, but in the aggregate the impacts can be substantial.

Discounting Agricultural Research Benefits over Time

Aggregate or market-level income (economic surplus) changes to producers and consumers resulting from agricultural research generally occur over several years. Therefore, the income changes can be discounted to account for the fact that income received sooner is worth more than income received later. Results are presented as internal rates of return (IRRs) to research expenditures or net present values (NPVs) (Box 11.2). IRRs should be considered only as rough approximations.

- First, IRRs may be based on projects and programs of various sizes. A high rate of return realized for a small research project may not carry over if the project is scaled up. Economists often use NPVs rather than rates of return to rank investments for that reason.
- Second, IRR calculations assume that returns can be reinvested over time at the calculated IRR (Alston et al. 2011; Rao, Hurley, and Pardey 2012). However, it may make more sense to assume that the returns can be reinvested in the future at the rate of return on alternative social investments. Rao, Hurley, and Pardey (2012) recalculate the rates of return to agricultural research for a large set of studies globally assuming the reinvestment rate is 3 percent. This modified IRR calculation reduces the average rate

BOX 11.2 Internal rate of return, net present value, and real social rate of return

Net present value (NPV) is the sum of discounted benefits and costs over time:

$$NPV = \sum_{t=0}^{T} \frac{(B_t - C_t)}{(1 + r)^t},$$

where *Bt* and *Ct* are benefits and costs in year *t*, *r* is the discount (interest) rate, and *T* is the time horizon. The internal rate of return (IRR) is the discount rate that reduces the NPV to zero:

$$\sum_{t=0}^{T} \frac{(B_t - C_t)}{(1 + IRR)^t} = 0$$

The IRR is a real rate of return (and does not include inflation). The real social rate of return includes the value of all (not just private) benefits and costs to society, including those resulting from health and environmental effects.

Source: Authors.

of return to agricultural research from 33 percent to 12 percent—still a decent return, but lower than the previous estimates.

- Third, IRRs can also be skewed upward when the benefits analysis focuses on a specific country and does not account for spill-ins of research knowledge from other sources (such as CGIAR). The resulting calculation of the return can be a legitimate estimate for the country in question, but would not apply to the world as a whole.

Perhaps in part because of the potential for bias, fewer studies over time have reported IRRs. Many of them estimate the NPV of agricultural research investments in which a real social rate of return of 3–5 percent is typically used to discount costs and benefits. NPV calculations still require proper accounting for research costs, and, when spill-ins of research knowledge occur, a careful effort must be made to attribute costs and benefits to specific research expenditures.

Looking Beyond Efficiency Benefits and Rates of Return

Agricultural research produces many types of technologies and institutional changes. The most common type of research to be evaluated in Africa is genetic improvement; however, other types of research such as pest management, conservation agriculture, and policy research have also been evaluated. Most of the assessments have focused on total income effects, but some have addressed poverty, nutritional, health, environmental, and other types of benefits.

When policymakers are interested in nonefficiency impacts, alternative and often complementary approaches are needed. Methods may include assessment of changes in poverty indexes (Moyo et al. 2007), calculations of changes in disability-adjusted life years (DALYs, a widely accepted measure of health outcomes) (Meenakshi et al. 2010; Nguema et al. 2011), and use of contingent valuation or choice experiments to place value on nonmarket benefits of improved technologies (Bonabona-Wabbi, Taylor, and Norton 2014; Vaiknoras, Norton, and Alwang 2015). In evaluating agricultural research benefits that are not productivity enhancing, the task of accurately and cost-effectively measuring what would have occurred without the research can be a challenge. Consequently there are fewer empirical results for these types of studies, although, as discussed below, the literature is growing.

Ex Ante Impact Assessment and Priority Setting

Ex post research impact assessment provides information that supports accountability in agricultural research, while ex ante assessment can support decisionmaking in program selection and funding. Structured priority-setting methods that include economic evaluation may lead to better judgments and expose poor ones. While these methods are no substitute for the judgments of scientists and administrators, they provide a mechanism for considering scientific and economic data that would otherwise be difficult to use (Alston, Norton, and Pardey 1995). Few agricultural research systems in Africa use structured priority-setting methods on a regular basis, although several have used them on occasion. Methods range from simple scoring activities to economic surplus analyses and mathematical programming (Ceesay et al. 1989; Teri, Mugogo, and Norton 1990; KARI 1991; Mills 1998; Mutangadura and Norton 1999; Thornton et al. 2000; Diagne et al. 2009; Manyong, Sanogo, Alene 2009; Wood and Anderson 2009).

Most ex ante impact analyses of agricultural research do not involve a full priority-setting analysis and process for a research system. They involve assessments of specific research topics or themes to help decisionmakers choose from among a limited set of research options or decide whether to continue supporting particular lines of research. Some of these analyses involve formal cost-benefit analysis (Rudi et al. 2010), but most are published in reports rather than refereed journals (see, for example, Norton and Philips 2011). Unless there is a unique twist to a method, journals seldom publish speculation, which is essentially what ex ante analysis represents. In fact, few fully fledged priority-setting analyses are published, except as the occasional book chapter. The purpose of ex ante or priority-setting studies is not publication, but improved decisionmaking.

Impact Evaluations of Agricultural Research in Africa

Several papers have summarized the results of previous agricultural research evaluations, including results for Africa. For example, Alston et al. (2000a, 2000b) summarized the results of studies that calculated rates of return to agricultural research and were completed by 1997. They also conducted a meta-analysis of the rates of return, and examined various factors that influenced those returns.[1] One such factor is the level of aggregation, because the lower the level of aggregation of research programs, the higher the variance in rates of return. Most research projects yield a zero rate of

1 The authors compared rates of return for studies that differ by econometric versus noneconometric approach, by research focus, by time period, by ex post versus ex ante analysis, by average versus marginal rate of return calculations, by real versus nominal rate of return calculations, and by level of aggregation of the programs evaluated.

return, and the estimated benefit from an aggregate research program is the average of the benefits from many projects with zero benefits, some with modest gains, and a few with high (and in some cases astronomical) gains. Many agricultural research evaluation studies assess the benefits of specific projects or technologies. Those studies seldom list the unsuccessful projects; hence, the level of aggregation is a critical factor when comparing studies on rates of return.[2] The overall mean rate of return for all regions in the Alston et al. (2000b) study was 99 percent, which was skewed upward by a few studies with very high estimated returns, as their median return was 48 percent.

Alston et al. (2000b) included in their study a summary of results from 48 studies for Africa. Those studies provided 188 estimates of rates of return to agricultural research. The estimated rates of return ranged from −100 to 1,490 percent, with a mean rate of return of 49 percent and a median return of 34 percent, which are a little lower than the average returns for all regions but are still high. Evenson and Gollin (2003) reported a similar 37 percent median rate of return in studies that evaluated economic benefits to all types of agricultural research in Africa.

Since the late 1990s, several additional evaluations of agricultural research in Africa have been conducted, including some studies of the environmental, poverty, and nutritional benefits of research. Four significant review studies that report agricultural research benefits for Africa have been published. Evenson and Gollin (2003) summarized the benefits of crop genetic improvement not only in Africa, but also around the world. Maredia and Raitzer (2006) presented evidence of the benefits of research undertaken by CGIAR and NARS partners in SSA. Renkow and Byerlee (2010) examined the impacts of CGIAR research in Africa and other regions, with the benefits categorized by type of research, such as genetic improvement, pest management, natural resource management, and policy analysis. Walker and Alwang (2015) report on a major effort to examine the inputs (research infrastructure), outputs (variety releases), outcomes (diffusion), and impacts of crop varietal improvement research in Africa between 1998 and 2011. They also describe country-to-country spillovers in adoption that are so important for many SSA crops and analyze the role of international agricultural research centers in supporting SSA research.

---

2 For example, Alston et al. (2000b) averaged results from studies that reported returns from individual projects, but did not include the reported zero returns for many of the projects mentioned in the same studies, hence biasing upward the overall results.

Impacts of Varietal Development in Africa

Evenson and Gollin (2003) found 92 million hectares were planted with modern varieties in SSA in 1998, representing 23 percent of the area devoted to the 10 crops considered (wheat, maize, rice, sorghum, millet, barley, lentils, beans, cassava, and potatoes). More than 1,150 varietal releases were made between 1965 and 1998. They estimated that the total genetic improvement contribution to yield growth over the period was 0.28 percent per year, which was less than half of the rate of growth for developing countries as a whole. They found significant adoption of improved maize varieties and hybrids in SSA, with 36 percent of the maize area planted with modern varieties in West and Central Africa, and 52 percent in East and Southern Africa in the late 1990s. Alene et al. (2009) updated the numbers for West and Central Africa for 2005, and found 60 percent of the maize area planted with modern varieties at that time.

Walker and Alwang (2015) update the Evenson and Gollin (2003) study, focusing on agricultural research and its impacts in SSA. Their findings show significant variability by crop and by country, but overall the stock of improved varieties in SSA has increased and continues to increase over time. More than 1,400 varietal releases occurred in SSA between 1998 and 2011 for the 20 crops on which they report. Using a more consistent methodology and investigating diffusion for twice as many crops as Evenson and Gollin (2003), they estimate that, as of 2011, 35 percent of SSA cropland (more than 107 million hectares) is planted to improved varieties. Adoption rates range from a high of about 90 percent of planted area for soybeans to around 5 percent for bananas and field peas. The volume identifies crop-specific constraints to wide adoption (for instance, 40-year-old groundnut varieties are still widely planted in West Africa, and sorghum varieties based on *Paramecium caudatum* types cannot compete with the dominant Guinean materials prevalent in the region). Fuglie and Marder (2015) found that diffusion of these varieties accounted for an overall average productivity gain on adopting areas of about 47 percent, and contributed to about 15 percent of the growth in food crop production in SSA between 1980 and 2010. By 2010, the higher productivity of improved food crop varieties had added US$6.2 billion to the yearly value of agricultural production in SSA.[3]

Many evaluations are part ex post and part ex ante in the sense that the research has been completed, and the varieties have begun to be, but have

---

3 All currency is in US dollars, unless specifically noted otherwise.

not yet fully been, adopted. Therefore, the person doing the analysis uses both ex post data and ex ante adoption projections. For example, Kalyebara et al. (2008) estimated that improved common bean varieties (*Phaseolus vulgaris*) were adopted on about half of the total bean area in East, Central, and Southern Africa over 17 years beginning in 1986, but they project benefits out to 2015. The benefits from this combination of ex post and ex ante analysis were estimated at $200 million (in 1986 dollars) on a research investment of $16 million.

Maredia and Raitzer (2006) reviewed impact studies of agricultural research on SSA in which CGIAR centers played a role. They found 52 studies, of which 34 involved improved crop varieties; of these, 16 studies assessed the benefits of the research at an aggregate level. Using only data from the most rigorous studies, they conducted a meta-analysis of the economic benefits of crop genetic improvement in eight crops (beans, cassava, maize, millet, potato, rice, sorghum, and wheat) and conservatively estimated $2.4 billion in economic benefits from 1978 to 2004 (in 2004 dollars).

Rusike et al. (2010) employed a combination of methods, including difference in differences, propensity score matching, and Heckman's treatment effects model, to assess the impacts of cassava research for development in Malawi. The normal selection bias associated with who adopted improved cassava varieties was compounded by issues of distinguishing the effects of relative price changes in cassava versus maize resulting from the country's structural adjustment program; the collapse of input, credit, and maize markets; changes in labor markets resulting from HIV/AIDS; and other factors. Assessing the impacts of agricultural research on development involves consideration of the whole innovation system with its input and output markets, extension service, farmers' organizations, and other groups in the value chain, in addition to the normal experimental work on improved planting materials. It involves careful attention to the pathways through which research eventually leads to impacts on development outcomes.

Rusike et al. (2010) tested hypotheses related to the effects of research on yields and on area planted with cassava, as well as market and institutional restrictions on scaling up the supply of cassava. They estimated the impacts of cassava research on caloric intake and food security, and evaluated the number of additional months that households were able to meet minimum caloric requirements from home-produced cassava and maize staples. The average treatment effect was about eight months or a 66 percent increase in months meeting the minimum. They found that, by 1995, yearly yields in the mostly cassava-growing and -consuming districts first exposed to the program were about 23 percent higher than they would have been in the absence of the program. Controlling for other observable factors, they found that the productivity of the area cropped to cassava was about 14 percent higher in those districts.

Recognizing that policymakers increasingly demand information on distributional impacts, a few studies have examined the impacts of crop genetic improvement on poverty reduction (Moyo et al. 2007; Alene et al. 2009; Larochelle et al. 2015; Zeng et al. 2015). Moyo el al. (2007) evaluated the total economic benefits of rosette virus–resistant peanut varieties developed by the International Crops Research Institute for the Semi-Arid Tropics (ICRISAT) in partnership with NARS and US universities, and then apportioned those benefits to poor people in Uganda based on the production and consumption of peanuts. Total benefits of $34–$62 million (in 2000 dollars) were estimated, and the poverty headcount index decreased by 0.5–1.0 percentage points in the study area under most assumptions. Alene et al. (2009) found a poverty reduction of 0.1 percent in 1981 to more than 1.26 percentage points in 2004 for improved maize varieties in West and Central Africa, with an average poverty reduction of 0.75 percentage points. This last figure represents an average of 740,000 people lifted out of poverty each year. The impacts were greatest in Benin, Ghana, and Nigeria, where maize accounted for more than 10 percent of total agricultural production.

Zeng et al. (2015) assessed the yield, income, and poverty impacts of improved maize varieties in Ethiopia developed jointly by the International Maize and Wheat Improvement Center and the Ethiopian NARS. Because maize varieties are often adopted on some, but not all, plots on individual farms, they used plot-level rathere than household farm-level estimates of varietal adoption, and allowed for heterogeneity in treatment effects in terms of yield gains across plots and farm households. The authors used an instrumental variables approach as the main identification strategy, and applied a backward derivation procedure in an economic surplus framework to identify the counterfactual price level from which the counterfactual income distribution was estimated. The authors assessed treatment effects, and computed poverty impacts as the differences between the observed and the counterfactual income distributions. They found that 42 percent of the producers were full adopters, 14 percent were partial adopters, and the rest were nonadopters. Adopting improved maize varieties in Ethiopia resulted in a 1.0–1.5 percent decrease in the overall poverty headcount ratio. A 2.4–2.6 percent decrease in the poverty depth index, and a 2.7–3.1 percent decrease in the poverty severity index (using the $1.25 a day poverty line).

Larochelle et al. (2015) examined the impacts of improved bean varieties on poverty and food security in Rwanda and Uganda. They conducted a treatment effect, economic surplus, and poverty analysis similar to that conducted by Zeng et al. (2015). Poverty in 2011 would have been about 0.4 and 0.1 percentage points higher in Rwanda and Uganda, respectively, in the absence of varietal improvement. A measure of household dietary diversity was developed to examine the effects of the improved bean varieties on food security among rural households. Food insecurity would have been substantially higher in Rwanda without the introduction of improved bean varieties, and noticeable impacts were also found in Uganda.

Limited work has been devoted to assessing the effects of varietal improvement on reducing yield variability, even as agricultural researchers have placed increased importance on innovations to reduce this variability. Yield stability is important to farmers, whose food security and livelihoods depend on it. Research that builds in resistance to insects and diseases, drought, or salinity can be especially helpful to poor farmers. While research could increase yield risk, especially if the genetic base is narrowed, evidence suggests that risk has decreased over time, as resistance to stresses has been built into improved varieties, at least for maize and wheat (Gollin 2006).

Additional evaluation of the benefits of yield stability is needed and will likely be forthcoming as adaptation to climate change increases in importance. Methods for economic evaluation of research that reduces yield variance have been developed and applied in ex ante analysis for drought-tolerant crop varieties in East and Central Africa (Kostandini, Mills, and Mykerezi 2011), but the methods have not been applied in ex post analysis. Part of the difficulty is that vulnerability measurement generally requires panel data, but panel studies of adequate size to make inferences to populations are extremely costly.

Several recent studies have assessed the nutritional benefits of consumption of biofortified crops (Low et al. 2007; Meenakshi et al. 2007, 2010; Gunaratna et al. 2010; Nguema at al. 2011). The challenge from an impact assessment perspective is to establish a causal link along an extended impact pathway from research on biofortification, through household-level adoption and subsequent impacts on productivity, income, and consumption. While strong evidence exists that increased consumption of micronutrients helps improve nutritional status, this linkage is only the end of a long causal pathway. As a result, few ex post studies exist to convincingly link research on biofortification to nutritional outcomes.

The HarvestPlus project has produced several studies, some of which evaluated the potential impacts of biofortification with provitamin A, iron, and zinc for several staple crops in Africa (for example, Meenakshi et al. 2010). Many of these studies are ex ante assessments and calculate DALYs. For example, Meenakshi et al. (2010) projected the benefits of vitamin A–enhanced cassava in the Democratic Republic of the Congo and in Nigeria, maize in Ethiopia and Kenya, and sweet potatoes in Uganda. Nguema et al. (2011) projected the benefits of vitamin A–t and iron-enhanced cassava in Nigeria and Kenya using a DALY approach and economic surplus analysis.

One useful feature of the DALY approach is that it facilitates analysis of cost-effectiveness, because improved varieties can be compared with other interventions, such as fortification and supplementation. A meta-analysis of nine studies of the benefits to malnourished children of quality protein maize found an average rate of height gain of 9 percent and an average weight gain of 12 percent (Gunaratna et al. 2010). Low et al. (2007) conducted an experiment with 850 households in Mozambique and reported a significant increase in vitamin A intake among young children who consumed orange-fleshed sweet potatoes. Microlevel assessments of nutritional benefits are necessary, but are not sufficient to establish a causal link between particular kinds of agricultural research and nutrition gains. There is clear need to broaden the analysis to the entire impact pathway.

Impacts of Pest Management Research in Africa

Two types of pest management research have been subject to economic impact assessment: classical biological control (CBC) and integrated pest management (IPM). CBC involves controlling a pest in its new environment by introducing a natural enemy from its original geographical environment. The most famous case of CBC was the introduction of a parasitic wasp (*Apoanagyrus lopesi*) into Africa from Paraguay to control the cassava mealybug (*Phenacoccus manihoti*) in more than two-dozen countries. Zeddies et al. (2001) estimated at least $9 billion in benefits (in 1994 dollars) for 27 countries discounted over 40 years. Coulibaly et al. (2004) estimated significant economic benefits to CBC for cassava green mites, mango mealybug, and water hyacinth. In fact, CBC has produced the largest economic benefits that have been measured ex post for agricultural research in Africa. The methods used to evaluate those benefits were simple cost–benefit analyses, but the results are credible because the pest-related losses and pesticide costs were easy to measure, and farm-level savings could be aggregated across all the farms whose exposure was eliminated by CBC. Because the parasitic wasps naturally multiply and require no active intervention by farmers, the issue of adoption measurement for this single-component technology was relatively simple.

Assessing the impacts of IPM is more difficult. Indeed, IPM is a package of technologies that includes components (such as improved varieties with insect and disease resistance, CBCs, and other types of biological control) and management practices (such as altered timing for planting, grafting on resistant rootstocks, use of pheromone traps, and reduced toxicity pesticides when pesticides are required). A specific IPM research program may include only a portion of these and other component technologies. Impact assessment in the context of IPM will help identify components and packages with the most potential for scaling up. It may also be used by research managers to identify fruitful lines of future IPM research or to help decide between investments in IPM research and other alternatives.

Several IPM evaluations in the past have evaluated farmer field schools (FFSs), which involve training programs for groups of around 25 farmers that take them through the whole crop season with weekly meetings. However, FFS is primarily a training program and not an agricultural research program, even though farmers do test some things on their own. Evidence is now clear that at a cost of $20–$60 per participant, FFS is a relatively expensive way to train large numbers of farmers, even if those who participate benefit economically (Feder, Murgai, and Quizon 2004; Ricker-Gilbert et al. 2008). The key to broader benefits from FFS is farmer-to-farmer spread—which should occur as FFS-trained farmers interact with their friends and neighbors—but there is little convincing evidence of this spread.

Economic evaluations of IPM research programs have found significant and large benefits in many countries around the world, but relatively few ex post impact studies have been conducted in Africa, except for individual component technologies, such as CBC for cassava mealybug (Zeddies et al. 2001); disease-resistant varieties, such as for rosette virus on groundnuts (Moyo et. al. 2007); and a host-free period for virus control in tomatoes (Nouhoheflin 2010). Evaluations that have been conducted are part ex post and part ex ante studies using economic surplus analysis (for example, Debass 2001).

Impacts of Natural Resource Management and Other Types of Research

Relatively few assessments have been conducted of the benefits of natural resource management (NRM) research related to agriculture in Africa or in other developing countries. Most NRM research in agriculture has focused on managing pests while minimizing pesticide use, reducing erosion, sequestering carbon in the soil, and conserving water. Examples include impact assessments of conservation agriculture (Nkala, Mango, and Zikhali 2011), zero tillage (Dalton, Yahaya, and Nabb 2014), and tree fallow (Ajayi et al. 2007) to reduce erosion, sequester carbon, and conserve water. Ajayi et al. (2007) estimated an NPV of $2–$20 million, with an IRR of 3–20 percent for tree fallow in maize in Zambia. Dalton, Yahaya, and Nabb (2014) estimated a 35 percent increase in net returns for no-till compared with conventional tillage in Nyoli, Ghana. And Nkala, Mango, and Zilhali (2011) assessed the impacts of conservation agriculture on productivity, household incomes, and food security in Mozambique. Farmers who used conservation agriculture were 53 percent more likely to experience an increase in productivity, but impacts on income and food security were not significantly different for those who used conservation agriculture and those who did not.

Most of these studies have measured the benefits of the technologies in terms of their contributions to income through agricultural productivity gains rather than environmental improvement, and most of the measured benefits occur on relatively small acreages. Part of the difficulty in assessing environmental benefits is that they are not priced in the market. Such methods as contingent valuation and choice experiments suffer from potential problems with hypothetical bias and limited geographical applicability. Choice experiments have recently grown in popularity in developing countries, including examples in Africa, and the method holds potential for future ex post evaluations (Bennett and Birol 2010). Other difficulties in assessing benefits of NRM are that biophysical measurement is difficult and costly, the environment is multifaceted, and many benefits occur at higher scales than plots or farms. Simulation modeling is one approach for addressing some of these issues.

Other types of agricultural research, such as policy analysis, have been evaluated as well, but relatively few of these studies have quantitatively estimated economic benefits, especially in Africa. Valuing policy research involves valuing information, and it is difficult to apportion credit for a policy change and to assess the counterfactual.

Benefits of Private Agricultural Research

Few studies measure the impacts of private agricultural research in Africa. Several studies do assess farm-level profit or the risk effects of specific technologies, such as seed and fertilizer generated and sold by private firms (for example, Regier, Dalton, and Williams 2012); however, no parallel set of published analyses calculates rates of return to research by private firms. Pray, Gisselquist, and Nagarajan (2011) report almost 1,300 cultivars registered from private firms in Kenya, Senegal, South Africa, Tanzania, and Zambia, with more than 100 of them in South Africa. They also report more than

$62 million (in 2008 dollars) spent on agricultural research in those countries, with more than $50 million spent in South Africa alone. Regier, Dalton, and Williams (2012) report data on increased yields and profits, and on reduced pesticide use associated with adoption of genetically modified crops in Burkino Faso and South Africa; however, they do not report on returns to research for the private firms that conducted the research. Nevertheless, it is clear that private investments in agricultural research for Africa have increased in recent years, and those investments are likely to have yielded social as well as private returns, particularly in South Africa.

Lessons for Evaluating the Impacts of Agricultural Research in Africa

The Need for Ex Post Evaluation

A body of evidence is developing on the benefits of agricultural research in Africa, despite the complexity and heterogeneity of its rainfed, smallholder agriculture. The evaluation studies usually rely on observational data, and increasingly employ improved techniques to identify causal linkages between research investments and their impacts. The critical challenge in such cases is to find a plausible strategy to identify the impact of adoption on observed changes at the field and household levels across a diverse agroecological and social landscape.

The best studies carefully evaluate their identification strategy and test its underlying assumptions. The adoption–impact relationship generally involves unobservable factors that estimation strategies must account for. Identifying the treatment effects requires a clear map of the impact pathway, which itself depends on the underlying theory of change. Investigating linkages along the pathway will expand the value of such assessments and help research managers understand why certain programs work, while others do not. This investigation may involve expanding the tools used for impact assessment; for example, RCTs can be used to evaluate alternative dissemination and education practices, even if they are less useful for evaluating agricultural research itself.

Nevertheless, most studies of the impacts of agricultural research are still at least partly ex ante because of a relative dearth of ex post evidence for many crops, types of livestock, countries, and types of research. Evidence of the impacts of agricultural research programs aimed at environmental, health, nutrition, and poverty objectives is scarce, despite the significant and increasing investments in these types of programs. Future funding for agricultural research will depend on filling the gap between the demand for and the supply of these types of evaluation studies. It is not necessary to evaluate all research projects, because evaluation costs money; however, impact assessment for a representative sample of research projects and improved technologies can demonstrate the return on investment, so as to maintain the flow of research resources. Evaluating a sample of various types of projects can also provide evidence about relative payoffs.

Methodological difficulties in measuring the benefits of certain types of research, such as NRM and policy studies, should not necessarily be taken as evidence that those areas are poor investments. Instead, they should stimulate research to overcome those difficulties, both for investigators and for data providers—such as the Agricultural Science and Technology Indicators initiative of the International Food Policy Research Institute—that can document the size of the research investments across topical areas.

Trade-offs between Cost and Credibility

The cost of conducting impact assessments of agricultural research differs substantially, depending on such factors as the depth and credibility of the analyses, the complexity of the interventions, the nature of the farming systems, the types of impacts assessed, and whether the evaluation is conducted ex ante or ex post. The least costly impact assessments (a few thousand dollars) are ex ante evaluations that (1) gather secondary price and quantity data on the commodities involved for a single geographic area, and (2) obtain expert opinions of costs, yield changes, and expected rates and timing of adoption of the technologies. Simple spreadsheets can be used to assess the market-level benefits and costs and to generate NPVs or IRRs for the research. Geographic price and technology spillovers can be included for a relatively small additional cost. The level of precision can be improved (at additional cost) if on-farm cost and return data are gathered in experiments with the new technologies as a substitute for expert opinions.

Once the interventions are completed (for example, the varieties are released), expert opinion on adoption can be replaced with expert panel elicitation or data from farm-household surveys. These surveys can be conducted once or over several years (with accompanying differences in costs). Developing a sample frame for such surveys is complicated, because a true measure of adoption should include evidence from all areas where adoption may have occurred. The research needs to begin with information about the geography of the potential spread of the technology.

The impact studies in Walker and Alwang (2015) used similar methodologies to generate samples representative of specific crop production in specific countries. That study's diffusion estimates were largely based on expert panels convened in each country; diffusion estimates from expert panels sometimes match the estimates from representative household surveys very well, but do not match well in others (Walker and Alwang 2015). The smaller and more diverse the farms and research interventions, the higher the associated costs of tracking diffusion will be (Alwang 2015). Tracing the impacts through a value chain and into factor and product markets adds additional costs, as does incorporating risk with respect to model parameters. Conducting an RCT may add little to substantial costs (thousands to millions of dollars), depending on the complexity of the technologies, the diversity of the agroecological and social environments, and the stage of the research. For many types of social science or NRM research, data-intensive simulation models or other approaches that do not involve farm-household–level surveys may be useful.

The question is often asked, what will be the cost to conduct an impact assessment of agricultural research in a specific country or region of SSA? The answer is that it depends on the breadth of the questions (for example, the types of impacts and interventions), and how much precision is desired. A recent tendency in the development literature is to improve precision in impact assessments by targeting narrow interventions that lend themselves to RCTs (often at a cost that exceeds $1 million). Few agricultural research interventions are simple or narrow; agricultural research evaluation budgets are typically small; and, for these and other reasons described above, few of the interventions have been evaluated with RCTs.

Impact evaluations of research targeting environmental improvement or policy change can involve valuing information that has direct effects off-farm. Even if such evaluations can be undertaken by NARS, the costs can be substantial.

The Importance of Multidisciplinary Interactions

Many impact evaluations of agricultural research require a basic understanding of the research itself, together with data on yields, input costs, and other factors. One way to acquire that understanding is to conduct multidisciplinary research that includes an impact assessment evaluator as part of the research team. Certain types of evaluation tools, such as RCTs, can be applied only if the research being evaluated is designed with impact evaluation in mind from the start. The lack of involvement early in the research process is one reason why there are so few impact evaluations. Therefore, embedding evaluators into agricultural research teams can be useful.

Using the Results of Impact Assessments

Impact assessments are useful not only for justifying research investments, but also for designing a research portfolio around the highest payoff activities. However, the difficulty in measuring the benefits of some types of agricultural research implies a need for caution. Funding agencies may be tempted to fund only activities with measurable impacts and, in the process, may skew the research program toward easily quantifiable, as opposed to high-payoff, research. Some of the highest-payoff research activities can be risky, with outcomes that are difficult to measure. In addition, many donors are interested in nonefficiency objectives, and measuring the trade-off between aggregate efficiency impacts and other impacts can represent an obstacle.

Ex Ante Impact Assessment and Priority Setting

Several lessons emerge from ex ante/priority-setting studies, including those completed in Africa.

- First, scoring is the method most commonly used for structured research priority setting, but it is also the easiest to abuse. It is popular because it facilitates a process in which stakeholders are involved in discussions and in weighting and ranking programs. Scoring is easy to abuse because, unless careful attention is devoted to defining research-system objectives, measuring impacts against those objectives, and deciding whose weights count, the results can be an odd ranking based on weighting apples against oranges.
- Second, the impacts of some types of research (such as policy research) on some objectives (such as environmental objectives) are inherently more difficult to measure than others, and yet must be, if all programs are to be compared.
- Third, ex ante impact assessment of major programs is useful, but structured priority-setting analysis can be costly and may best be reserved for strategic planning, possibly every five years. Mutangadura and Norton (1999) is a good example of a study that would be difficult to duplicate every year, because it involved hundreds of impact assessments for multiple commodities, types of research, regions, and farm types. Nevertheless, such studies can be very useful for periodic decisionmaking.

- Fourth, the presence of both geographical spillovers in agricultural research and regional research programs means that donor priorities often need to be set at levels higher than for a specific country, and regional priorities must be reconciled with country-level priorities.
- Fifth, priority setting requires specific economic analysis skills, which must be institutionalized or purchased when a study is conducted. Priority setting at CGIAR centers often draws on internal capacity for analysis, but country-level studies often involve outside consultants when major priority-setting analyses are completed. Internal ex ante and ex post impact assessment capacity needs improvement in many research systems.

Where feasible, a useful mechanism for setting agricultural priorities is to project the NPV of research benefits by program, making use of economic surplus and geographic information system (GIS) tools. In some cases, trade-offs in meeting efficiency versus nonefficiency objectives can be assessed in this framework. However, the increased emphasis over time on project-specific as opposed to program-core funding has reduced the use of this approach for priority setting, although the techniques are still useful for market-level economic evaluations of some projects. Projecting impact pathways for technologies and policy changes is useful in almost all cases. For programs and objectives for which economic surplus is difficult to measure, alternative measures—even if in physical terms—can be used, and then the economic value sacrificed can be projected for programs that shrink when others expand.

Building Capacity for Impact Assessment

Impact assessment requires economic evaluation skills, an ability for evaluators to work jointly with biological scientists and understand the basics of the research being evaluated, financial resources for the evaluations, and a solid plan for collecting and managing data. Many evaluations in NARS and in international agricultural research centers fail on the last point. Few systems systematically collect and store the necessary data for either ex ante or ex post analysis. The first step in a workable data system is to maintain a close working relationship between the evaluators and other scientists, so that the appropriate data are properly collected.

Conclusion

The number of ex post impact assessments of agricultural research has grown over time. In addition to economic gains from productivity growth, the types of benefits being assessed have broadened in the past decade to include nutritional improvement, poverty reduction, and NRM. Estimated returns to agricultural research in Africa—often in the range of 20–40 percent for rates of return, or 8–18 percent if calculated with a modified IRR formula—suggest that research is a high-payoff development intervention, even if the total number of evaluations is relatively small. Many research benefits accrue to consumers as a result of the (lower) output price effects associated with the additional production that results from the new technologies. Increasing evidence is also confirming the substantial poverty-reducing and nutritional benefits of agricultural research.

Pressure is building to improve the credibility of agricultural research evaluations to provide solid evidence of the effectiveness of alternative investments (such as in health and education). As a result, increased attention has been devoted over time in research evaluations to identifying a causal relationship between agricultural research and outcomes. With observational data, this identification relies on statistical techniques, often with untestable underlying assumptions. As a result, a relatively small percentage of all agricultural research programs has been subjected to rigorous economic impact evaluation. Evaluation methods that are popular for other types of interventions, such as RCTs, hold less promise for agricultural research because of the complexity and simultaneous nature of many research interventions, the lag lengths involved with agricultural research, and the need to conduct on-farm testing on the plots of willing (and, hence, nonrandom) farmers. Nevertheless, once new agricultural practices are in the diffusion process, they may be amenable to RCTs.

Because of the continual need for impact evaluations, agricultural research institutions may employ in-house evaluation experts. Such experts may be involved at all stages of the research process, or they may conduct evaluations when the need arises without prior involvement in the research. Use of such experts can increase the odds that the evaluators will understand the research being evaluated, may improve the quality of the data collected by the research system, and may facilitate midstream adjustments to a research project when the need arises. However, in-house experts also may become captive to the interests of the scientists or administrators—or at least it may appear that way—reducing the credibility of the impact analysis. While the use of outside experts can reduce this problem, it may be at the expense of a full understanding of the research or the possibility of timely midstream redirection. A combination of both internal and external impact evaluation may be the solution for some programs.

Ex ante economic evaluations of component technologies are as common as or more common than ex post evaluations; however, few of them are used for systemwide research priority setting. Priority-setting exercises can be expensive for agricultural research systems, in terms of both time and money; hence, they are most useful for strategic planning every few years. Economic surplus analysis, GIS, and trade-off analysis are perhaps the preferred methods for such priority setting.

Many, if not most, evaluations of agricultural research and development do not include quantitative impact assessment. Instead, they involve qualitative evaluations of program impacts or assessments of whether research programs accomplish their objectives. As assessment methods improve, and as funding sources demand more quantitative evidence of impact, the number and quality of impact assessments of agricultural research are likely to increase.

References

Ajayi, O., F. Place, F. Kwesga, and P. Mafongoya. 2007. "Impacts of Improved Tree Fallow Technology in Zambia." In *The Impact of Natural Resource Management Research: Studies from the CGIAR,"* edited by H. Waibel and D. Zilberman. Wallingford, UK: CABI Publishing.

Alene, A., A. Menkir, S. Ajala, B. Badu-Apraku, A. Olanrewaju, V. Manyong, and A. Ndiaye. 2009. "The Economic and Poverty Impacts of Maize Research in West and Central Africa." *Agricultural Economics* 40: 535–550.

Alston, J., M. Andersen, J. James, and P. Pardey. 2011. "The Economic Returns to U.S. Public Agricultural Research." *American Journal of Agricultural Economics* 93 (3): 1257–1277.

Alston, J., C. Chan-Kang, M. Marra, P. Pardey, and T. Wyatt. 2000b. *A Meta-Analysis of Rates of Return to Agricultural R&D*. Research Report 113. Washington DC: International Food Policy Research Institute.

Alston, J., M. Marra, P. Pardey, and T. Wyatt. 2000a. "Research Returns Redux: A Meta-Analysis of the Returns to Agricultural R&D." *Australian Journal of Agricultural and Resource Economics* 44 (2): 185–215.

Alston, J., G. Norton, and P. Pardey. 1995. *Science Under Scarcity: Principles and Practice for Agricultural Research Evaluation and Priority Setting*. Ithaca, NY: Cornell University Press.

Alwang, J. 2015. "Implications for Monitoring Progress and Assessing Impacts." Chapter 21 in *Crop Improvement, Adoption and Impact of Improved Varieties in Food Crops in Sub-Saharan Africa*, edited by T. Walker and J. Alwang. Wallingford, UK: CABI Publishing.

Bennett, J., and E. Birol. 2010. *Choice Experiments in Developing Countries*. Northhampton, MA: Edward Elgar Publishers.

Bonabona-Wabbi, J., D. Taylor, and G. Norton. 2014. "Willingness to Pay to Avoid Consumption of Pesticide Residues in Uganda: An Experimental Approach." *African Journal of Agricultural and Resource Economics* 9 (3): 226–238.

Ceesay, S., E. Gilbert, B. Mills, J. Rowe, and G. Norton. 1989. *Analysis of Agricultural Research Priorities in The Gambia.* Project Paper 16. Canberra: Australian Centre for International Agricultural Research; The Hague: International Service for National Agricultural Research.

Coulibaly, O., V. Manyong, S. Yaninek, R. Hanna, P. Sanginga, D. Endamana, A. Adesina, M. Toko, and P. Neuenschwander. 2004. *Economic Impact Assessment of Classical Biological Control of Cassava Green Mite in West Africa.* Cotonou, Benin: International Institute for Tropical Agriculture.

Dalton, T., I. Yahaya, and J. Nabb. 2014. "Perceptions and Performance of Conservation Agriculture Practices in Northwestern Ghana." *Agriculture, Ecosystems, and Environment* 187: 65–71.

Datta, S., and S. Mullainathan. 2012. *Behavioral Design: A New Approach to Development Policy.* Center for Global Development Policy Paper 16. Washington, DC: Center for Global Development Policy.

Debass, T. 2001. "Economic Impact Assessment of IPM CRSP Activities in Bangladesh and Uganda: A GIS Application." MSc thesis, Virginia Tech, Blacksburg, VA.

Diagne, A., P. Kormawa, O. Youm, S. Keya, and S. N'cho. 2009. "Priority Assessment for Rice Research in Sub-Saharan Africa." Chapter 8 in *Prioritizing Agricultural Research for Development*, edited by D. Raitzer and G. Norton. Wallingford, UK: CABI Publishing.

Duflo, E., M. Kremer, and J. Robinson. 2008. "How High Are Rates of Return to Fertilizer? Evidence from Field Experiments in Kenya." *American Economic Review Papers and Proceedings* 98 (2): 482–488.

Evenson, R., and D. Gollin, eds. 2003. *Crop Variety Improvement and Its Effects on Productivity.* Wallingford, UK: CABI Publishing.

Feder, G., R. Murgai, and J. Quizon. 2004. "Sending Farmers Back to School: The Impact of Farmer Field Schools in Indonesia." *Review of Agricultural Economics* 26: 45–62.

Fuglie, K. and J. Marder. 2015. "The Diffusion and Impact of Improved Food Crop Varieties in Sub-Saharan Africa." Chapter 17 in *Crop Improvement, Adoption and Impact of Improved Varieties in Food Crops in Sub-Saharan Africa*, edited by T. Walker and J. Alwang. Wallingford, UK: CABI Publishing.

Gollin, D. 2006. *Impacts of International Research on Inter-temporal Yield Stability in Wheat and Maize: An Economic Assessment.* Mexico City: International Maize and Wheat Improvement Center (CIMMYT).

Gunaratna, N., H. De Groote, P. Nestel, K. Pixley, and G. McCabe. 2010. "A Meta-Analysis of Community-Based Studies on Quality Protein Maize." *Food Policy* 35 (3): 202–210.

Hareau, G., G. Norton, B. Mills, and E. Peterson. 2005. "Potential Benefits of Transgenic Rice in Asia: A General Equilibrium Analysis." *Quarterly Journal of International Agriculture* 44: 229–246.

Imbens, G., and J. Wooldridge. 2009. "Recent Developments in the Econometrics of Program Evaluation." *Journal of Economic Literature* 47 (1): 5-86.

Kalyebara, R., D. Andima, R. Kirkby, and R. Buruchara. 2008. *Improved Bean Varieties and Cultivation Practices in Eastern-Central Africa: Economic and Social Benefits.* Cali, Colombia: Centro Internacional de Agricultura Tropical.

KARI (Kenya Agricultural Research Institute). 1991. *Kenya's Agricultural Research Priorities to the Year 2000.* Nairobi.

Kostandini, G., B. Mills, and E. Mykerezi. 2011. "Ex Ante Evaluation of Drought-Tolerant Varieties in Eastern and Central Africa." *Journal of Agricultural Economics* 62 (1): 172 –206.

Larochelle, C., J. Alwang, G. Norton, E. Katungi, and R. Labarta. 2015. "Impacts of Adopting Improved Bean Varieties on Poverty and Food Security in Uganda and Rwanda." Chapter 16 in *Crop Improvement, Adoption and Impact of Improved Varieties in Food Crops in Sub-Saharan Africa,* edited by T. Walker and J. Alwang. Wallingford, UK: CABI Publishing.

Low, J., N. Osman, M. Arimond, F. Zano, B. Cunguara, and D. Tschirley. 2007. "A Food-Based Approach Introducing Orange-Fleshed Sweet Potatoes, Increased Vitamin A Intake, and Serum Retinol Concentrations in Young Children in Rural Mozambique." *Journal of Nutrition* 137: 1320–1327.

Manyong, V., D. Sanogo, and A. Alene. 2009. "The International Institute of Tropical Agriculture's Experience in Priority Assessment of Agricultural Research." Chapter 4 in *Prioritizing Agricultural Research for Development,* edited by D. Raitzer and G. Norton. Wallingford, UK: CABI Publishing.

Maredia, M., and D. Raitzer. 2006. *CGIAR and NARS Partner Research in Sub-Saharan Africa: Evidence of Impact to Date.* Rome: Science Council Secretariat.

Meenakshi, J., N. Johnson, V. Manyong, H. De Groote, J. Javelosa, D. Yanggen, et al. 2007. *How Cost-Effective Is Biofortification in Combating Micronutrient Malnutrition? An Ex Ante Assessment.* HarvestPlus Working Paper 2. Washington, DC: International Food Policy Research Institute.

Meenakshi, J., N. Johnson, V. Manyong, H. De Groote, J. Javelosa, D. Yanggen, et al. 2010. "How Cost-Effective Is Biofortification in Combating Micronutrient Malnutrition? An Ex Ante Assessment." *World Development* 38 (1): 64–75.

Mills, B. 1998. *Agricultural Research Priority Setting: Information Investments for the Improved Use of Research Resources*. The Hague: International Service for National Agricultural Research; Nairobi: Kenya Agricultural Research Institute.

Moyo, S., G. Norton, J. Alwang, I. Rhinehart, and C. Deom. 2007. "Peanut Research and Poverty Reduction: Impacts of Variety Improvement to Control Peanut Viruses in Uganda." *American Journal of Agricultural Economics* 89 (2): 448–460.

Mutangadura, G., and G. Norton. 1999. "Agricultural Research Priority Setting under Multiple Objectives: An Example from Zimbabwe." *Agricultural Economics* 20: 277–286.

Nguema, A., G. Norton, M. Fregene, R. Sayre, and M. Manary. 2011. "Expected Economic Benefits of Meeting Nutritional Needs through Biofortified Cassava in Nigeria and Kenya." *African Journal of Agricultural and Resource Economics* 6 (1): 70–86.

Nkala, P., M. Mango, and P. Zikhali. 2011. "Conservation Agriculture and Livelihoods of Smallholder Farmers in Central Mozambique." *Journal of Sustainable Agriculture* 35: 757–779.

Norton, G., and J. Alwang. 2012. "Frontiers in Impact Assessment of Agricultural Development Interventions." International Track session paper presented at the annual meeting of the American Applied Economics Association, Seattle, Washington, August 8.

Norton, G., and E. Phillips. 2011. *Potential Benefits of the Integrated Breeding Platform for Wheat, Sorghum, Cassava, and Rice Report to the Generation Challenge Program*. Mexico City: International Maize and Wheat Improvement Center (CIMMYT).

Nouhoheflin, T. 2010. "Assessing the Economic Impacts of Tomato Integrated Pest Management in Mali and Senegal." MSc thesis, Virginia Tech, Blacksburg, VA.

Pray, C., D. Gisselquist, and L. Nagarajan. 2011. *Private Investment in Agricultural Research and Technology Transfer in Africa*. Conference Working Paper 13 from the ASTI/IFPRI–FARA Conference "Agricultural R&D: Investing in Africa's Future: Trends, Challenges, and Opportunities." Accessed March 2013. www.asti.cgiar.org/pdf/conference/Theme4/Pray.pdf.

Rao, X., T. Hurley, and P. Pardey. 2012. *Recalibrating the Reported Rates of Return to Food and Agricultural R&D*. Staff Paper P12-8. St. Paul: University of Minnesota, Department of Applied Economics.

Regier, G., T. Dalton, and J. Williams. 2012. "Impact of Genetically Modified Maize on Smallholder Risk in South Africa." Paper presented at the 16th Conference of the International Consortium on Applied Bioeconomy Research, Ravello, Italy, June 24–27.

Renkow, M., and D. Byerlee. 2010. "The Impacts of CGIAR Research: A Review of Recent Evidence." *Food Policy* 35: 391–402.

Ricker-Gilbert, J., G. Norton, J. Alwang, M. Miah, and G. Feder. 2008. "Cost-Effectiveness of Alternative IPM Extension Methods: An Example from Bangladesh." *Review of Agricultural Economics* 30: 252–269.

Rudi, N., G. Norton, J. Alwang, G. Asumugha. 2010. "Economic Impact Analysis of Marker-Assisted Breeding for Resistance to Pests and Post-Harvest Deterioration in Cassava." *African Journal of Agricultural and Resource Economics* 4 (2): 110–122.

Rusike, J., N. Mahungu, S. Jumbo, V. Sandifolo, and G. Malindi. 2010. "Estimating Impact of Cassava Research for Development Approach on Productivity, Uptake and Food Security." *Food Policy* 35: 98–111.

Teri, J., S. Mugogo, and G. Norton. 1990. *Analysis of Agricultural Research Priorities in Tanzania.* The Hague: International Service for National Agricultural Research.

Thornton, P., T. Randolph, P. Kristjanson, S. Omamo, and J. Ryan. 2000. *Priority Assessment for the International Livestock Research Institute, 2000–2010.* Impact Series 6. Nairobi, Kenya: International Livestock Research Institute.

Vaiknoras, K., G. Norton, and J. Alwang. 2015. "Farmer Preferences for Attributes of Conservation Agriculture in Eastern Uganda." *African Journal of Agricultural and Resource Economics.* June: 158–168.

Walker, T., and J. Alwang. 2015. *Crop Improvement, Adoption and Impact of Improved Varieties in Food Crops in Sub-Saharan Africa.* Wallingford, UK: CABI Publishing.

Wood, S., and J. Anderson. 2009. "Strategic Priorities for Agricultural Development in Eastern and Central Africa: A Review of the Institutional Context and Methodological Approach for Understanding a Quantitative, Sub-Regional Assessment." Chapter 11 in *Prioritizing Agricultural Research for Development*, edited by D. Raitzer and G. Norton. Wallingford, UK: CABI Publishing.

Zeddies, J., R. Schaab, P. Neuenschwander, and H. Herren. 2001. "Economics of Biological Control of Cassava Mealybug in Africa." *Agricultural Economics* 24 (2): 209–219.

Zeng, D., J. Alwang, G. Norton, B. Shiferaw, M. Jaleta, and C. Yirga. 2015. "Ex Post Impacts of Improved Maize Varieties on Poverty in Rural Ethiopia." *Agricultural Economics* 46: 158–173.

Chapter 12

MONITORING AND EVALUATION TO STRENGTHEN THE PERFORMANCE OF AGRICULTURAL R&D

Howard Elliott and John Lynam

Over the past decade, donors have made a concerted effort to improve aid effectiveness. This intent was codified by the Development Assistance Committee of the Organisation for Economic Co-operation and Development in the Paris Declaration of 2005, under which development agencies agreed to manage their aid budgets on the basis of five principles, two of which included managing development funds for results and mutual accountability (Chapter 6, this volume). The focus on results-based management and accountability further emphasized evaluation methods and monitoring systems in contractual arrangements associated with the investment of international public funds in Africa south of the Sahara (SSA), including investment in such public goods as agricultural research. The focus on performance monitoring in agricultural projects was reinforced by the entry of the Bill & Melinda Gates Foundation into agricultural development in SSA and by the integration of the private sector's "bottom line" into project development and results frameworks.

The emphasis on accountability within the timeframe of project investment shifted the focus in performance monitoring from ex post impact assessment to monitoring and evaluation (M&E) methods, and thus from economic methods to methodologies coming from the expanding disciplinary field of evaluation. This disciplinary differentiation created boundaries between M&E frameworks incorporated in donor projects, where the focus was direct benefits and project targets and impact assessment associated with measuring both first- and second-order economic benefits, such as real income gains to consumers and producers, environmental protection, and institutional learning (Chapter 11, this volume).

Because most donors, except the World Bank, were not funding national agricultural research capacity and development projects, the focus on developing M&E methods for the relatively new area of agricultural research

The authors are grateful to Leonard Oruko for collegial comments on draft versions of this chapter.

shifted donor investments to Africa's subregional organizations (SROs), the Comprehensive Africa Agriculture Development Programme (CAADP), and CGIAR. This is where M&E capacity was principally built, with a resulting shift from incorporating M&E into projects, to developing M&E frameworks for institutions. Developing M&E frameworks to improve institutional management and learning, as opposed purely to accountability, was a natural outcome of this shift. M&E approaches have always been a compromise between the needs of the funders to know "Did it work?" versus the needs of the implementers to learn "How did it work?" This is the balance between accountability and learning.[1]

Donor support to agricultural research in SSA essentially comes from development budgets. Over time, the overall development objectives against which investments are measured have expanded; this applies equally to agricultural research. The initial focus on efficiency gains in smallholder farming systems that resulted in the large production increases under the Green Revolution has been systematically extended over the past several decades to include reduced rural poverty; sustainable natural resource management (NRM); and, most recently, improved child and maternal nutrition. Research planning and the design of M&E systems become much more complex with this increasing array of objectives. Holding agricultural research accountable for these development objectives at the same time that the timeframe to demonstrate results has telescoped in the shift from ex post impact assessment to results-based management has significantly moved agricultural research toward the development end of the R&D spectrum. This shift has moved research off station; has expanded the scope of research into postharvest, marketing, and support services; and has put a premium on institutional partnerships (Chapter 13, this volume). In the process, this has put more emphasis on learning.

This chapter begins with a discussion of the performance monitoring challenge in African agricultural research with respect to monitoring change in smallholder agricultural systems, the time lag in agricultural research, the context for technology adoption, and causality in agricultural research impact pathways. Next, four M&E cases illustrate the need to balance the objectives

---

1 Some of this balance is achieved when internal project-level monitoring processes track moves "along the crawl space" of projects, while the program or institutional processes track higher-order moves along the impact pathway, which does not exclude rigorous evaluation, such as randomized control trials. Structured experiential learning and real-time performance data with internal feedback to decisionmaking can use within-project variations in design as their own counterfactual, and thus reduce the incremental cost of evaluation (Pritchett, Samji, and Hammer 2013).

of accountability and learning made possible by M&E: (1) monitoring inputs versus outputs, (2) monitoring research process versus outputs, (3) monitoring research outputs versus broader innovative processes, and (4) monitoring the research process versus the development process. Thereafter follow discussions of the evolution of evaluation tools, the most-used logic models and their perceived strengths in dealing with the impact pathways from research outputs to development outcomes, and evaluation methodology and the efficacy of pilot studies versus formal evaluation designs. The chapter concludes with the search for accountability in investment in agricultural research and development (R&D).

The Challenge of Performance Monitoring in African Agricultural Research

Monitoring change in African agricultural systems is both complicated and data intensive because of the heterogeneity of agriculture in Africa, which creates market inefficiencies and increases the costs of adaptive research and extension. The InterAcademy Council (IAC) (2004) recognized this heterogeneity, noting that Africa would need thousands of mini-revolutions rather than a single Green Revolution. While the IAC identified four predominant systems that, in aggregate, were the likely levers for growth, they were in fact composed of many smaller systems operating within local constraints of markets and policies, reflecting the "small-country problem" in SSA. The small scale and heterogeneity of "development domains" (areas characterized by similar agroecologies, population densities, and market access), and the inherent weakness of public services, complicate planning of agricultural research and thus the M&E processes that support it.[2] Impact pathways were less predictable, and monitoring change in farming and market systems is commensurately more labor and data intensive than in the more homogeneous Green Revolution Asia.

Research has raised agricultural productivity in SSA, but at a rate well below that of other developing countries (Chapters 2 and 3, this volume). To achieve the 6 percent agricultural growth rates postulated by CAADP, Africa would have to achieve productivity-based growth of 3 percent, matching rates achieved by Brazil and China, which have both invested more heavily in

2 The International Food Policy Research Institute's addition of population and market factors to the traditional crop "recommendation domain" highlights the importance of public services and the complexity of planning agricultural research (Johnson et al. 2011).

research and rural services than Africa has. Even with doubling investment in research, ending discriminatory policies against agriculture, and doubling irrigated areas, individual countries would still encounter the small-country problem of high costs of policy and regulatory change to serve small development domains. Cooperation with neighbors, organizing technological spillovers, and building adaptive research capacity will be critical in improving the effectiveness of agricultural R&D (Chapter 14, this volume).

The lag between investment in agricultural research and actual impact on farm productivity is particularly difficult to specify. Pardey and Beddow (2013) argue that, for the United States, it is the accumulation of research results over the long run that accounts for differences in observed productivity among states: scientists build on the work of previous research going back much longer than the 10–12 years commonly estimated for the development and release of a crop variety, and up to 50 percent of the benefits can be attributed to spill-ins from outside. Crop varietal technology timelines illustrate the lengthy innovation process typical of agriculture, and the complementary roles of the public and private sectors. Moreover, the timeline between research discovery and commercial interest is longer than is often assumed. In developed countries, a large share of basic research is conducted in universities and public research institutes, whereas the private sector is predominantly investing in developmental research. In Africa, the public sector—made up of both national and international research agencies—must undertake the whole research chain, from basic to developmental research, which in more developed countries is largely in the hands of the private sector.

The heterogeneity of African agricultural systems, the large range of commodities in these systems, and the relatively small development domains involved often overwhelm limited research budgets. In Africa in the past two decades, the relative weakness of the private sector and underfunded public systems define three needs: (1) to catch up in the area of basic science that underpins key commodities, (2) to build capacity to access knowledge widely, and (3) to create linkages with the wider innovation system.

What Gets Monitored?

Project Versus Institutional Funding

A recurring preference of donors for project rather than long-term program or institutional funding has concerned analysts since the early 1980s (Ruttan 1982). The advantages of long-term program or institutional support include lower transaction costs, the ability to plan and complete long-term path-breaking projects, and the opportunity to develop human and institutional capacity around challenges. On the other hand, contracts, short-term programs, and competitive grants create flexibility to change programs (not just at the margin) and assert individual donor interests. Such approaches require special attention to the continuity of the scientific core and to capacity building, which are not secured by other budgets.

The original CGIAR, with core support directed to addressing major global challenges, typified the type of institutional innovation that could bring results. The erosion of the core-funded model, with its periodic External Program and Management Reviews, gave way to a proliferation of project-based activities, new initiatives, and enabling structures and mechanisms in which individual donors carried out their separate reviews, often as subcomponents of larger programs. This situation meant that donors ended up doing their own M&E by research project and theme, with donor-specific criteria. The resulting burden on national systems was recognized in the Paris Declaration on Aid Effectiveness (OECD 2005), which recommended greater coordination among donors and the provision of budgetary support, rather than project aid.

At the regional level, agricultural productivity programs, such as the East and West African agricultural productivity programs (EAAPP and WAAPP, respectively), were created in the World Bank under multidonor trust funds that provided loans to participating countries for a mixture of national and regional activities. SROs became involved, essentially managing competitive grant mechanisms. Given the importance of CGIAR in regional networks and the new intention to align with African agendas, M&E tools at all levels have been rapidly sought to structure a results framework around consensus on an African agricultural development agenda.

Accountability and Learning in M&E

The common belief that accountability and learning are necessarily competing objectives is disputed by Guijt (2010). However, project formats usually require accountability statements about predefined goals, and a specification of the activities and interim results that lead to their achievement (for example, in "results-based management"). Theories of change and the learning that is embedded in them can become overly rigid, as project contracts often require proof of deliverables and set milestones that negate the adaptive management that is so essential in research programs. Often there is a fundamental disconnect between the rhetoric about the need for learning and the reality of procedures that funding agencies require. Whether it occurs because of an inability to deal with complex realities, a preference for the easily measurable over more difficult-to-measure objectives, or human and institutional capacity constraints, the tendency for accountability to prevail over learning is commonly acknowledged.

- Learning is necessary for practical improvement, strategic adjustment, and rethinking driving values. Nongovernment organizations (NGOs) are often in the lead in promoting learning through evaluation. ActionAID International (2011) is one among several NGOs that seeks "triple loop learning": (1) Are we doing things well? (2) Are we doing the right things? (3) Were our assumptions right? There needs to be clarity about who is expected to learn, for what purpose, and at what level the learning is aimed, which is then incorporated into the design of the M&E system.
- Accountability has many forms: to managers or funders (upward accountability), to civil society (social accountability), to partners (mutual accountability), and to researchers themselves (strategic accountability, in the sense of "Did we act as effectively as possible?").

Guijt (2010) argues that M&E must adapt expectations of accountability and learning to contextual characteristics. In ordered situations, cause and effect are clear and primarily linear, and impact pathways can be analyzed within a results-oriented system. In complex situations, such as research on smallholder systems in Africa, learning is needed to recognize emerging patterns, to respond, and to adapt. For example, the attempts to define an impact pathway for agricultural approaches to health and nutrition outcomes have identified multiple interactions, in part that derive from three separate goals, each potentially requiring its own dedicated policies, instruments, and intervention strategies. This recognizes not only that agricultural research is a "blunt instrument" for achieving policy goals, but also that in adding a goal, such as nutrition, a new instrument is needed. The skills required for compliance with donors, for strategic accountability, and for learning are convergent, so capacity building and the design of M&E systems should go beyond compliance to ensure that all three needs are met.

Causality in Impact Pathways

The evolution of evaluation approaches in research is associated largely with CGIAR, particularly during a period when it was challenged to prove its contribution to development goals, and when donors were withdrawing their

FIGURE 12.1 Types of assessments and evaluations on R4D results chains

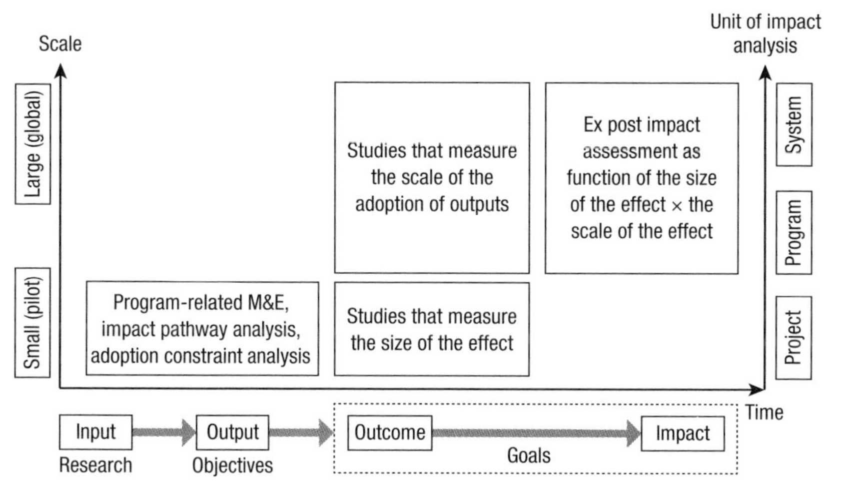

Source: Maredia (2013).
Note: M&E = monitoring and evaluation.

support for research at the national level. CGIAR's Standing Panel on Impact Assessment (SPIA) provides a useful classification of different types of assessments and evaluations on research for development (Figure 12.1).

Until CGIAR reform in 2008, evaluation of investment in agricultural research in CGIAR centers was based almost wholly on periodic external reviews in the form of ex post Impact Assessment (epIA). Such assessments were based on where research outputs were known to have been widely adopted and had impacts. These impacts were modeled through economic surplus methods that depended on translating cost reductions and production increases at the farm level into market impacts on price and the distribution of net benefits between consumers and producers. Only those parts of the research portfolio that had demonstrated impact were assessed, and epIA gave an estimate of the returns to investment on that particular line of research. Such rate-of-return studies had little to say about the market and institutional context within which the technology diffused, how the technology was disseminated, the effect of the technology on ecosystem services, or the potential for reproducing such impacts under other contexts.

In sum, there was little potential for deeper learning about causality within the impact pathway from such epIAs. Moreover, by the time of the impact study, the research program had often moved on to other research lines. Such assessments essentially provided an ex post measure of the returns on investment, but they provided little information on how investments in agricultural research could best achieve the development objectives that framed donor funding strategies. It was at this point that the expanding field of evaluation, which had been applied widely in the health and education sectors and was being applied internally in donor programs, was integrated into agricultural research programs.

The demands from investors in international public goods, such as agricultural research, were that aid recipients must demonstrate contributions to development objectives from their funding. Since investments in agricultural research had a longer time lag than most other investment, it would be necessary to monitor more immediate outcomes that could be causally linked to these broader development objectives—what became known within CGIAR as "intermediate development objectives." For agricultural research in Africa, this has had a number of effects.

- First, research programs have had to be much more explicit about their impact pathways and the chain of causality between the production of research outputs, their dissemination, and the impacts on a possible range of intermediate outcomes from gender equality to enhanced ecosystem services.

- Second, in an African context, this has expanded the scope of research from a sole focus on productivity-enhancing technologies, to research on improved market efficiency and smallholder access to markets, innovations in crop or livestock insurance, more cost-effective dissemination methods using information and communications technologies, or farmers' organizations that improve access to capital and technologies by the rural poor or women in rural households. Such research in SSA recognizes that innovations are needed in market, institutional, and farmers' organizations to support adoption and effective use of more traditional production technologies.

Finally, enhanced accountability for impact has necessitated the development of more effective partnerships with other actors in the agricultural sector in order to achieve broad-scale development outcomes, which in turn has increased the focus on research within a larger agricultural innovation system (Chapter 13, this volume). M&E systems are developed on the basis of such increased understanding of impact pathways. Moreover, Maredia (2013) notes that in moving down the impact pathway, the assessment of results in terms of intermediate development outcomes becomes less a case of analyzing "causal attribution" and more one of analyzing "causal contribution."

What Needs to Be Monitored?

Impact pathways for agricultural research, especially in an African context, are highly context dependent, have long timeframes, and require functional organizational partnerships where many institutions have relatively weak capacity. With the expansion of agricultural research into institutions, policy, land use, and the environment, there is an expanding range of processes that research institutes develop and deploy to meet development objectives. This expansion results in significant complexity in designing monitoring systems that meet accountability needs along the impact pathway, but at the same time provide information for assessment, program adaptation, and learning. Such monitoring systems for agricultural research may be conceived in terms of different stages or levels. The discussion that follows is only suggestive of the design issues in developing such a monitoring system.

Monitoring Inputs Versus Outputs: The Case of Plant Breeding

Agricultural research is in many ways a design process that combines "engineering" methods with scientific creativity. This qualitative and uncertain nature of agricultural research has made it difficult to measure the performance of the agricultural research process itself. Much of the focus in such monitoring systems has been on the inputs into the process. Agricultural Science and Technology Indicators data collection is a very good and representative example of such monitoring, as the data collection focuses on primary inputs of scientific personnel, operating funds, capital investments, research infrastructure, and training.

What has been more difficult is to relate such inputs, including organizational alternatives, to the performance of research institutes and the overall efficiency of the research process. The clearest performance indicator has been publications, as measured in terms of journal impact and the number of citations. However, knowledge production does not easily translate into relevant technologies and management systems, nor does it necessarily translate into impact on intermediate development outcomes. In fact, many development agencies suggest a trade-off in publishable research and what is increasingly termed agricultural research for development. The problem thus remains how to monitor the research process itself as the first stage in an impact pathway.

Plant breeding would be a tractable example of monitoring at this stage. Numerous design choices go into a breeding program, particularly the number of breeding projects (the number of agroecologies that require separate adaptation), the number of traits, whether populations or elite lines are to be developed, the breeding methodology employed (for example, recurrent selection versus broadening parents in the crossing program), whether to employ marker-assisted selection (as is increasingly becoming the case), and the number and location of testing sites. All of these choices have cost and resource implications and essentially determine the inputs that go into a breeding program. The question becomes whether the program could be more cost-effective and in relation to what performance criteria.

Optimally, the performance criterion would be the number of adoptable varieties; however, there is a significant time lag between investment decisions in breeding programs and a sufficient understanding of actual farmer adoption, so more intermediate performance criteria are needed. These could include the number of varieties released under national varietal release regulations, the results in the national yield performance trials, the results in the multilocational variety trials, or trait performance in the selection process. All of these factors are used in varietal evaluation (with increased precision from the first to the last listed), but with increasing time lags in terms of feedback into breeding program design and adaptation. Developing a monitoring system involves inherent trade-offs among precision of outcomes, time lags, data costs, and the need to make adjustments in the breeding program—that is, to answer the question of whether the program is performing well.

Monitoring Process Versus Outputs

To deal with the lag problem in adaptive management of breeding programs, innovations have been in process. Participatory breeding and participatory varietal selection have been designed to telescope the lag time between varietal development and farmer evaluation. The potential trade-off has been in how locally adapted such varieties will be, which may be overcome by incorporation into multilocational trials. To a significant extent, the results of the process provide the outcome indicator, and the focus is on making the process as cost-effective as possible. In this way, monitoring shifts from a focus on outputs to more of a focus on process variables; at the same time, it extends the research to a closer interface with development agents and outcomes—what is more broadly termed "client-oriented research." Research moves primarily on farm, which in turn requires a specification of context and representativeness, particularly given the heterogeneity of African farming systems. In this regard, the research moves to another stage in the impact pathway, with significant differences in how the monitoring process is designed.

Monitoring Research Outputs versus Innovation Process

As previously discussed, there has been an evolution of systems approaches to the changing objectives of national systems and funding agencies, the need for scientists to design results to fit in wider knowledge systems, and the need for institutions to work together in a broader innovation systems framework (Elliott 2008; Chapter 13, this volume). The convergence of policy, science, and institutional perspectives on innovation systems evolved as follows:

1. **Policy and Systems Perspective**
   National agricultural research institutes →
   National agricultural research systems →
   Agricultural knowledge and information systems →
   Agricultural productivity programs →
   National agricultural innovation systems
2. **Science Perspective**
   Crop improvement →
   Cropping systems →
   Natural resource management →
   Climate change/sustainable intensification
3. **Institutions and Change**
   Client-oriented research →
   Farmer-to-farmer dissemination →
   Demand-driven advisory services →
   Integrated agricultural research for development/innovation platforms

Evaluation systems have grown up around benchmarking systems and measuring these evolutions (Spielman and Birner 2008; Spielman and Kelemework 2009a, 2009b; Ragasa, Abdullahi, and Essegbey 2011). The target for such analysis would be institute directors and ministers of agriculture who wonder whether they are doing the right things, whether they are making the correct assumptions, and whether they are linking to the rest of the national agricultural innovation system.

Monitoring the Research Process versus the Development Process

The emphasis on development outcomes has fundamentally shifted the monitoring of agricultural research, driven by the introduction of "results frameworks." The research process is expected to identify scenarios of progress along impact pathways leading to development outcomes. Since national agricultural research systems (NARSs) and CGIAR are both dependent on donors (and their funding comes from development budgets, rather than budgets for scientific cooperation), it is understandable that the agenda has moved toward the development end of the research-for-development spectrum. The challenge for research is to respond to the concerns of development (and "speak their language") without abandoning the primary role of researchers. This is discussed below in the context of designing results frameworks and defining intermediate development outcomes along the impact pathway.

The move to "results-based management" (RBM) has been led by donors who have to reassure their funders, policymakers, or constituents that public money is being spent responsibly. RBM shifts the justification of the investment in agricultural research from ex post impact evaluation to direct monitoring of the results emanating from the investment. For donors, these results should be measured in terms of impact on development outcomes. For agricultural R&D, the difficulty is the timeframe over which such results are expected to be generated. Moreover, the evaluation tools that have traditionally been used, such as logical framework ("logframe") analysis, outcomes mapping, and participatory impact pathway analysis, have been designed, essentially, for development programs. The approaches all involve some defined investment or intervention that is expected to yield gains to some target group, while identifying the policy and institutional assumptions under which that gain can be achieved with some stated probability of success. Most critically, within RBM, if a constraint is binding and there is no budgetary or program plan within the original project to change it, the project must include within itself the means of removing the constraint.

RBM systems are being instituted in most agricultural research organizations that depend on donor funding, particularly the Alliance for a Green Revolution in Africa, the Forum for Agricultural Research in Africa (FARA), the SROs, and CGIAR. CGIAR is often a vehicle through which research breakthroughs come into practice in the NARSs (Chapter 3, this volume). CGIAR's chief executive officer recently noted (CGIAR 2013b):

> Starting today, we are presenting and discussing the Theories of Change, Impact Pathways and Intermediated Development Outcomes for the CGIAR Research Programs (CRPs), with our donors and our partners. What we are talking about here are clear, development outcomes that have been reached by consensus. The Programs, together with their partners, can be held accountable for delivering these outcomes. Program level outcomes tied to outcomes at CGIAR System

> Level . . . are in turn linked to the new Sustainable Development Goals. Together with solid monitoring and reporting, these outcomes constitute the foundation, the building blocks for the CGIAR's results based performance management system.

The Essentials of a Results-Based Management System

An RBM system must be designed for a specific user (or partnership) working within a described context if it is to serve both learning and accountability objectives. However, many elements are common to an RBM, as is illustrated by CGIAR's template for CRPs (summarized in Appendix Table 12A.1). The template is an appropriate example, because it is designed for partnerships involving CRPs, NARS organizations, and partners in advanced research institutes, and it highlights some of the complexities and responsibilities inherent in an RBM system.

Whose Agenda Is Being Monitored and Evaluated?

The influence of donors on the M&E agenda is often determinant. In response to their own constituents' demands for justification, donors have pulled the whole planning, monitoring, learning, accountability, and evaluation system toward accountability for final outcomes.[3] At the same time, donors have prescribed processes, funded particular technical partners, and recommended institutional arrangements—sometimes with limited evidence for their prescriptions.

The introduction of the theory of change (TOC) as the latest refinement of logic models is an attempt to force planners to be more explicit about their assumptions and evidence for presumed links between efforts and results. The original logframe analysis outlined a development hypothesis by which inputs were transformed into outputs, outputs contributed to strategic objectives, and strategic objectives contributed to goals. Other variations of the planning tool argued that activities contributed to outcomes, outcomes contributed to purposes, and purposes contributed to goals. The approaches all required that planners specify the assumptions under which the development pathway would

3 The order in which planning, M&E, learning, and accountability are presented is significant. While the temptation to create an acronym "PMEAL" is present, most practitioners assert that planning, monitoring, and learning come before evaluation and accountability and represent the interest of the scientists and implementers of development for efficiency and outcomes.

logically lead to its goals. Where planners identified missing resources, skills, or institutions, the project would be required to include the additional resources or actions needed to ensure the result.

A contemporary approach, called "outcome mapping," worked backward from desired goals in a participatory mode to ensure the necessary skills, processes, and institutions for success were in place. Since these logical exercises became an integral part of project documents, their targets often became rigid deliverables, where missed milestones were treated as failures, rather than as learning opportunities.

In a more adaptive mode, many agricultural research institutions use participatory impact pathway analysis (PIPA), which develops the logic model and impact pathway in participation with stakeholders. PIPA separates the performance management stages of inputs-into-outputs from the final evaluation. Monitoring and learning are continuous and separated from "evaluation," which takes place after a project's completion. The participatory learning process permits correction of the impact pathway and even revision of the expected goals themselves. Participants derive outcome targets and milestones, which are regularly revisited and revised as part of the project's M&E framework. This goes beyond the traditional use of logic models by engaging stakeholders in a structured participatory process, promoting learning, and providing a framework for "action research" on processes of change (Douthwaite et al. 2008). To do this, the monitoring indicators must be SMART—that is, specific, measurable, attainable, relevant, and time-bound or trackable (Chapter 11, this volume). Thus, stakeholders know which predictions or hypotheses are being confirmed and which are being corrected.

Understanding the Data Requirements of a Full Results-Based Management Approach

Recent contributions to conceptualizing the need for detailed TOCs and attempts to measure them come from the World Agroforestry Centre, which reviewed attempts to scale up agroforestry to sustainably increase production and maintain environmental services (Coe, Sinclair, and Barrios 2014, 73):

> Evidence suggests that this will not be achieved by wide scale promotion of a few iconic agroforestry practices. Instead, three key issues need to be addressed. First, fine-scale variation in social, economic and ecological context and how this creates a need for local adaptation. Second, the importance of developing appropriate service delivery mechanisms, markets, and institutional contexts, as well as technologies. Third,

> appropriate research design, within the scaling process, that enables co-learning amongst research, development and private sector actors. This requires a new paradigm that builds on previous integrated systems approaches, but goes further, by embedding research centrally within development praxis.

Achieving development outcomes with agricultural technologies, particularly in an African context, requires fine-scale targeting and adaptation, where results critically depend on context, which in turn influences the design of the dissemination or scaling program. Understanding the adoption of new techniques and the impact on household welfare requires sex-disaggregated, household-survey data across institutional, economic, and ecological contexts within the target region or population. Developing a systematic household-survey capacity to monitor change and impact on household welfare requires efficient sampling frames, standardized survey instruments, and the ability to resurvey. Moreover, household panel survey designs usually have to be supplemented with surveys of marketing agents, service providers, and agroclimatic conditions in order to locate the household within its socioeconomic context. Such surveys are costly within an African context, particularly if conducted independently for every research program. This has led to efforts to develop standardized national household surveys within a panel survey design under the Living Standards Measurement Studies–Integrated Surveys on Agriculture. However, these efforts have been instituted only in a few countries, and essentially shift the public-goods nature of the costs to national statistical agencies, often supported by development partners.

A renewed commitment to agricultural statistics and information may be an important force for improved knowledge and attention to agriculture. Regular collection of intercensus data on agriculture will be enhanced, but the flood of information on markets, trade, and prices made possible by mobile telephony will be a potential game changer for M&E. Remote sensing will be another technology harnessed for agricultural development. As discussed above, household surveys will be a third tool. The adoption of the Global Strategy to Improve Agriculture and Rural Statistics (GSARS) (WB–FAO–UN 2010) aims to produce a minimum set of baseline data, improve mainstreaming of agricultural statistics in national statistical systems, and build capacity. With support from the Bill & Melinda Gates Foundation and the UK Department for International Development, tailored strategies are being developed for individual countries. Open data, combined with information from statistical databases, may eventually provide practical solutions for farmers. As the coordinator of GSARS cautions, however, statistics is an inferential science that requires representative samples, not just more data (Guillaume-Gentil 2015). Harnessing the data revolution for planning and performance management will require careful design.

Targeting and Scaling Out Interventions

A recent study by the CRP on Climate Change, Agriculture and Food Security deals with the conceptual issue of targeting and scaling out interventions through an iterative mapping process. The authors developed guidelines for evaluating and prioritizing potential interventions (Herrero et al. 2014, Abstract):

> There are real needs and opportunities for well-targeted research and development to improve the livelihoods of farmers while at the same time addressing natural resource constraints. The suitability and adoption of interventions depends on a variety of bio-physical and socioeconomic factors. While their impacts—when adopted and out-scaled—are likely to be highly heterogeneous, not only spatially and temporally but also in terms of the stakeholders affected. In this document we provide generic guidelines for evaluating and prioritising potential interventions through an iterative process of mapping out recommendation domains and estimating impacts. As such, we hope to contribute to the inclusion of such important considerations when agricultural innovations are targeted and scaled out.

Targeting and scaling out are key components of an integrated ex ante assessment process, as well as fundamental to implementing downstream adaptive research and scaling programs. "Two underlying questions are addressed by this framework: which data are required for targeting and scaling out, and how can the data be integrated to assess different impacts of a range of interventions" (Herrero et al. 2014, 8). The data used in this ex ante assessment provide the basis for establishing the indicators by which progress can be monitored.

The complexity (and cost) of such an exercise is evident in the steps necessary for discerning how scalable specific practices might be for improving food security, NRM, and livelihoods, and for mitigating the impacts of climate change. The steps involve identifying

1. the characteristics of the intervention that may affect its use and adoption in agricultural systems;

2. the recommendation domain for the products of research and where these are likely to be applicable;
3. the groups of people who are likely to be affected by the output of the technology/intervention; and
4. the nature of the impacts, in terms of their type, their magnitude, and the associated trade-offs at different temporal and spatial scales, and among the different types of impacts.

The detailed knowledge of population structures and behavior, as well as the ability to link this knowledge to land quality and potential, underlines the data intensity of such an exercise. It is definitely beyond the capability of most NARSs and requires collaboration with advanced research institutes.

Approaches to Developing an M&E Capacity within African NARSs

The development of an M&E capacity within African NARSs cannot be separate from the development of sufficient well-trained graduates at the MSc and PhD levels with the technical skills and field experience in agriculture demanded by employers. This demand comes from the banks, private agribusiness, universities, research institutes, and donor agencies that hire away the best candidates while lamenting the inability of NARSs to retain staff. Since evaluation can be taught to agricultural scientists, as well as economists and agricultural economists, the pool begins with a mix of agricultural skills and experience. The Regional Universities Forum for Capacity Building in Agriculture (RUFORUM) is an example of a regional approach to train high-quality postgraduates who are demanded by the market, and M&E is part of their MSc program in research methods.

A second assertion is that a great deal of capacity development is already occurring in Africa. However, it is poorly coordinated among national educational programs, donor support to higher education, and private-sector demand. Moreover, donor-sponsored training tends to target bilateral projects and initiatives. Several African countries have been leaders in moving planning and evaluation forward. With the support of the Rockefeller Foundation and the former International Service for National Agricultural Research, the Kenya Agricultural Research Institute (KARI) was an early adopter and codeveloper of formal approaches to research planning adapted to local conditions (authors' personal experience).

Third, efforts by CGIAR to establish RBM in its CRPs will have the effect of transferring skills to the SROs and NARSs; they must speak the same language if they are to develop the partnerships needed to achieve intermediate development outcomes. Many senior African leaders started their careers in planning and evaluation within NARSs, moved to SROs to work on regional M&E standards, and subsequently joined international organizations, donor agencies, or international research centers. This career trajectory should not be lamented; rather, it should be treated as part of the incentive structure for high-quality planning and evaluation in NARSs.

Finally, capacity development for M&E is a continuing investment. The focus of evaluation changes over the life cycle of an intervention, the users of the information are different, and the tools are continuously evolving. All this militates in favor of efforts to agree on common concepts and definitions, and compatible data collection.

The following sections outline two innovative experiences with M&E related to a larger agenda of policy and technical change. The first case documents the long-term use of rigorous M&E by the International Potato Center (CIP) to prove that orange-fleshed sweet potato (OFSP) can be a successful and economic means of addressing vitamin A–deficiency blindness in children. Studies have successively proved the effectiveness of biofortification, evaluated efficiency in reaching end users, and monitored efforts to reach policymakers and agents of change. The second case looks at the data-driven, evidence-based approach of the Ethiopian Agricultural Transformation Agency (ATA), and the role that rigorous M&E has played in the early stages of its implementation of a longer-term transformation agenda.

Evolving the Evaluation Focus with the Success of the Intervention

CIP runs a multistakeholder program: the Sweetpotato for Profit and Health Initiative (SPHI), with the goal of reaching 10 million households across 17 SSA countries over 10 years, through the widespread uptake of sweet potatoes to reduce malnutrition among children under the age of five years. A major program under SPHI applies the lessons of a HarvestPlus project (HarvestPlus 2012) and is now an initiative called Reaching Agents of Change (RAC). In turn, RAC's M&E plan is a useful example of a well-articulated RBM framework. CIP and Helen Keller International will reach out to regional and national agencies who will act as agents of change for OFSP.

Working back from its goals of reducing vitamin A deficiency and food insecurity, particularly of women and young children, RAC sets up specific objectives related to (1) generating new investments to scale up the adoption of OFSP in five countries, and (2) building the capacity of implementation agencies to design and implement technically strong and cost-effective interventions focused on (or including) OFSP. A third specific objective deals with the management of RAC itself. Its TOC is based on knowledge of what works and how change is expected to occur. Working back from a vision of success, it poses three questions:

1. What advocacy activities are necessary to generate investments by governments, donors, and NGOs to scale up OFSP?
2. What materials and capacities do implementing agencies need to deliver strong interventions?
3. What resources, capacities, and activities will the project need to manage for results?

The M&E plan is structured with biweekly activity reports, quarterly monitoring of achievements toward milestones, and a special narrative on systematic, corrective actions taken based on lessons learned. Comparing experiences across Burkina Faso, Ghana, Mozambique, Nigeria, and Tanzania, the RAC project will generate lessons on what works across different contexts and what corrective measures are needed in individual countries (Mbabu et al. 2015).

The example of OFSP illustrates several important considerations in designing an M&E system: (1) clarity in specifying goals, (2) a decent TOC that allows for adaptive management, (3) data collection instruments that identify needs for corrective action in real time, and (4) the involvement of local implementers in managing for results. The RAC program promotes policy advocacy at one end and technical skills development at the other. RAC's M&E plan identifies the levels of responsibility for implementation and arms the actors with the skills and tools to manage within each level. Information on activities carried out is recorded biweekly, whereas quarterly reports are filed with the M&E officer.

The OFSP case highlights the way that M&E and impact evaluation methodology have adapted to the task: first, randomized controlled trials to establish proof of concept; second, less costly evaluation of the phase of reaching end users; and currently, detailed M&E of the TOC associated with reaching agents of change. The RAC project is being piloted and monitored in two countries (CIP 2012).

Monitoring and Evaluation in a Results-Based Management System

The ATA has a much broader agenda than agricultural research—notably, the transformation of Ethiopia's agricultural economy. The goal is an agricultural sector that is more technologically intensive, creates more value up and down the value chain, contributes more to social goals, and ultimately represents a smaller share in gross domestic product through its contribution to growth elsewhere. ATA's strategy is to focus on transformation clusters that are regionally and geographically identified for their industrial and export potential, and to address bottlenecks in support systems and value chains. ATA was able to base its planning on baseline studies carried out by the International Food Policy Research Institute and the Ethiopian Development Research Institute (EDRI) under the Ethiopian Strategy Support Program (ESSP). The studies and ongoing support enable ATA to identify priority points of entry along value chains, as well as key systemic bottlenecks, while building toward a broader transformation of policies and institutions.

ATA is unique among similar entities in Africa because of (1) its political support and organizational status; (2) the analytical base on which it was founded; and (3) its structure as a data-driven, results-based learning organization. ATA serves as secretariat to an Agricultural Transformation Council chaired by Ethiopia's prime minister and comprising key ministries, has a reporting role to the Ministry of Agriculture, and plays an important role in building capacity within the ministry and regional bureaus of agriculture. ATA has developed internal monitoring, learning, and evaluation systems for reporting on its own deliverables in advancing more than 100 proposed solutions to overcome prioritized constraints in value chains, systemic bottlenecks, and cross-cutting issues.[4] The deliverables are time denominated and, as such, impose a strong internal discipline. ATA reports quarterly to the Transformation Council regarding key issues and the status of each deliverable (not started, significant issues, slight issues, on track, or completed), all of which is supported by analysis and suggested actions and is included in ATA's annual report.

ATA has been judicious in managing its dual relationship with the Transformation Council and Ministry of Agriculture. Being mandated to identify and address bottlenecks, it is not an implementing agency per se.

4 A "deliverable" is an activity that has not yet become an outcome or an impact, such as a strategy for a particular commodity or goal.

Nevertheless, it does work with other agencies and departments in building capacity and providing analytical and reporting strength. ATA is accumulating a track record in overcoming near-term technical and logistic constraints, while moving forward in addressing the more difficult policy and institutional constraints needed for the long-term transformation of Ethiopia's agricultural economy. These constraints lie in trade policy, financial systems, cooperatives, and extension. ATA's success has originally been in addressing priorities identified by the prime minister—recognizing that cooperatives and extension play an important national role in the political structure—while at the same time being part of the transformation agenda.

Even though ATA's experience is too recent to evaluate, some lessons are emerging. First, building capacity in the Ministry of Agriculture is proving more difficult than originally expected and will require additional attention. Second, while it is still too soon to evaluate the impact of ATA's strict monitoring of its deliverables on the broader transformation agenda, its discipline in tracking implementation does afford some credibility, which carries over to its longer-term strategic analyses. ATA's ability to call on ESSP for research—for example, on policies relating to the importation of wheat versus enhanced domestic procurement efforts, while maintaining market stability—puts ATA at the heart of policy and institutional transformation. Third, the Transformation Council's link to the prime minister's office is a valuable asset. It is recognized, however, that ATA's data-driven, evidence-based approach, led by the Bill & Melinda Gates Foundation and supported by other donors, will be challenging to replicate in other countries.

Creating the Necessary "Architecture" for Monitoring and Evaluation

With the 2013 memorandum of understanding (MOU) between the African Union (through CAADP) and CGIAR, a new "architecture" is creating the prospect of bringing order to the complexity of a global system. The elements of this new architecture, if implemented, include

1. aligned strategies at the regional, subregional, and national levels through an MOU between CAADP and CGIAR;
2. coordinated CGIAR center activities through multiactor and multistakeholder CRPs;
3. a fund council representing the collective will of the donors (constituent CGIAR centers receive some overhead through CRPs, although

inadequate support to the core remains a potential weak spot in the system);

4. funding from Africa's national governments, which have committed to meet the CAADP target of investing 10 percent of their national budgets in agriculture;
5. compatible tools for planning, monitoring, and evaluation, eventually following similar results frameworks supported by FARA and the SROs at the national level;
6. regional programs, such as EAAPP and WAAPP, that have pioneered the way for the World Bank to fund regional activities through loans to participating countries and to engage the SROs in program management; and
7. independent advice from the International Science and Partnership Council (ISPC), which has been monitoring the experience with impact pathways. .

Making the Architecture Work for Decisionmaking, Measurement, Learning, and Evaluation

This chapter has shown how M&E has evolved over the past three decades, along with planning tools that address the changing concerns of policymakers, funding agencies, and implementers of agricultural research. Their needs, respectively, are for tools that (1) reduce the risk of decisionmaking; (2) provide accountability to government and funding agencies that money was well spent; and (3) enable implementers to identify problems, flexibly adjust activities, and readjust goals in real time.

The institutional architecture is largely in place: the CGIAR centers are well located throughout the African continent through regional and subregional offices, while the CRPs target agreed-upon problems with strong participation from NARSs, advanced research institutes and development-oriented partners, NGOs, and agencies. Through the African Union, New Partnership for Africa's Development, and CAADP, there is a policymaking forum linked to the heads of state and a technical structure through FARA and the SROs for collaboration among NARSs. Many countries, however, lack strong articulation of a national agricultural innovation system to bring research, higher education, extension, and the private sector together. CAADP compacts seek to close these gaps while committing governments to a meaningful funding target to support

APPENDIX TABLE 12A.1 CGIAR's essential requirements of a results-based management system

Requirements	Notes (abbreviated from source document)
1. Statement of strategic goals (1 page)	Strategic goals of the CGIAR Research Program (CRP) should be reachable within 12–15 years and show contribution to system-level objectives. Note that increasing productivity or income cannot be assumed to automatically lead to poverty alleviation or food security, and that productivity or income growth cannot be maintained in the long term without effective management of natural resources and effective policies.
2. Impact pathway, theory of change (TOC), key hypotheses (4 pages)	• Highlight the TOC and key hypotheses upon which the TOC is based, and which the CRP will test in its own M&E system. • Synthesize the major elements of the CRP's high-level impact pathways and their supporting assumptions and TOCs.
3. Justification of international comparative advantage (1.5 pages)	Explain why the CRP has a clear comparative advantage. This is equivalent to demonstrating that no institutions are better placed to produce the necessary results, given the impact pathway of the CRP.
4. Intermediate development outcomes (IDOs), targets, and Indicators (3 pages)	For each intermediate IDO, specify the target population and geographical area concerned and quantify the targets. Indicate the level of uncertainty associated with those targets and, if necessary, plans to refine them. Robust indicators of progress toward these targets must be specified.
5. Flagship projects and clusters of activities (15 pages)	The guidelines are specific to CGIAR and call for organization of flagship programs and clusters of activities underneath the flagships. This organizes work in the most effective manner to produce the outcomes, outputs, and IDOs and monitor progress. The flagships must show the linkages the CRP has built with other CRPs, with a description of the type of research undertaken and its sustainability.
6. External partnership strategy and intellectual property issues (6 pages)	• Explain the CRP's strategy concerning external partnerships (non-CGIAR partners) from discovery to scaling-up phases. • Include a typology of your partners and their functional role in the CRP, including governance. • Explain how intellectual property issues are managed through these different partnerships.
7. M&E and risk management strategy (5 pages)	• Describe the CRP's approach to risk management and mitigation. • Describe how the CRP tracks its own progress within clusters of activities and subcluster levels, how adjustments/alignments are implemented when progress is not as expected, and how this approach fits with the overall results-based management implemented at the CRP level.
8. CRP governance (1.5 pages)	• Describe how the CRP management structure draws upon resources and talents across CGIAR centers and beyond the CGIAR system. • Indicate how the recommendations of the external review of management and governance in CRPs will be implemented by this CRP.
9. Budget request (3 pages)	• Explain the budget needed to produce the outputs, research outcomes, and IDOs. • Indicate the budget allocated to partners outside of CGIAR over time.

Source: CGIAR (2013a).

Note: M&E = monitoring and evaluation.

the system. The growing divide in human resource research capacity among the region's countries must be reversed (Chapter 8, this volume).

African countries have not been slow to adopt new practices in planning, monitoring, and evaluation when they serve national needs. KARI was an early adopter of formal priority setting before many countries in other regions of the world. Ethiopia's ATA is pioneering harmonized data for coordination of the system. The migration of African expertise among leading countries and institutions at the regional and subregional levels has helped spread the concepts and the development of tools. Nevertheless, the uniform development of the data, tools, and practices among countries, which would make M&E a powerful tool, is lacking. Donors continue to have their preferred partners under various alliances or bilateral programs for historical reasons. The recent Science Agenda for Agriculture in Africa argued that the African Union should take up the challenge to ensure that no country is left behind in having sufficient scientific capacity (FARA 2013). Its adoption by the heads of state in Malabo is encouraging. It is hoped that they will address the inadequacy of data for planning and monitoring Africa's agricultural growth and development, which includes linkages with R&D and national-level data sharing to draw institutions of agricultural knowledge together in a system.

References

ActionAID Interntional. 2011. "Alps: Accountability, Learning and Planning System: 2011 Update." Accessed March 2014. www.actionaid.org/sites/files/actionaid/alps2011_revised06july2011.pdf.

CGIAR. 2013a. "Guidance for CRP Second Call. Draft version 2." Accessed March 2014. http://library.cgiar.org/bitstream/handle/10947/2896/Guidance_for_CRP_Second_Call_-_DRAFT_-_Version_2.pdf?sequence=1.

CGIAR. 2013b. "Results Based Management for the CGIAR Research Program Portfolio." Accessed July 2013. www.cgiar.org/consortium-news/results-based-management-for-the-cgiar-research-program-portfolio/.

CIP (International Potato Center). 2012. *Reaching Agents of Change: Monitoring and Evaluation Plan 2011–2014*. Nairobi, Kenya.

Coe, R., F. Sinclair, and E. Barrios. 2014. "Scaling Up Agroforestry Requires Research 'in' Rather than 'for' Development." *Environmental Sustainability* 6: 73–77.

Douthwaite, B., S. Alvarez, G. Thiele, and R. Mackay. 2008. *Participatory Impact Pathway Analysis: A Practical Method for Project Planning and Evaluation*. ILAC Brief 17. Rome: Institutional Learning and Change Initiative.

Elliott, H. 2008. "The Evolution of Systems Thinking Towards Agricultural Innovation Systems." Annotated PowerPoint presentation to IFPRI International Conference, "Advancing Agriculture in Developing Countries through Knowledge and Innovation," Addis Ababa, Ethiopia, April 7–9. Accessed March 2014.

FARA (Forum for Agricultural Research in Africa). 2013. *Science Agenda for Agriculture in Africa (S3A): "Connecting Science": A Science Agenda for the Transformation of Agriculture in Africa.* A Report of an Expert Panel. Accra, Ghana. Accessed April 2016. www.asti.cgiar.org/sites/default/files/pdf/s-t-partnerships/S3A.pdf.

Guijt, I. 2010. "Accountability and Learning." In *Capacity Development in Practice.* London: Earthscan.

Guillaume-Gentil, A. 2015. "Decision Support Systems: Open and Big Data." SPORE 175 (April–May). Accessed April 2016. http://spore.cta.int/images/magazines/Spore-175-EN-WEB.pdf.

HarvestPlus. 2012. Disseminating Orange-Fleshed Sweet Potato: Findings from a HarvestPlus Project in Mozambique and Uganda. Washington, DC.

Herrero, M., A. Notenbaert, P. Thornton, C. Pfeifer, S. Silvestri, A. Omolo, and C. Quiros. 2014. *A Framework for Targeting and Scaling-Out Interventions in Agricultural Systems.* CCAFS Working Paper 62. CGIAR Research Program on Climate Change, Agriculture, and Food Security (CCAFS). Copenhagen, Denmark. Accessed March 2014. http://ccafs.cgiar.org/publications/framework-targeting-and-scaling-out-interventions-agricultural-systems#.Uzp0516gJFQ.

InterAcademy Council. 2004. *Realizing the Promise and Potential of African Agriculture: Science and Technology Strategies for Improving Agricultural Productivity and Food Security in Africa: Executive Summary.* Amsterdam.

Johnson, M., S. Benin, X. Diao, and L. You. 2011. *Prioritizing Regional Agricultural R&D Investments in Africa: Incorporating R&D Spillovers and Economywide Effects.* Conference Working Paper 15 from the ASTI/IFPRI–FARA Conference "Agricultural R&D: Investing in Africa's Future: Analyzing Trends, Challenges, and Opportunities," Accra, Ghana, December 5–7. Accessed April 2013. www.asti.cgiar.org/pdf/conference/Theme4/Johnson.pdf.

Maredia, M. 2013. "Evaluating Impacts of Agricultural Research: Lessons and Challenges Based on the Experience of the CGIAR." Presentation on behalf of the Standing Panel on Impact Assessment of the CGIAR. Cairo. March 15, 2014.

Mbabu, A., H. Munyua, G. Mulongo, S. David, and M. Brendin. 2015. *Learning the Smart Way: Lessons Learned by the Reaching Agents of Change Project.* Nairobi, Kenya: International Potato Center.

OECD (Organisation for Economic Co-operation and Development). 2005. *The Paris Declaration on Aid Effectiveness (2005) and the Accra Agenda for Action (2008).* Accessed March 2014. www.oecd.org/dac/effectiveness/34428351.pdf.

Pardey, P., and J. Beddow. 2013. *Agricultural Innovation: The United States in a Changing Global Reality*. Chicago Council on Global Affairs. Accessed April 16. www.aspeninstitute.org/sites/default/files/content/docs/ee/ChicagoCouncil_AgriculturalInnovation_2013_0.pdf.

Pritchett, L., S. Samji, and G. Hammer. 2013. "It's All about MeE: Using Structured Experiential Learning ("e") to Crawl the Design Space." Center for Global Development Working Paper 322. Accessed April 2016. www.gsdrc.org/document-library/its-all-about-mee-using-structured-experiential-learning-e-to-crawl-the-design-space/.

Ragasa, C., A. Abdullahi, and G. Essegbey. 2011. *Measuring R&D Performance from an Innovation Systems Perspective: An Illustration from the Nigeria and Ghana Agricultural Research Systems*. Conference Working Paper 14 from the ASTI/IFPRI–FARA Conference "Agricultural R&D: Investing in Africa's Future: Analyzing Trends, Challenges, and Opportunities," Accra, Ghana, December 5–7. Accessed July 2013. www.asti.cgiar.org/pdf/conference/Theme3/Ragasa.pdf.

Ruttan, V. 1982. "Institutional and Project Funding of Research." Chapter 9 in *Agricultural Research Policy*. Minneapolis: University of Minnesota Press.

Spielman, D., and R. Birner. 2008. *How Innovative Is Your Agriculture? Using Innovation Indicators and Benchmarks to Strengthen National Agricultural Innovation Systems*. ARD Discussion Paper 41. Washington, DC: World Bank.

Spielman, D., and D. Kelemework. 2009a. *Measuring Agricultural Innovation System Properties and Performance: Illustrations from Ethiopia and Vietnam*. IFPRI Discussion Paper 851. Washington, DC: International Food Policy Research Institute.

———. 2009b. "Measuring Agricultural Innovation System Properties and Performance: Illustrations from Ethiopia and Vietnam." Paper presented at the International Association of Agricultural Economists Conference, Beijing, August 16–29.

WB–FAO–UN (World Bank, Food and Agriculture Organization of the United Nations, and United Nations). 2010. *Global Strategy to Improve Agriculture and Rural Statistics*. Economic and Sector Work Report 56719-GLB. Washington, DC. Accessed July 2015. www.fao.org/docrep/015/am082e/am082e00.pdf.

PART 5

Rationalizing and Aligning Institutional Structures

Chapter 13

INTEGRATING AGRICULTURAL RESEARCH INTO AN AFRICAN INNOVATION SYSTEM

John Lynam, Joseph Methu, and Michael Waithaka

In the 1970s and 1980s, national agricultural research institutes (NARIs) were created by consolidating disparate agricultural research units across various ministries into autonomous parastatals, which in many ways had the unintended side effect of isolating these new entities. Clear institutional boundaries reinforced the disarticulation between research and extension, and served to limit the interaction of NARI-based scientists with farmers (although this was partly addressed by farming-system research programs of the 1980s). All this occurred before the market liberalization of the 1990s, which left little scope for interaction with the private sector. This lack of organizational connectivity has continued to the present, even as the private sector has expanded, extension has moved to more pluralistic systems with greater connectivity, and farmers have significantly enhanced communications options through cellular phones. NARIs have tended to interact primarily with other agricultural research organizations, whether through programs initiated by subregional organizations (SROs) or by CGIAR research networks (Chapters 2 and 15, this volume), which has often limited the relevance of the research, the effective testing of improved technologies, and efficient deployment.

To ensure the effectiveness of agricultural research, NARIs in Africa south of the Sahara (SSA) need to be more outwardly focused and develop better linkages with principal actors in the agricultural sector. This is necessary for new knowledge generated by the NARIs to be effectively used, and relies on innovative capacity across the rural economy. Such a focus on innovation within the broader agricultural sector, primarily through improved market integration, communication, and institutional linkages across the different actors in the sector, has been formalized as agricultural innovation systems (AIS). Within the context of AIS, agricultural research is only one of a number of contributors to the rural innovation process; interactions among all actors are critical to ensuring and sustaining agricultural innovation across the sector.

This chapter explores the potential for organizing African agricultural research within an AIS framework. The first section evaluates the principal issues involved within the specific context of the region. The second section presents a concrete example of the approach in the form of a case study. Thereafter, the principal methodologies for bringing about this kind of organizational transformation are reviewed. Finally, existing examples of the region's use of the AIS approach are reviewed to explore lessons learned to date.

Agricultural Research within an Agricultural Innovation System

Agricultural research is but one of a number of complex contributors to AIS, where the fundamental idea is to improve linkages between research institutions and other actors in the pursuit of agricultural growth and structural transformation. Within the context of an AIS, rather than a rigidly defined division of labor that governs interinstitutional relationships, stakeholders interact or "cluster" around a particular problem, often defined as an "innovation platform." This interaction creates the conditions for effective problem solving, which in an African context most often revolves around improving smallholder productivity or incomes. This improvement can be realized through new technologies or improved access to markets, inputs, credit, or insurance.

Innovation operates through the application of new knowledge, enhanced organizational capacity, or changes in policies or institutions, all of which require effective coordination. Such coordination can be provided by fully functioning and interlinked markets. However, in African rural economies, markets are often underdeveloped, and coordination must be provided by other actors. The objective is to ensure the resolution of the problem, which in essence means being more responsive to farmers' needs and providing the conditions for interventions to be effective. This requires changes in how agricultural research institutes view their roles, how they are organized, and how they link with other actors within the AIS.

Extension as the Early Source of Innovation

Prior to market liberalization in the 1990s, extension systems were seen as the unique source of rural innovation in African agriculture, offering a single organizational model in the form of the training and visit (T&V) system (Benor, Harrison, and Baxter 1984). The World Bank provided loans to more than 50 developing countries to adopt the T&V system during 1975–1998

(Anderson, Feder, and Ganguly 2006). It was a top-down system, delivering a limited number of well-proven extension messages to communities through lead farmers. Moreover, links to agricultural research were managed at the national level and were rarely effective. T&V was both labor intensive and expensive; structural adjustment programs and the resulting controls on budgetary deficits essentially ended World Bank loans for T&V. The system ultimately succumbed to most of the design problems facing large, publicly funded extension programs: "scale, inadequate interaction with the agricultural research systems, inability to attribute benefits, weak accountability, and lack of political support" (Anderson, Feder, and Ganguly 2006, 2).

However, no fiscally conservative model existed to take its place, so a period of experimentation with extension methods ensued from the late 1990s and into the new century. Such experimentation was aided by the rapid expansion of nongovernmental organizations (NGOs) as a result of democratization and the strengthening of civil society in a large number of African countries. Many of these extension methods were based in rural areas and, to compete for funding, had to offer innovative approaches to improving farmers' incomes. Many countries started with organic farming approaches, but also experimented with group approaches that built social capital.

A similar, but more widely disseminated, approach was farmer field schools (FFSs). This method was originally developed in Asia through the extension of integrated pest management and was introduced in East Africa in 1995 (Davis et al. 2012). The approach uses experiential learning in a group format, not only to familiarize farmers with new production techniques, but also to build social capital to empower them to continue to innovate. One study evaluating three countries in East Africa (Davis et al. 2012) found that FFSs were successful in improving both the productivity and the incomes of participating farmers, and that this applied particularly to female farmers and farmers with limited education. FFSs never evolved into a national agricultural extension system, however, partly because agricultural extension is one of the principal services to be included in district-level decentralization, so units could choose the extension methodology they deemed most appropriate. Similarly, the expanding private sector—often through the development of agro-dealer networks—also became a principal source of extension advice, particularly in terms of input management. These trends, together with the range of actors providing advisory services, led to the adoption of multistakeholder, pluralistic extension systems, where government was only one provider of such services.

Decentralization, multiple actors providing extension services, and the need for sustainable financing of extension motivated an experimental design of an advisory services system in Uganda. The National Agricultural Advisory Services (NAADS) program was piloted in selected districts in 2001 and was designed around three principal elements: (1) developing structures (farmers' forums) to articulate demand for advisory services, (2) meeting that demand through the provision of contracts to private advisory service suppliers, and (3) transitioning financing from public funds to farmers paying for services. NAADS invested significantly in training farmers' groups as part of their participation in the program. Although there was quality assurance for agricultural service providers, at least one of two impact studies (Okoboi, Kuteesa, and Barungi 2013) found that poor-quality service provision was a principal reason for the lack of adoption of new technologies and limited evidence of improved crop productivity or agricultural incomes. The other impact study (Benin et al. 2011)—although not finding evidence for technology adoption—did find impacts on participant farms in increased crop and livestock productivity and agricultural incomes. In terms of financing NAADS, both farmers' payments for services and contributions from local government remained insignificant. Possibly the two major constraints in the implementation of NAADS were the lack of effective capacity in agricultural service provision across the country and the problem of elite capture in awarding service contracts.[1] Building effective rural innovation capacity at scale has thus remained a challenge in Africa.

Agricultural extension remains an area of continuing experimentation in terms of methods, principal focus on expanding links to the private sector, financial sustainability, and the use of information and communications technologies (ICTs). ICTs have expanded rapidly, involving such areas as rural radio (for example, Shamba Shape Up in East Africa), extension and market advice over cell phones, and the use of video (for example, Digital Green). These techniques are very cost-effective in reaching large numbers of farmers. Such platforms can sensitize farmers to production problems, market opportunities, or improved inputs, but large questions remain as to their ability to address location-specific needs and farmers' ability to translate such information into changes in management practices, especially without the critical step of learning by doing found in more intensive methods, such as FFSs.

The evolution in extension methods and organizational models provides many of the preconditions for a more robust AIS and for enhanced rural innovation capacity. The experience with farmers' groups, improved social capital,

1 The president of Uganda called for the dissolution of NAADS in 2014, which would require parliamentary approval. The future of the program remains uncertain.

experimentation with methods, improved links with the private sector and other actors, the decentralization of capacity, and the testing of alternative financing models are all indicative of a more demand-responsive, flexible, and outward-looking capacity in providing advisory services. Nevertheless, this ongoing evolution will require the integration of new sets of skills, a better blending of new ICT approaches with learning by doing, and in the end better connectivity to agricultural research and sources of improved agricultural practices.

An Evolving Role for Agricultural Research

As discussed above, the move to organize research within an AIS comes at a time when agricultural extension is being reorganized using diverse methods and approaches. In addition, a decade after market liberalization, the private sector is investing in agriculture, thereby strengthening supply chains, developing input markets, and expanding agroprocessing capacity. These changes collectively provide both an opportunity and a challenge in terms of effectively coordinating the integration of smallholders into markets, improving their productivity, and increasing their marketable surplus, which is essentially the basis of most investment plans under the Comprehensive Africa Agricultural Development Programme (CAADP). Improving adoption of advanced technologies is directly linked to improved access to input, credit, and output markets, which requires approaches beyond traditional extension methods toward strengthening the capacity to innovate. This dynamic institutional context and the increased emphasis on results and accountability (Chapter 12, this volume) have expanded the roles of agricultural research, beyond just the production of scientific knowledge and development of technologies, into building innovation capacity.

Linking the extension of enhanced technologies to improving market access requires better links with the private sector, more effective collective action and farmers' organizations, and innovative extension approaches and new skills. These requirements emphasize value-chain approaches in organizing innovation platforms, but at the same time they require something of an "honest broker" to integrate competing private-sector actors. NARIs, especially given the distribution of their research stations and status as autonomous parastatals, are in a position to play that role, given their internal capacity. As an example, the Kenya Agricultural Research Institute (KARI) has organized its research and outreach around priority value chains (Miruka et al. 2012). In taking this approach, KARI requires new skills in facilitating, organizing, and coordinating linkages throughout the innovation platform, which in turn requires capacity beyond that needed to conduct agricultural research and development (R&D) programs (see the discussion of KARI in the last section of this chapter).

The need for new, "soft" (that is, facilitation) skills and improved capacity in such areas as facilitation, business development, farmers' organizations, and communication raises the question of where such capacity should be developed within the sector. Extension would be a logical option, but would require a significant shift in skills and knowledge, especially considering that most extension personnel have limited (usually diploma-level) training. Some NGOs have the needed skills, and many could take on a brokering role between farmers' organizations and private and public entities; however, their ability to deploy the necessary capacity would be limited geographically and by their dependence on external funding. With increasing accountability, agricultural research institutes are being nudged to take on this role. The question then becomes whether capacity should be built only in the larger systems. It can be argued that smaller research systems already function in a brokering role by accessing external sources of technology and adapting it to local conditions; that adaptation role could easily be broadened from a narrow focus on adaptive research to one of coordinating innovation platforms.

Research within an Agricultural Innovation System under Capacity Constraints

As previously noted, a fully functioning AIS in SSA is still under development. As Sumberg (2005, 24) argues, the intent is not an integrated system, but an interacting one, where "greater interaction or feedback between . . . actors makes a system more dynamic, which is manifest by the properties of robustness, flexibility, and the ability to generate and respond to change." Using an AIS as a basis for institutional reform has shifted the debate away from an entity's internal capacity to undertake relevant research, to its capacity to interact and link with other, functionally different actors in the agricultural sector. However, these two capacities are quite different in terms of skill sets, disciplinary mixes, and mobility, yet they must interact and complement each other within a research institution. As Horton (2012, 316) notes, research within an AIS requires "new competencies related to communication, participatory planning, facilitation of teamwork, and learning-oriented evaluation. Conventional capacity development has concentrated on developing the knowledge and skills of individuals, but research organizations that perform effectively in innovation systems also require changes in policies, management systems, and incentives." This requires a rebalancing of competencies and changes in internal management and financial allocations to link internal research programs with external application—namely, innovation.

Positioning agricultural research in a more proactive role within an AIS thus requires leadership and a fundamental decision to improve the organization's capacity to interact with other actors in the sector. Some research institutes adopt the position that the capacity to foster innovation exists in other organizations, so they can play a more traditional role. This is what Reddy, Hall, and Sulaiman (2011) refer to as "knowledge generation and adaptation," as distinct from "knowledge application." For those institutes where developing innovation capacity requires a more proactive role, the issue becomes where to train staff in these new skills, how to organize such capacity, and what percentage of resources to devote to it. For supporting staff, this would require short courses, but for program leadership, in most cases degree-level training is required.

The Regional Universities Forum for Capacity Building in Agriculture (RUFORUM) has launched a PhD program in rural innovation, which provides both theory and practical skills, including soft skills, in this emerging field. The degree will be offered by Makerere University, Egerton University, and Sokoine University, and will draw on a student body from East and Southern Africa (Chapter 9, this volume). Nevertheless, for African conditions, few models exist to draw from regarding how to organize such programs—although Spielman, Ragasa, and Rajalahti 2012 review methodologies and various institutional arrangements in improving organizational interaction and linkages and Moock (Chapter 10, this volume) evaluates the potential of regional networks in building such capacity. Moreover, experience with the internal program organization of research institutes is minimal, possibly apart from the work by Mbabu and Hall (2012) in Papua New Guinea.

Balancing Context and Scale in an Innovation Process

Research within an AIS is applied quite differently across the developing world, depending on how well a country's agricultural markets are integrated and linked to global agricultural markets, and how well the focal points for problem identification—that is, the matrix of professional, trade, commodity, and farmers' organizations—are developed (Lynam 2012). In countries such as Chile or Thailand, where these conditions are met, research tends to (1) focus more on the use or agroprocessing end of the value chain; (2) be more effectively integrated with private-sector application; (3) be more basic, so as to feed into private-sector product development and give the industry a competitive edge when competing in the world market; and (4) focus on broader-based transformation across the industry, defined either by commodity or input, thus achieving some scale economies in application. This research is often organized around a cluster approach, where "cluster-based policy aims at removing the imperfections of innovation systems by enabling them to function more efficiently and avoid coordination failures" (Theus and Zeng 2012, 396). Clusters form the platform for problem identification and innovation, driven by a dynamic market environment.

Such market conditions are not met in SSA. Innovation tends to be focused more at the production end of the value chain, where there is significant public-sector participation and uncertain, often tenuous links to an emerging private sector. Given the heterogeneity facing farmers in agroecological, market, and institutional conditions, understanding context is critical to the innovation process (Hounkonnou et al. 2006; Reddy, Hall, and Sulaiman 2011) and to the adaptation and application of knowledge. At issue within an African context are how the innovation process is coordinated across the different actors, and how responsive it is to context. There is no "market for innovations" (Sumberg 2005) to provide such coordination; at the farmer level, the innovation process is usually facilitated, often involves enhanced community-level innovation capacity (such as Triomphe 2012), and is often organized within a value-chain platform. The heterogeneity of the farmer context gives rise to the question of how to achieve sufficient scale in a facilitated innovation process to cover the organizational, coordination, and transaction costs.

In many ways, ensuring effective approaches to scaling both research and innovation processes in SSA is the core design issue for deploying both organizational capacity and programmatic investments. For large investors in agricultural development in the region (such as the Bill & Melinda Gates Foundation), this has become a critical design requirement for investments. The approach to scaling the dissemination of input-based technologies tends toward a relatively narrow set of technologies distributed through existing or potential input markets. From an AIS perspective, problem solving within a heterogeneous environment most often leads to localized solutions with potential for farmer-to-farmer diffusion, but limited scope for spatial diffusion. Neither a relatively narrow spatial scope to localized innovations nor wide market distribution but irregular adoption of a narrow set of improved inputs has the potential to meet the needs of African smallholder farmers.

This challenge has led to other approaches to attaining scale, which primarily focus on developing better linkages and improved competencies of actors within the AIS following two principal, but reinforcing, pathways (Hounkonnou et al. 2006). One approach is to build enhanced social capital in rural communities through more effective farmers' organizations, community empowerment, or methods that enhance farmers' innovation capacity. Methods include FFSs and the codesign of innovations (Triomphe 2012). The second pathway relies on organizational and institutional change to support the broader, more effective application of such methods, thereby targeting systemwide capacity and the formation of efficient organizational partnerships.

For African national agricultural research systems (NARSs), undertaking research within an AIS requires developing an efficient adaptive research capacity with links to innovation platforms that operate at a sufficient scale. An efficient adaptive research capacity explores agricultural heterogeneity through systematic site selection, characterization, appropriate trial design, integrated analysis, and extrapolation. Thus, ad hoc empiricism is replaced with information platforms that build systematically over time within cost-efficient, hierarchical designs. Such capacities do not exist in most NARSs. Adaptive research is costly, but NARSs have the potential to develop adaptive research networks across field capacities through NGOs, extension providers, and CGIAR centers, where a systematic design could substantially improve the value of the information generated. Such capacity can in turn be linked to facilitated innovation platforms that coordinate actors in the agricultural sector around particular problems, value chains, or agroecological zones. With decentralization of many public services, a relevant scale for coordinating actors and managing heterogeneity is the district level, especially given that NARSs operate through a distributed system of research stations. Nevertheless, successfully operating innovation platforms at this scale would require an effective organizational structure to reach out to and link farmers. Thus, information and farmers' organizations are essential building blocks for achieving scale in the context of agroecological and socioeconomic heterogeneity.

Innovation Systems and Agricultural Development in Africa South of the Sahara

Structural changes are occurring in both national and global food and agricultural systems. They include the integration of agriculture into global and regional markets, and the emergence of consumers as key drivers of technological change. Singly and in combination, these changes are fostering the development of AIS in Africa.

Recent moves to integrate African markets have been spearheaded by subregional economic blocs, with the goal of reducing tariffs and easing crossborder trade. In addition, urbanization, the revolution in ICTs, and the emergence of a middle-income class are changing consumer preferences and demand. All have contributed to an expansion in agroprocessing and intraregional trade in higher-value products, including vegetable oils, processed milk, and milled flours, such as for maize, and canned vegetables and fruit juices. All of this requires restructuring of traditional value chains, where quality and timeliness of supply are critical to their effective functioning. The example of the seed potato value chain is presented in the Box 13.1.

BOX 13.1 Actors in a seed potato value chain

A project on developing seed potato value chains in Burundi, Kenya, and Uganda is being developed with support from the Association for Strengthening Agricultural Research in Eastern and Central Africa. Many actors are directly and indirectly involved, collectively making up an agricultural innovation system. Some actors operate at the *macro level*, meaning they help to provide an enabling environment within which the value chain can operate. Actors at the *meso level* provide services to the value chain and include research institutions that develop technologies, such as new seed varieties and other research outputs. *Microlevel* actors are directly involved in adding value to and moving the commodity along the pathway from production to consumption. *Product innovations* occur during the input and production segments of value chains; *process innovations* occur during the transformation phase, and *systems-level transformations* occur during the sales and distribution phase. Development projects have traditionally invested heavily in product innovations, but have paid little attention to either process- or systems-level innovations.

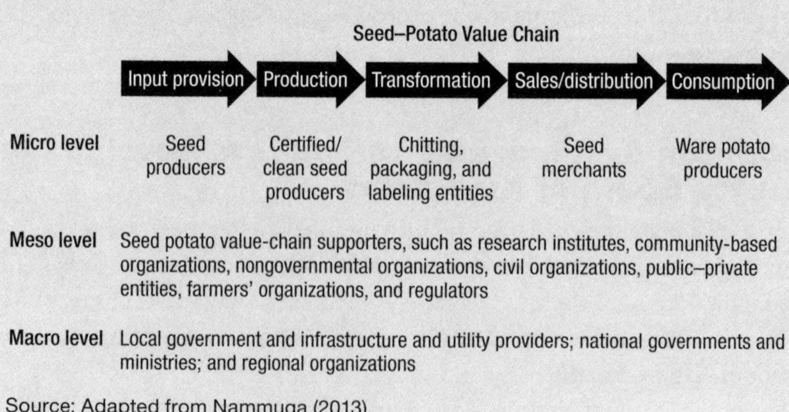

Source: Adapted from Nammuga (2013).

A development strategy that adopts an agricultural innovation framework represents a shift from a mode of simple technology delivery to one of strengthening the ability and capacity of actors in the agricultural sector to innovate. Five main agricultural innovation systems elements that support this are (1) technology aggregation, (2) knowledge and skills, (3) markets, (4) financing, and (5) an enabling environment.

Increased investments are needed to support innovations at the process and systems levels, especially in terms of adding value and linking farmers to markets. Such investments should target reforms in policies and legislation, institutional structures, and infrastructure development. Moreover, capacity-building interventions targeting individuals, institutions, and organizations are needed at the lower end of the chains (that is, at the farm and small-scale trade levels), which is inherently complicated, given the breadth and diversity involved. Capacity-strengthening interventions should aim to enhance the ability of the different value-chain actors to improve their interlinkages, as strong organizational linkages are an important ingredient for innovation.

Capacity development at the individual level should support farmers' ability to adapt available technologies to their production environments and conditions. At the organizational level, farmers need support in developing effective organizations that enable them to profitably acquire inputs and information; access ways of collectively marketing their produce; and develop the necessary management capacities in the areas of human and financial resources, as well as ICTs

Institutional linkages, the incentives to innovate, and greater responsiveness to consumer demand rely on well-functioning markets. Effective integration of smallholders into higher-value supply chains relies on such factors as higher-potential agroecologies, good access to road infrastructure, distance to principal urban markets, and location of processing infrastructure. Only a small minority of African farmers are located in such favorable market contexts. The majority of smallholder farmers face a number of constraints in terms of access to and integration into markets; availability of inputs, especially fertilizer; incentives for collective action; functionality of services, such as extension and credit; and, in summary, appropriate incentives for adoption of new technologies. Farmers thus tend to focus on the production of basic food staples, with a significant portion maintained for subsistence.

AIS approaches adopt a more holistic approach to resolving such interconnected production, organizational, and market constraints, but with an inherent bias toward those farmers and locations where market potential is more developed. A basic question is how AIS approaches can be adapted to the particular conditions of African agricultural contexts in order to reach the greatest number of farmers.

Innovation Methodology at Different Scales

Building on Systems Methods

AIS practice in Africa is still evolving. It has its roots in a number of integrated systems methodologies that have been developed over the past couple of decades. In Latin America and Asia, such systems focus on upstream research that supports agroenterprise clusters higher in the value chain. In contrast, AIS approaches in Africa focus on facilitating farm-level innovation, but are usually linked to commercial or market opportunities. AIS approaches do not exclude objectives related to food security or farming-system resilience, but methods for addressing these objectives are not as well developed, and coordinating their development and application at scale will be more difficult compared with value-chain approaches.

Agricultural innovation methods build on the trend since the 1990s toward more integrated approaches, particularly in achieving impact with R&D in natural resource management (NRM) (Hagmann et al. 2002). NRM-related R&D was based on systems approaches, was context specific, relied on tools and methods rather than physical technologies, involved farmer participation and adaptive management approaches, and was facilitated by a range of institutional actors. Since 2000, CGIAR has moved to formalize these methods and approaches and to integrate productivity and NRM research and application (Campbell and Sayer 2003; CGIAR Science Council 2003). While this effort has resulted in a set of principles and approaches, it has not had widespread application and, thus, is lacking an evolving community of practice.

Within CGIAR, integrated natural resource management (INRM) was overtaken by the development of challenge programs, and in Africa by the response to the 1990s market liberalization process. In attempts to link smallholder farmers to markets, the slow response of the private sector to the withdrawal of state agencies from input supply and output marketing resulted in an expansion of facilitated approaches. These approaches usually involved interventions at various points in value or supply chains and, as such, usually involved a commodity focus. Market access thus provided the incentive environment for farmers to adopt improved technologies. Nevertheless, given the commodity orientation of such value-chain approaches, new technology targeted productivity increases, thus moving away from INRM's focus on systems and resource management.

All of these elements came together in the design of CGIAR's SSA Challenge Program in what was termed Integrated Agricultural Research for Development (IAR4D). The design of this program in 2003 built on the foundations of INRM in linking productivity and NRM research; at the same time, however, it added research on markets. All of these elements were deemed necessary for developing smallholder agriculture and achieving what has more recently been termed "sustainable system intensification." The intent was to improve the integration of agricultural research and, hence, the impact on development outcomes, which required crossing traditional institutional boundaries, particularly between research and extension. The characteristics of IAR4D were defined by Hawkins et al. (2009, 3) as follows:

- IAR4D is about change or innovation as an outcome, not just about information, knowledge, or technology as a product;
- IAR4D places "research" as one of the components contributing to the development process, rather than its pivotal point;
- IAR4D focuses on processes and performance rather than just products (technologies, policies)—or, to put it another way, improved processes are the product.

As such, IAR4D was completely integrated into evolving ideas in the development of AIS, as was exemplified by the SSA Challenge Program (which is discussed in more detail later in this chapter).

Marrying Theory with Organizational Change and Methodological Practice

AIS/IAR4D practice continues to evolve and, hence, has succeeded where INRM effectively stopped. Much of this work is carried out by CGIAR centers or northern universities (Hounkonnou et al. 2012). The research, per se, focuses more on methodology and adaptive research than on traditional applied research. AIS methods are thus a bridge between developing technologies or managing components (such as varieties or soil fertility practices), and applying, adapting, or integrating them into farming systems. This is partly because the methods focus on identifying problems at the farming-system or community level, and, to maintain momentum in the innovation process, an initial testing of potential "on-the-shelf" technology options is required. However, the intent is to combine the farm-productivity dimension with institutional and organizational innovations that significantly change the possibilities and incentives available to farmers. Organizational innovations require the involvement of other actors in the agricultural sector, often at (but not limited to) the district level. As previously discussed, such organizational linkages are facilitated through innovation platforms (Nederlof, Wongtschowski, and van der Lee 2011).

In a recent systematic review of capacity strengthening in agricultural research, Posthumus, Martin, and Chancellor (2012, 1) found that "at the level of national agricultural research systems (NARSs), investments need to be made in strengthening relationships between research, extension, higher education, civil society, the private sector and farmer organisations to enhance innovation." The AIS approach shifts the focus of capacity strengthening from internal organizational reforms and human capital to improvements in organizational linkages. Spielman, Ragasa, and Rajalahti (2012, 277) note that "there is a case for both market and nonmarket approaches to improving demand articulation and organizational interfaces [in agricultural research systems]. They include investment in formal mechanisms that provide stakeholder input to research organizations, more participatory mechanisms that bring researchers and farmers together to solve problems, innovation platforms that address larger, more complex challenges with diverse actors, commercialization programs that move research into the marketplace, and financing mechanisms that encourage collaborative research." That is, NARI approaches to improving institutional linkages can be implemented incrementally, adding increasing complexity and building on experience and learning over time.

The core AIS concept, which dynamically involves core agricultural sector actors in the research process, has been tested in a few countries, but the results suggest a low degree of integration of research within the larger system. Spielman and Kelemework (2009, 6–8 and 22) evaluated interactions across four principal categories of actors (that is, domains) within the Ethiopian AIS (knowledge and education, bridging institutions, business and enterprise, and an enabling environment), and found that the interactions were primarily within individual domains (for example, private-sector actors). In terms of interactions across domains, "while respondents from all domains were satisfied with linkages with bridging institutions (in this case, linkages with public extension services), they were largely dissatisfied with their linkages with collaborators in all other domains, particularly the knowledge and education domain (in this case, public research organizations and institutes of higher learning)." This lack of interaction in the area of research was confirmed in Ghana and Nigeria (Ragasa, Abdullahi, and Essegbey 2011, 17), where in

Ghana "less than 30 percent of researchers in [research] institutes and faculties reported being involved in research–extension linkage committee (RELC) activities. Half of these organizations said that less than 10 percent of their researchers were involved in RELCs. A survey of 237 agricultural researchers suggests that 87 percent were not involved in RELCs." This trend was reflected in interactions with other domains—including farmers, with whom almost a quarter of researchers had not interacted in the past year.

An AIS perspective highlights the relatively autonomous functioning of NARSs in Africa. Moreover, as Mbabu and Ochieng (2006, 8) note, "many publicly funded agricultural organizations in Africa—such as agricultural research organizations, universities, extension services, and farmers' organizations—are facing a crisis of confidence among key stakeholders arising out of the failure to deliver the desired development impact." In an African context, the production of technologies by NARIs is not sufficient to produce the increases in smallholder productivity that are essential for economic growth. This is not a call for yet another major reform of African NARSs, but rather for developing the necessary leadership, organizational incentives, and shared vision, recognizing that joint action produces synergies far greater than can be achieved independently. Yet, significant transaction costs are involved in facilitating these organizational linkages and the financial and logistical resources required to support such mechanisms. In addition, new (primarily soft) skill sets are needed to facilitate these linkages, and whether these skill sets should be developed within NARIs or provided by other, more specialized organizations depends on context.

The Centrality of and Capacity in Adaptive Research

To develop the foundation to improve organizational linkages and function within an AIS, NARIs need to build their relative comparative advantage—producing new knowledge and technologies. However, research stations do not provide an effective locus for developing and managing organizational linkages. Adaptive research trials are a principal activity around which such linkages can be initiated. They are not only essential for dealing with the heterogeneity and scalability issues, as discussed above, but also provide a mechanism for interactions with farmers; developing a dialogue with extension providers; testing private-sector inputs, particularly varieties and fertilizer blends; and providing links to market agents and credit agencies. If properly designed, adaptive research trials can be integrated into more complex organizational mechanisms, such as innovation platforms, or can act as an initial bridge to wider interactions with other actors in the agricultural sector.

The N2Africa program has been initiated with demonstration and adaptive research trials as its core activity (Giller et al. 2013, 165):

> N2Africa focuses on the delivery and dissemination (D&D) of the best available nitrogen N2-fixing legume technologies . . . . Monitoring and evaluation (M&E) seek to understand why certain technologies work best for particular farmers, and feedback loops through adaptive research seek to refine and improve the technologies through addressing those problems that emerge. Thus, the emphasis is on improving N2-fixing legume technologies, solving problems encountered in the field, understanding how to tailor technologies to different farms and farming systems and using this understanding to refine D&D.

Understanding how to efficiently adapt technologies to different farming systems and agroecologies is a necessary first step in facilitating innovation and adoption by farmers. Where more complex technologies are involved, especially working with several components in a farming system, these trials could evolve into more participatory technology design and codesign with farmers (Triomphe 2012). This would require trade-offs between systematic data collection and location-specific adaptive management and learning, but the trade-offs could be managed as long as the researchers and farmers were clear about the objectives of the trials.

Adaptive research trials can thus become the basis for articulating demand, with researchers interacting with farmers to diagnose problems, then developing a location-specific understanding of the principal productivity constraints and yield responses to technological and management interventions. If done systematically—for example, through agroecological and socioeconomic stratification of the target area and population—NARIs could develop their understanding of the heterogeneity of yield responses and the potential for adoption, while enhancing their interactions with farmers over time. The basis for scaling the results would also exist to the extent that an organizational structure for farmers was in place. Alternatively, close linkages with extension services would allow recommendations to be targeted more precisely. Moreover, variations in farmers' circumstances could be incorporated as a basis for managing heterogeneity, rather than relying on national- or even regional-level recommendations (such as those used for fertilizer combinations and rates).

A systematic adaptive research network could not only feed information back into the research design process, but also feed critical information forward to rural credit and insurance schemes and to market agents. Production credit and insurance rely on understanding productivity responses under temporal and spatial variability. Moreover, agroenterprise investors need to understand the farm-level costs of production, farmers' supply responses, and seasonal variability. Location matters in their investments in terms of logistical and raw material supply costs. Thus, research can offer critical insights to an innovation platform, in essence supplying the public goods that justify public investment. Such capacity comes at some cost, however, in terms of operating budgets, which tend to be the most insecure, in terms of both the amount and the time of release, given that adaptive research is time sensitive if the resulting information is to be accurate. Therefore, an AIS is based as much on product and information flows as it is on organizational linkages and processes. In effect, these two aspects complement each other.

Emerging Experience with Agricultural Innovation Systems in Africa South of the Sahara

Effectively positioning national agricultural research in SSA within an AIS presents a number of constraints. A functional AIS relies on the following characteristics:

1. a sufficient array of institutions that are networked and have the necessary capacity;
2. deep entrepreneurship and innovation capacity, usually driven by well-functioning markets;
3. facilitated or self-organizing organizational platforms that permit effective networking; and
4. finances that cover the transaction costs inherent in the networking process.

Innovation primarily arises from interactions among the different actors within these networks—for example, among trade associations, farmers' organizations, and agricultural research institutes (Hall, Dijkman, and Sulaiman 2010a). However, the region's agriculture sectors are generally characterized by underdeveloped input and output markets; weak or nonexistent service delivery organizations (particularly in terms of rural credit and insurance, but also extension services in many areas); an emergent private sector; and a large smallholder sector that is only just developing a commercial orientation.

Given these constraints, the African experience with AIS approaches tends to focus on developing the innovation capacity of smallholder farmers, is facilitated by external agencies, is primarily led by international research organizations, and is developed in the context of multicountry projects. The fodder adoption project is a good example, whereby a consortium of CGIAR centers used an innovation systems approach to support innovation in the use of fodder technologies by smallholders. This was a facilitated approach focusing on developing innovation capacity by (1) strengthening weak ties among actors, (2) filling organizational gaps, (3) strengthening the supply system for fodder seeds, and (4) interacting with policymakers to improve policies (Ayele et al. 2012).

Strengthening networks and filling organizational and market gaps are characteristic of AIS approaches in Africa and lead to the question of how these approaches can be appropriately scaled. A review of the experiences of four African programs that use an AIS approach follows.

Research into Use

The Research into Use (RIU) program of the UK Department for International Development was specifically designed to put the scientific results of the Renewable Natural Resources Research Strategy (RNRRS) program into practical use by farmers at a relevant scale. The program's focus was on the actual application of knowledge, rather than the generation of knowledge (Hall, Dijkman, and Sulaiman 2010b; Clark 2013). RIU was launched in 2006 and ran for five years—a period too short to demonstrate significant farm-level impact at scale, especially given changing objectives after a midterm review. The RIU program's initial design was technology led, essentially applying on-the-shelf technologies resulting from the program. However, over the course of the RIU program, the activities were framed within an innovation systems perspective, to maximize learning from the program and ensure private-sector participation (Clark et al. 2012; Clark 2013).

The organization of the RIU program in Africa involved two program strands comprising seven country programs and a competitive "best bets" grant program.[2] The national programs also had two prongs: developing a national innovation coalition and facilitating the development of innovation platforms, essentially organized around specific commodity value chains. The national innovation coalition was primarily designed to support capacity strengthening in the innovation process, especially in terms of facilitating the development of innovation platforms and assisting farmers with communicating their concerns. The innovation platform aspect initially had a technology

2 Best bets refers to technological options determined to have the best potential to move on to farmer testing and adoption.

focus, but evolved into commodity value-chain platforms, in many cases focusing on farmers capturing an increasing proportion of the value-added in the supply chain.

The innovation platforms tended to be organized at the district or state level, although in Malawi they were organized at the national level. The composition of the innovation platforms also varied significantly, with much broader participation of different actors in Nigeria, but much greater involvement of farmers' groups and NGOs in Malawi and Rwanda, as well as much more focus on production, the market, and adding value. Research institutes also participated (especially in Nigeria), but they did not lead the facilitation of innovation platforms. The dominance of farmers' groups in Malawi and Rwanda often resulted in too much of a bias toward farmers' interests at the expense of viable business models in the supply chain (Gildemacher and Mur 2012).

The RIU program explicitly separated rural innovation from research, following a more linear model of first funding research under the RNRRS program and then applying that research in the RIU program. In effect, the application was driven by the available research products. In an evaluation of the RIU program, Gildemacher and Mur (2012, 166) note that, at least initially, the program went "against the principles of needs-driven research, and of making use of multiple sources of innovation. Rather than starting with the open question of needs and then engaging in a wide search for possible solutions from different sources, pre-conditions were set that reduced the chances of effective innovation." The evaluation assessed the different innovation platforms and best bets in terms of farm-level impacts and the development of innovation capacity. Although the five-year timeframe limited the potential for drawing conclusions, the evaluation focused on five cases and found that in one case there was significant impact but limited innovation capacity (Nigeria), while in another there was significant innovation capacity but limited impact, at least to date (Rwanda). This suggests that where preconditions of markets and effective institutions are in place and agroecologies do not vary (as in Nigeria), technology is adopted at scale (as in the Green Revolution). Such conditions are still relatively rare in SSA, however, and building innovation capacity linked to multiple sources of technology provides an entry point into broader-based agricultural development, although as Gildemacher and Mur (2012) note, what remains lacking is a framework to do this at a larger scale.

The SSA Challenge Program

The SSA Challenge Programme (SSA-CP) was a CGIAR response to developing a research program across CGIAR centers focusing on smallholder development in the region. The design of the program built on the methodological synthesis in INRM and attempted to integrate the increasing work on smallholder market access following market liberalization in the 1990s. The conceptual framework became known as IAR4D. The original intent was to directly link to the evolving conceptual thinking in AIS using a demand-driven farmer approach to establish system entry points and set the research agenda. However, the CGIAR Science Council found this open-ended process antithetical to ensuring high-quality science. A succeeding round of interactions between program management and the CGIAR Science Council resulted in a program design that would test the relative effectiveness of IAR4D using a randomized experimental design applied within three well-defined benchmark sites in West, East, and Southern Africa (Lynam, Harmsen, and Sachdeva 2010). The central methodology employed within IAR4D was the facilitation of innovation platforms.

The organization of the platforms varied somewhat across East, West, and Southern Africa, but was essentially focused at the subdistrict level because it facilitated the experiment's design by providing a framework for randomization. The platforms drew on representatives of farmers' groups, key market agents, and local government officials. A hierarchical approach to farmers' involvement enabled a cascade approach to building capacity and providing services to farmers' groups.

The innovation platforms were facilitated by university and extension personnel working with CGIAR centers, with a continuing question of how formalized the platforms should be, particularly, to interact with local government. In East and Southern Africa, the entry point for the innovation platforms was access to markets, where resolving that issue would lead to farmers' demand for new technologies. In West Africa, where the benchmark site was dominated by Nigeria, markets for principal crops were not viewed as a constraint, so the entry point focused on access to improved production technology. In all cases, the programs sourced on-the-shelf technology. Integrating productivity-enhancement techniques with improved NRM proved challenging within the time constraints of the program. These different entry points led to significant differences in the mix of participants in the platforms.

The role of agricultural research institutes in facilitating innovation capacity has been debated in AIS literature, particularly in terms of whether research institutes would bias innovation platforms to their own interests, and whether technology should take the lead as an entry point into the rural innovation process. CGIAR centers acted as "honest brokers" in facilitating the development of the platforms, with a research interest in evaluating how to achieve development outcomes more cost-effectively. The brokering role was critical in creating a neutral platform for identifying problems and developing potential solutions. Given the rural locus of the innovation process and the role that improved technologies play in rural innovation, such agricultural research institutes as CGIAR centers are well placed to facilitate the development of innovation platforms, because they have the necessary skills and operating budgets. However, only the larger NARIs would likely be in a position to develop the necessary skills and operating capacities. Alternative organizations that have both of these capacities are relatively limited in an African context, although in many cases, international NGOs are moving into this existing gap. The experience of the SSA-CP suggests that research institutes can play a leadership role in an AIS, but such facilitation is not necessarily limited to research institutes.

In an African context, innovation platforms should focus on cost-effectively facilitating rural innovation capacity at a sufficient scale, but also at a sufficiently contextualized local level. Finding this balance and creating the public funding that supports the inherent transaction costs involved should be a priority. Another priority is the need to design innovation platforms and build the capacity to facilitate their development. The SSA-CP had these priorities as a principal research objective, although it was not possible to vary scale because of the experimental design. As with many evaluations of organizational innovations, there was an inherent difficulty in adequately specifying the counterfactual. Moreover, the SSA-CP was critically time constrained, and funding was insufficient for the program to complete an adequate implementation period to evaluate effectiveness, within about two years.

Initial findings, however, suggest increased adoption of crop management technologies in innovation platform-facilitated communities compared with a control community. This was not true for soil management or postharvest technologies (Pamuk, Bultea, and Adekunle 2014). Moreover, bottom-up innovation platforms, as measured by variation in priorities across communities, tended to be more successful in facilitating adoption compared with top-down innovation platforms. Unfortunately, most of the African programs that are testing an AIS approach are project based, and securing multiyear funding has been a major constraint to effective implementation.

The African Highlands Initiative

In many ways, the African Highlands Initiative (AHI) developed the methodological underpinnings and practice of the principals underlying INRM. Its advantage in comparison to similar work in Africa is that it operated for a relatively long time by current project standards (from 1995 until 2008), after which components of the program were absorbed into the World Agroforestry

Centre's East Africa Regional Program. The evolution of the program over this period incorporated the conceptual thinking surrounding IAR4D and the AIS. The AHI organized its activities around benchmark sites across five countries, with an experiment in scaling out impact into other sites at the end of the program.

AHI's principal focus was NRM within the high-population-density areas of the East African highlands. In its first phase, the research targeted principal farming systems, so it had to be combined with improved productivity. The second phase expanded the scope and scale of research to watersheds, which involved an understanding of community-level collective action and developing broader governance and decisionmaking mechanisms. The work went on to look at how to integrate these institutional innovations at the district level. Much of the research resulted in developing methods at different scales and understanding interactions among technological change, market access, farmer investment in the natural resource base, institutional and governance innovation at the watershed and then the district level, and links to policy reform (see German et al. 2012 for a review of the program's work). Given the relatively long-term involvement in each of the benchmark sites, the final phase was designed to test whether both the soil and land management approaches could be scaled out to other districts.

AHI worked from farm to watershed to district levels, thereby integrating technical change, watershed management, and policy. However, scaling out technical innovations and scaling up institutional and policy innovations beyond the benchmark sites and districts proved to be more difficult and time constrained (Opondo et al. 2012). The AHI did this by developing interlinked innovation platforms at the watershed and district levels. An evaluation of this stage (Amede 2012) suggested that the watershed-level innovation platforms functioned well in the scaling out of technical innovations and sustainable land management techniques. However, they performed not as well at the district level, primarily because district government participation introduced incongruent interests. Unfortunately, this key issue was not tested further. The critical constraint, however, appeared to be the need for institutional change, district-level buy-in, and expanded participation in the district-level innovation platform by other soil- and land-management stakeholders.

Reform of Kenyan Agricultural Research

The dominance of international and regional actors in the testing of AIS approaches in Africa demands an assessment of the uptake of these ideas by NARSs and NARIs. Kenya provides an ongoing example of this process because it has undergone systematic policy reform within the agricultural sector, including research and extension. This reform has been seen as an integral aspect of economic transformation, starting with the Strategy for Revitalizing Agriculture (2004–2014), which was recently superseded by the Agricultural Sector Development Strategy (2010–2020). The 2010–2020 strategy is aligned with the CAADP process and provides the framework for finalizing new agricultural research and agricultural extension policies. The CAADP Compact resulted in the Medium-Term Investment Plan for 2010–2015 (Government of Kenya 2010), which prioritized investments across CAADP's four pillars; yet only 1 percent of the overall investment budget was earmarked for research and extension (Pillar 4). The problem of underfunding is recognized in the 2012 National Agricultural Research System Policy (Government of Kenya 2012, 15): "Government funding has been directed primarily to maintaining the core functions of public research institutes. Over recent years, foreign assistance has taken an increased share in funding both core functions and stand-alone projects. . . . The overall funding base for agricultural research remains fragile and unsustainable."

Nevertheless, the strategic focus of the 2010–2020 strategy and 2012 research policy was to commercialize smallholder agriculture, generate a yearly growth rate in the sector of 7 percent, and transform the sector to one that is essentially market driven. This would require significant increases in smallholder productivity. The investment gap for agricultural research is being filled by the World Bank–supported Kenya Agricultural Productivity and Agribusiness Program (KAPAP) and for extension by KAPAP and the Swedish government. At the same time, the Government of Kenya (2010, 13) notes the significant challenges in meeting these goals: "Asia's Green Revolution took place within the context of irrigated specialized agriculture, stabilized prices, public provision of subsidized inputs, assured markets for farm outputs, and cheap credit. In contrast, Kenya must achieve a largely market-led agricultural transformation within a context of mostly rainfed and highly diversified smallholder agriculture, high-cost agricultural input and output marketing, volatile prices, inefficient land, labor and credit markets, and a vibrant but relatively low-capacity private sector." These binding constraints on agricultural growth have informed the policy process through the recognition of the importance of organizational, institutional, and technological innovation in transforming smallholder agriculture.

The policy focus on commercializing smallholder agriculture has resulted in a priority-setting process based on commodities and an implementation process based on value chains. Thus KARI's approach to adaptive research is value chains for agricultural products, with a particular focus on partnerships, markets, and gender (Miruka et al. 2012). In research policy, value-chain and AIS approaches are directly integrated: "Some entities are embracing the concept of an 'Innovation Systems Framework' within which the agricultural product value chain and the Integrated Agricultural Research for Development (IAR4D) are included. This new approach seeks to (1) promote vertical coordination and horizontal integration within and among commodities, (2) fill in the missing nodes in the value chain continuum, and (3) improve linkages among research, education and extension. In Kenya, successes recorded in some areas such as horticulture, tea, banana, and dairy production show that in commercial commodities where the stakeholders are systematically consulted, the research agenda can be better focused and deliver impact" (Government of Kenya 2012, 11).

KARI's adoption of the commodity value-chain approach has led it to develop further capacity in marketing and, more recently, in facilitating local innovation platforms essentially based on commodity value chains (Makini et al. 2013). Scope is addressed through the future development of these innovation platforms regionally and nationally, also within a value-chain framework. Moreover, the move toward an AIS was complemented by policy changes in agricultural extension through the National Agricultural Sector Extension Policy of 2012. Extension was devolved to the counties with the intention of improving service provision by allowing farmers to choose the provider. Under KAPAP, extension focuses on particular value chains in pilot counties through farmers' "common interest groups." The elements of an AIS are in place, but county capacities are just being established, and the extension system has yet to be fully scaled out. KARI is also being reorganized under the new research policy, and government commitment to fund research and extension remains elusive.

Kenya's move toward a more fully integrated AIS has been led by an interconnected policy reform process embedded in larger political reforms under the new constitution, facilitated by both CAADP and KAPAP. The eventual impact, however, will depend on implementation. Some argue that a commodity value-chain approach stalls reform in commodity parastatals (Poulton and Kanyinga 2013). There is also the risk that relying on commercial, market-driven approaches will not have the requisite impact on balanced growth, food security, and poverty reduction within an agricultural sector that has high poverty rates, constraints on farm size, and a significant area in marginal agroecologies. The question of how to build these objectives into an evolving AIS in SSA remains a critical issue.

Conclusion

AIS approaches have shifted the capacity-strengthening debate on agricultural research in Africa from how NARIs are internally organized, to how they effectively link to other actors in the agricultural sector and how the new knowledge they generate is applied within a smallholder context. The AIS approach sets the objective of increasing smallholder productivity in the context of market-driven rural development, given high market-access costs, evolving input markets delivered at relatively high prices, incipient rural credit and insurance markets, and a reforming extension capacity. NARIs provide necessary but insufficient inputs into the rural innovation process; to ensure the adoption of new technologies and management techniques, they must interact with an expanding private sector, an increasing array of farmers' and civil society organizations, and reforming public-sector service providers. Market liberalization and democratization have been the principal drivers in this dynamic process, and agricultural research has an opportunity to move from a state of near isolation to developing essential organizational linkages for rural innovation and smallholder development.

A functional, national AIS is currently only emergent within a few African countries, based on pilot and methodological work by primarily international entities and a desire by a number of principal donors to structure their funding around innovative approaches. The Kenyan case suggests one course through a holistic and coordinated policy reform process, in part reflecting opportunities available through a new constitution. However, the impacts of such reforms are only achievable through effective implementation requiring financial resources.

Organizing research within an AIS requires developing new capacities and skill sets, reviewing internal organization, and expanding field-level operating capacity. In summary, organizing research around an AIS requires increased operating budgets, but government funding for research remains limited and highly dependent on external sources. The chicken-and-egg question remains of how to demonstrate impact to justify increased funding, when increased funding at critical junctures is necessary to produce those impacts. The challenge in moving this agenda forward will be in progressively demonstrating pilot-level impact to justify funding for expanded implementation. The failure to provide sufficient funding to adequately test the AIS approach in SSA-CP, even with initial positive results, is yet another example of an opportunity squandered and the perception that the approach is ineffective.

From a methodological perspective, two principal limitations are inhibiting the full development of the potential and effectiveness of AIS approaches.

The first is the need to move beyond innovation platforms organized around value chains. The value-chain approach has natural organizational logic, but it neglects noncommercial, mostly marginal agroecological zones, the enterprise complexity of smallholder farming systems, the balancing of food security and poverty objectives, and the longer-term investments needed to ameliorate degradation of the natural resource base. Second, the AHI demonstrated that it is possible to improve NRM, but it requires very different techniques and a much longer time period. Moreover, the more recent focus on sustainable intensification in smallholder systems requires more of a farming-systems perspective rather than a focus on a single commodity. A commercially viable cash crop can provide a suitable entry point in the innovation process, but it needs to be integrated into the rest of the farming system, particularly with the objectives of improved resource use efficiency and sustainable resource management, especially of soil, water, and trees.

Finally, given the budgetary constraints under which these approaches will be deployed, cost-effectiveness will depend on balancing heterogeneity, context, and scale. Given that innovation platforms involve facilitation with high transaction costs, spreading these costs at a scale that does not neglect farmer and market heterogeneity will be a critical balancing act. Understanding context heterogeneity more rigorously using spatial analysis tools and being able to allocate scarce organizational resources in relation to that understanding is one way of improving program efficiency. This will also involve more systematic approaches to adaptive research, as well as a better understanding of how to effectively build innovation capacity in rural settings. The most important factor in moving AIS approaches forward, however, will be in testing and comparing implementation options in terms of innovation platforms, innovation brokers, and scale. This will increase program costs and fuel the investment versus evaluation conundrum, whereby on one hand donors want to limit resources to just supporting implementation of programs, but on the other they want to increase the evidence of impact. African AIS are currently straddling this particular fence.

References

Amede, T. 2012. *Project Evaluation Report on Enhancing the Adaptive Management Capacities of Rural Communities for Sustainable Land Management in the Highlands of Eastern Africa: Devolution of SLM Scaling Up Roles and Responsibilities.* Project Evaluation Report. Ottawa: International Development Research Centre. Accessed October 2013. http://hdl.handle.net/10625/50308.

Anderson, J., G. Feder, and S. Ganguly. 2006. *The Rise and Fall of Training and Visit Extension: An Asian Mini-Drama with an African Epilogue.* World Bank Policy Research Working Paper 3928. Washington, DC: World Bank.

Ayele, S., A. Duncan, A. Larbi, and T. Tan Khanh. 2012. "Enhancing Innovation in Livestock Value Chains through Networks: Lessons from Fodder Innovation Case Studies in Developing Countries." *Science and Public Policy* 39 (3): 333–346.

Benin, S., E. Nkonya, G. Okecho, J. Randriamamonjy, E. Kato, G. Lubadde, M. Kyotalimye, et al. 2011. *Impact of Uganda's National Agricultural Advisory Services Program.* Washington, DC: International Food Policy Research Institute.

Benor, D., J. Harrison, and M. Baxter. 1984. *Agricultural Extension: The Training and Visit System.* Washington, DC: World Bank.

Campbell, B., and J. Sayer, eds. 2003. *Integrated Natural Resource Management: Linking Productivity, the Environment and Development.* Wallingford, UK: CABI Publishing in association with CIFOR (Center for International Forestry Research).

CGIAR Science Council. 2003. *Towards Integrated Natural Resource Management: Evolution of NRM Research within the CGIAR.* Rome: CGIAR Science Council Secretariat.

Clark, N. 2013. "Putting Research into Use (RIU): Technology Development for the Poor Farmer in Low Income Countries." *The World Financial Review.* Accessed October 2013. www.worldfinancialreview.com/?p=824.

Clark, N., A. Frost, I. Maudlin, and A. Ward. 2012. *Technology Development Assistance for Agriculture: Putting Research into Use in Low Income Countries.* London: Routledge.

Davis, K., E. Nkonya, E. Kato, D. Mekonnen, M. Odendo, and R. Miiro. 2012. "Impact of Farmer Field Schools on Agricultural Productivity and Poverty in East Africa." *World Development* 40 (2): 402–413.

German, L., J. Mowo, T. Amede, and K. Masuki, ed. 2012. *Integrated Natural Resource Management in the Highlands of Eastern Africa: From Concept to Practice.* Abingdon, Oxon, UK: Earthscan.

Gildemacher, P., and R. Mur, eds. 2012. *Bringing New Ideas into Practice: Experiments with Agricultural Innovation. Learning from Research into Use in Africa (2).* Amsterdam: Royal Tropical Institute; Edinburgh: Research into Use. Accessed October 2013. www.kitpublishers.nl/net/KIT_Publicaties_output/ShowFile2.aspx?e=2054.

Giller, K., A. Franke, R. Abaidoo, F. Baijukya, A. Bala, S. Boahen, K. Dashiell, et al. 2013. "N2Africa: Putting Nitrogen Fixation to Work for Smallholder Farmers in Africa." In *Agro-Ecological Intensification of Agricultural Systems in the African Highlands,* edited by B. Vanlauwe, P. van Asten, and G. Blomme. London: Routledge.

Government of Kenya. 2010. *Medium-Term Investment Plan: 2010–2015: Agricultural Sector Development Strategy.* Nairobi.

———. 2012. *National Agricultural Research System Policy.* Nairobi.

Hagmann, J., E. Chuma, K. Murwira, M. Connolly, and P. Ficarelli. 2002. "Success Factors in Integrated Natural Resource Management R&D: Lessons from Practice." *Conservation Ecology* 5 (2): 29.

Hall, A., J. Dijkman, and R. Sulaiman. 2010a. *Research into Use: Investigating the Relationship between Agricultural Research and Innovation.* UNU-MERIT Working Paper Series 2010-44. Maastricht, The Netherlands: Maastricht Economic and Social Research Institute on Innovation and Technology, United Nations University.

———. 2010b. "Research into Use: An Experiment in Innovation." *LINK Look*. Maastricht, The Netherlands: Maastricht Economic and Social Research Institute on Innovation and Technology, United Nations University. Accessed March 2011. www.innovationstudies .org/images/stories/linklookmarch2010.pdf?phpMyAdmin=Gla9117azp0fOuSptKR Au4Hzb3e.

Hawkins R., W. Heemskerk, R. Booth, J. Daane, A. Maatman, and A. Adekunle. 2009. *Integrated Agricultural Research for Development (IAR4D). A Concept Paper for the Forum for Agricultural Research in Africa (FARA) Sub-Saharan Africa Challenge Programme (SSA-CP).* Accra, Ghana: Forum for Agricultural Research in Africa.

Horton, D. 2012. "Organizational Change for Learning and Innovation." In *Agricultural Innovation Systems: An Investment Sourcebook*. Washington, DC: World Bank.

Hounkonnou, D., D. Kossou, T. Kuyper, C. Leeuwis, E. Nederlof, N. Röling, O. Sakyi-Dawson, et al. 2012. "An Innovation Systems Approach to Institutional Change: Smallholder Development in West Africa." *Agricultural Systems* 108: 74–83.

Hounkonnou, D., D. Kossou, T. Kuyper, C. Leeuwis, P. Richards, N. Röling, O. Sakyi-Dawson, et al. 2006. "Convergence of Sciences: The Management of Agricultural Research for Small-Scale Farmers in Benin and Ghana. *Netherlands Journal of Agricultural Science* 53 (3/4): 343–367.

Lynam, J. 2012. "Agricultural Research within an Agricultural Innovation System: Overview." In *Agricultural Innovation Systems: An Investment Sourcebook*. Washington, DC: World Bank.

Lynam, J., K. Harmsen, and P. Sachdeva. 2010. *Report of the Second External Review of the Sub-Sahara Africa Challenge Programme (SSA-CP).* Rome: CGIAR Independent Science and Partnership Council.

Makini, F., G. Kamau, M. Makelo, W. Adedunle, G. Mburathi, M. Misiko, P. Pali, et al. 2013. *Operational Field Guide for Developing and Managing Local Agricultural Innovation Platforms.* Nairobi: Kenya Agricultural Research Institute.

Mbabu, A., and A. Hall, eds. 2012. *Capacity Building for Agricultural Research for Development: Lessons from Practice in Papua New Guinea.* Maastricht, The Netherlands: Maastricht

Economic and Social Research Institute on Innovation and Technology, United Nations University.

Mbabu, A., and C. Ochieng. 2006. *Building an Agricultural Research for Development System in Africa.* ISNAR Discussion Paper 8. Washington, DC: International Food Policy Research Institute.

Miruka, M., J. Okello, V. Kirigua, and F. Murithi. 2012. "The Role of the Kenya Agricultural Research Institute (KARI) in the Attainment of Household Food Security in Kenya: A Policy and Organizational Review." *Food Security* 4: 341–354.

Nammuga, P. 2013. Unpublished ASARECA project reports. Association for Strengthening Agricultural Research in Eastern and Central Africa, Entebbe, Uganda.

Nederlof, S., M. Wongtschowski, and F. van der Lee, eds. 2011. *Putting Heads Together: Agricultural Innovation Platforms in Practice.* Royal Tropical Institute Development Policy and Practice Bulletin 396. Amsterdam: KIT Publishers. Accessed April 2016. repub.eur.nl/pub/77629/vellema_etal-2011_ugandainnovationplatform_kit-nederlof_etal_.pdf.

Okoboi, G., A. Kuteesa, and M. Barungi. 2013. *The Impact of the National Agricultural Advisory Services Program on Household Production and Welfare in Uganda.* African Growth Initiative Working Paper 7. Washington, DC: The Brookings Institution.

Opondo, C., J. Mowo, F. Byekwaso, L. German, K. Masuki, J. Wickama, W. Mazengia, et al. 2012. "Institutional Change and Scaling Up." Chapter 6 in *Integrated Natural Resource Management in the Highlands of Eastern Africa: From Concept to Practice*, edited by L. German, J. Mowo, T. Amede, and K. Masuki. Abingdon, Oxon, UK: Earthscan.

Pamuk, H., E. Bultea, and A. Adekunle. 2014. "Do Decentralized Innovation Systems Promote Agricultural Technology Adoption? Experimental Evidence from Africa." *Food Policy* 44: 227–236.

Posthumus, H., A. Martin, and T. Chancellor. 2012. *A Systematic Review on the Impacts of Capacity Strengthening of Agricultural Research Systems for Development and the Conditions of Success.* London: EPPI-Centre, Social Science Research Unit, Institute of Education, University of London.

Poulton, C., and K. Kanyinga. 2013. *The Politics of Revitalizing Agriculture in Kenya.* Future Agricultures Consortium Working Paper 59. Accessed April 2016. www.future-agricultures.org/publications/research-and-analysis/working-papers/doc_download/1731-the-politics-of-revitalising-agriculture-in-kenya.

Ragasa, C., A. Abdullahi, and G. Essegbey. 2011. *Measuring R&D Performance from an Innovation Systems Perspective: An Illustration from the Nigeria and Ghana Agricultural Research Systems.* Conference Working Paper 14 from the ASTI/IFPRI–FARA Conference "Agricultural R&D: Investing in Africa's Future: Analyzing Trends, Challenges, and Opportunities,"

Accra, Ghana, December 5–7. Accessed January 2015. www.asti.cgiar.org/pdf/conference/Theme3/Ragasa.pdf.

Reddy, V., A. Hall, and R. Sulaiman. 2011. *The When and Where of Research in Agricultural Innovation Trajectories: Evidence and Implications from RIU's South Asia Projects.* UNU-MERIT Working Paper Series. Maastricht, The Netherlands: Maastricht Economic and Social Research Institute on Innovation and Technology, United Nations University.

Spielman, D., and D. Kelemework. 2009. *Measuring Agricultural Innovation System Properties and Performance: Illustrations from Ethiopia and Vietnam.* IFPRI Discussion Paper 851. Washington, DC: International Food Policy Research Institute.

Spielman, D., C. Ragasa, and R. Rajalahti. 2012. "Designing Agricultural Research Linkages within an AIS Framework." In *Agricultural Innovation Systems: An Investment Sourcebook. Module 4: Agricultural Research within an Agricultural Innovation System.* Washington, DC: World Bank.

Sumberg, J. 2005. "Systems of Innovation Theory and the Changing Architecture of Agricultural Research in Africa." *Food Policy* 30: 21–41.

Theus, F., and D. Zeng. 2012. "Agricultural Clusters." In *Agricultural Innovation Systems: An Investment Sourcebook.* Washington, DC: World Bank.

Triomphe, B. 2012. "Co-Designing Innovations: How Can Research Engage with Multiple Stakeholders?" In *Agricultural Innovation Systems: An Investment Sourcebook.* Washington, DC: World Bank.

Chapter 14

CROSSBORDER COLLABORATION IN AFRICAN AGRICULTURAL RESEARCH: THE POLITICAL ECONOMY OF TECHNOLOGY SPILLOVERS

Johannes Roseboom

Determining the best way to organize and fund agricultural research is a topic that has received considerable attention in recent years (GFAR 2011), particularly in relation to Africa south of the Sahara (SSA) (Roseboom 2011). Although national governments hold the primary responsibility for the organization and funding of agricultural research, these issues have inherent international dimensions, because most agricultural research challenges extend beyond national borders. The resulting incongruence between internationally relevant problems and nationally driven solutions leads to wasteful duplication of effort, as well as underinvestment. A key concept in the discussion of these issues is what economists call "technology spillovers," whereby the benefits of advances in knowledge and technology developed in (and paid for by) one jurisdiction spill over into another.

This chapter first presents an overview of current thinking about agricultural research spillovers; it goes on to summarize recent attempts to quantify the potential for agricultural technology spillovers globally, across SSA's subregions, and between member countries of the subregional organizations (SROs) of national agricultural research systems (NARSs) in SSA[1]; it then provides an assessment of the implications of this information for regional collaboration in agricultural research. The remaining discussion focuses on benchmarking some of the key characteristics of crossborder collaboration in agricultural research in the United States, the European Union (EU), and SSA. The chapter closes with a summary of the main conclusions.

The author would like to thank participants of the writers' workshop organized by Agricultural Science and Technology Indicators in July 2013 for their feedback on an early draft of this chapter and various reviewers for their comments at various stages.

1 The SROs are the Association for Strengthening Agricultural Research in Eastern and Central Africa (ASARECA), the Centre for the Coordination of Agricultural Research and Development in Southern Africa (CCARDESA), and the West and Central African Council for Agricultural Research and Development (CORAF/WECARD).

Technology Spillovers

It is a commonly held notion that—in addition to local investment in agricultural research—technology spillovers play an important role in explaining local changes in agricultural productivity. Nevertheless, few agricultural productivity studies have actually tried to capture this effect quantitatively, and those that have often attribute a sizable share (in many cases more than half) of the measured productivity increase to such technology spillovers (Alston 2002). These results suggest that agricultural productivity studies that do not account for technology spillovers substantially overestimate the contribution of local research. A better understanding of how agricultural technology spillovers occur should help to improve the design of more effective agricultural research systems.

Byerlee and Traxler (2001) distinguish three types of research spillovers:

1. *knowledge-related spillovers*, involving knowledge generated elsewhere, such as a new discovery or research method, that is applied in developing a new technology;
2. *technology-related spillovers*, which occur when an actual technology is transferred or adapted to a new environment; and
3. *price-related spillovers*, involving the adoption of a more efficient technology that affects the price of a commodity in locations both where the technology was adopted, and—through market effects—where it was not adopted.

Research spillovers are usually cast as spillovers across locations, but they can also occur across stages of research—that is, between basic, applied, and adaptive research (Box 14.1, Figure 14.1). The spillover from basic to applied

BOX 14.1 The different stages of research

Basic research aims to increase understanding of fundamental principles; it is curiosity driven, without any concrete (economic) application of the knowledge in sight. In contrast, *applied research* uses the knowledge generated through basic research to develop technologies and solutions that target concrete problems and opportunities. *Adaptive research* aims to modify an existing (prototype) technology or solution to local circumstances.

Source: Author.

FIGURE 14.1 Knowledge and technology spillovers at different stages of the research process

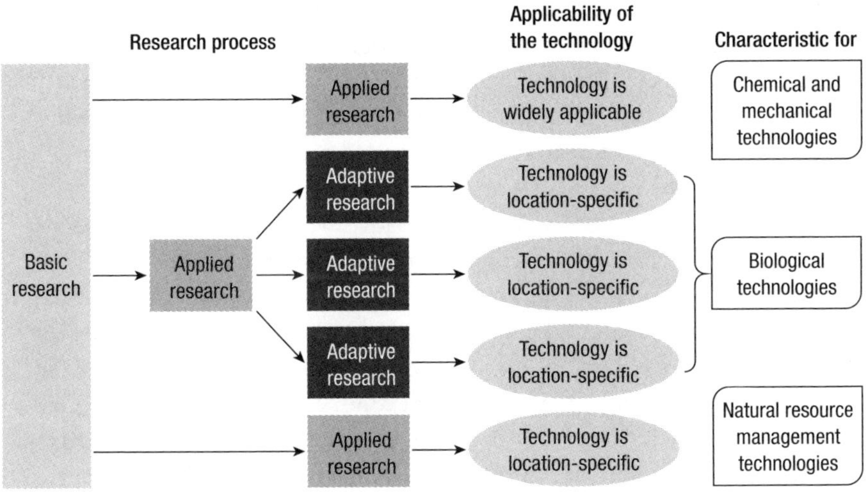

Research process
Applicability of the technology
Characteristic for
Basic research
Applied research
Applied research
Adaptive research
Adaptive research
Adaptive research
Applied research
Technology is widely applicable
Technology is location-specific
Technology is location-specific
Technology is location-specific
Technology is location-specific
Chemical and mechanical technologies
Biological technologies
Natural resource management technologies

Source: Author.

research is mostly knowledge spillover, whereas the spillover from applied to adaptive research is technology spillover. The dominant direction of the spillover flows from basic to applied research, and from applied to adaptive research, but feedback loops also occur (for example, information from adaptive research trials feeding back into the applied research agenda). Moreover, these spillovers between different phases in the research process are usually *intended*. Not highlighted in Figure 14.1 are *unintended* technology spillovers between applied research programs—for example, a technology developed for irrigated rice production spilling over to rainfed rice production.

Knowledge spillovers from basic to applied research can have huge and long-lasting impacts by affecting multiple domains of applied research; take, for example, the discovery of DNA. Nevertheless, such scientific discoveries are rare and unpredictable. In today's highly interconnected world, the diffusion of basic research results is almost instantaneous, whereas in the past it could take centuries to spread across the world. The more critical bottleneck nowadays is whether countries have the scientific capacity to absorb and apply this knowledge. All countries in the world are aware of what can be done with modern biotechnology, but only the larger and richer ones actually have the capacity to deploy that knowledge.

The spillover of agricultural technologies from applied to adaptive research varies by technology. Some (such as mechanical and chemical technologies) can be used worldwide with little or no adaptation to local circumstances, while others (particularly biological technologies) can require significant adaptation to local circumstances. Moreover, in certain instances, it is not a matter of adapting an existing technology; rather, it relates to conducting applied research to develop an original technology for a specific location (for example, natural resource management technologies).

A gamut of combinations of applied and adaptive research exists in the conduct of agricultural research to generate both widely applicable and location-specific technologies. Whether a prototype technology should undergo adaptive research depends (a) on the additional yield per hectare generated by the adaptive research program, (b) on the area under production affected by the adapted technology, and (c) on the costs of conducting the adaptive research program. Adaptive research only makes sense when the additional benefits it generates (a × b) can at least pay for its additional costs (c).[2]

This stylized representation of the research process assumes a world without national borders and markets, and with a single supranational entity deciding on the optimal design of the agricultural research system. Introducing national borders, markets, and competing approaches significantly complicates the research process and brings price-related research spillovers into play.

First, depending on the factors affecting the supply and demand of a particular commodity, producers will pass on some or all of the research benefits to consumers in the form of lower prices. The more inelastic (that is, fixed) the demand, the larger the share of the benefits consumers will receive through lower prices. The impact of price-related spillovers in four "stylized" markets is presented in Table 14.1. By adopting the perspective of a national planner, these price-related spillovers affect the optimal allocation of research resources. Not only does the absolute size of the research benefits matter (relative to the cost of generating them), but also to whom the benefits will accrue. For example, investing in research that targets foreign markets with fairly low demand elasticity does not make much sense from a national perspective because the research benefits will go to foreign consumers. Moreover, insight into who benefits from research helps to determine who should pay for it. For example, in commodity markets with high demand elasticity, it is more opportune to tax producers for the costs of the research, because they have a lot to

2 Byerlee and Maredia (1999) point to a common mistake that adaptive research programs make in assuming there will be no yield improvement without adaptive research. This is not necessarily so.

TABLE 14.1 **Price-related spillovers under different scenarios**

Demand	Local market	Foreign market
Low elasticity	Research benefits are passed on to local consumers in the form of lower prices	Research benefits are passed on to foreign consumers in the form of lower prices
High elasticity	Research benefits are mainly captured by local producers	Research benefits are mainly captured by local producers

Source: Author.

gain through the potential to increase their sales; however, the same cannot be said in commodity markets with low demand elasticity.

Second, a further distinction can be made between adopters and non-adopters of technology. Nonadopters are not directly affected by technology changes, but are indirectly affected by lower prices induced by improved technologies; hence, they may lose market share to technology adopters (particularly in markets with low demand elasticity). Therefore, innovation processes not only generate winners (technology adopters and consumers), but also losers (technology nonadopters).

A third aspect is that the introduction of multiple national jurisdictions (rather than a single supranational one) causes a fragmentation of applied and adaptive research into parallel, national research efforts targeting the same commodity, agroecology, production system, or problem. This leads not only to research duplication, but also to less ambitious research agendas founded on decisionmaking processes that account only for the *national* benefits of agricultural research investments.

At the same time, because of the duplication in research effort, national (and state and provincial) jurisdictions create a substantial amount of *unintended* technology spillover between parallel research programs in different jurisdictions targeting the same commodity, agroecology, production system, or problem. Because countries differ by size and stages of economic development, unintended technology spillovers tend to run from large to small countries and from technologically more advanced to less advanced countries—a factor that is often captured in the form of yield differences (technology spillovers generally run from countries with higher yields to those with lower yields). However, countries that operate in isolation, have perfect technology proximity (that is, high similarity in agricultural production and agroecology), and are equal in size and technology advancement have little to gain from each other in terms of these unintended technology spillovers. They can be expected to conduct very similar research, resulting in a high degree of research duplication.

A fouth dimension requiring consideration is the economies of size of the research itself. Byerlee and Traxler (2001) present a conceptual framework incorporating this aspect together with market size and spill-ins in order to determine the optimal size of national research investment in a given market. They distinguish the following strategies, outlining increasingly more ambitious, complex, and costly research agendas through which countries can take advantage of emerging innovation opportunities:

1. *Spontaneous diffusion of improved technologies* without the benefit of local research and development (R&D)—as argued above, spontaneous spill-ins vary considerably across technologies (the higher the spill-in, the less the need for additional action)

2. *Direct transfer of technologies* after testing and screening by local R&D programs for suitability to local environments

3. *Adaptive transfer of technologies*, whereby final technologies from elsewhere are subject to local *adaptive* research before local release (for example, the use of imported varieties as parents in local breeding programs)

4. *Comprehensive applied research*, whereby imported knowledge from basic research conducted elsewhere is used in local applied research programs to produce homegrown technologies

5. *Comprehensive basic and applied research* that uses imported knowledge and includes the ability to conduct basic and pretechnology research

Byerlee and Traxler (2001) argue that these increasingly complex research capacities often lead to discontinuities in the research production function. For example, the transition from Strategy 2 to Strategy 3 involves the addition of a crossing program and early generation selection, which is considerably more complex and expensive to undertake than is simple testing of imported varieties. Depending on the volume of research benefits that can be captured by a country (mostly defined by market size), countries will adopt a more ambitious research strategy.

Because of limited market size, small countries probably best settle for Strategies 1 and 2, or Strategy 3 in the case of a technology that affects a major commodity or multiple commodities. Strategies 4 and 5 are largely out of reach for small countries because of the high costs of such activities relative to the limited volume of national production that could benefit. These countries are very much net beneficiaries of technology spillovers, but at the same time have to accept technologies that are not necessarily a perfect match for their circumstances. Nevertheless, despite their less ambitious research agendas, small countries tend to invest proportionally more in agricultural research than large countries (Pardey, Roseboom, and Anderson 1991).

Exploring SSA's Technology Spillover Potential

The presence of agricultural technology spillovers has mostly been addressed in the context of agricultural productivity studies that try to establish a link between research investments and advances in productivity. All have struggled with how to capture the spillovers and how to attribute observed productivity changes to research conducted locally, to the spillover of technology from elsewhere, or to a combination of both of these factors. Alston (2002) discusses how different studies have dealt with these problems and shows that the assumptions imposed can result in significant differences both in the volume of research benefits and in the attribution of those benefits to an individual's or organization's own research and that of others. How best to capture technology spill-ins in agricultural productivity studies is still subject to extensive discussion and experimentation.

Another line of research, pioneered at the International Food Policy Research Institute (IFPRI), looks at the potential for technology spillovers across locations based on similarities in production, agroecology, and other factors. The result is policy-relevant, geospatial "domains," such as agroecological zones, that take into account large amounts of data on rainfall, sunshine, altitude, soil type, infrastructure, population, production, and so on.

Global and Regional Technology Spillovers

Pardey et al. (2007) present a global picture (based on 156 countries) of technology proximity based on similarities among countries and groups of countries in terms of agroecology and output mix. They calculated a measure of proximity for both factors on a scale of 0, indicating no similarity, to 1, indicating complete similarity (Table 14.2).[3] Together, the agroecological and output proximity scores result in a technology proximity score. The results of their study reveal very low technology proximity between low-income and high-income countries, which is unfortunate. High-income countries are technologically more advanced and account for around two-thirds of global

3 For details on the methodology used to measure similarity, see Pardey et al. (2007).

TABLE 14.2 Agroecological and output proximity of countries clustered by income level

Income bracket	Agroecological proximity				Output proximity			
	High income	Higher middle-income	Lower middle-income	Low income	High income	Higher middle-income	Lower middle-income	Low income
High income	1.00	—	—	—	1.00	—	—	—
Higher middle-income	0.81	1.00	—	—	0.95	1.00	—	—
Lower middle-income	0.56	0.69	1.00	—	0.74	0.71	1.00	—
Low income	0.06	0.13	0.44	1.00	0.38	0.38	0.64	1.00

Source: Pardey et al. (2007).

public and private spending on agricultural R&D. Pardey et al. (2007) argue that the structural differences in agroecology and agricultural output very much curtail the potential for technology spillovers from high-income to low-income countries. Another concern is that in recent years the public agricultural research agenda of high-income countries has shifted away from productivity enhancement to more qualitative aspects, such as health and ecological issues. These are not necessarily the research priorities of low-income countries that are still struggling with food security. In other words, low-income countries are very much on their own when it comes to developing better agricultural technologies.

Using the same methodology and dataset, Pardey et al. (2007) also focus more closely on Africa (Table 14.3). Results reveal that North Africa has very low agroecological proximity to SSA; its output proximity to SSA is slightly better, but is weaker than its proximity to the rest of world. These results

TABLE 14.3 Agroecological and output proximity among African subregions and with the rest of the world

Subregion	Agroecological proximity					Output proximity				
	North Africa	East Africa	Southern Africa	West Africa	Rest of the world	North Africa	East Africa	Southern Africa	West Africa	Rest of the world
North Africa	1.00	—	—	—	—	1.00	—	—	—	—
East Africa	0.00	1.00	—	—	—	0.41	1.00	—	—	—
Southern Africa	0.13	0.75	1.00	—	—	0.53	0.70	1.00	—	—
West Africa	0.01	0.85	0.81	1.00	—	0.21	0.52	0.40	1.00	—
Rest of the world	0.27	0.32	0.29	0.36	1.00	0.73	0.60	0.71	0.31	1.00

Source: Pardey et al. (2007).

suggest that the potential for technology spillovers between North Africa and SSA is low, and that North Africa would be better off seeking research collaboration elsewhere.

The agroecological proximity between East, West, and Southern Africa stands out as being relatively high. Surprisingly, however, output proximity is substantially weaker. In particular, West Africa stands out as having a relatively low output-proximity score, both with the other subregions and with the rest of the world, which reduces the potential for technology spillovers. Southern Africa includes South Africa, which is an important stronghold of agricultural research capacity in the region. Individually, South Africa's agroecological proximity to the other subregions is weak (less than 0.2); its output proximity to the other subregions is also lower than that of Southern Africa, but less dramatically so. This finding dampens high expectations regarding South Africa's potential lead role in agricultural innovation in the region.

Pardey et al. (2007) also point to the fact that geographic proximity does not necessarily translate into technological proximity. It may be that a country has more in common in terms of its technology needs with countries in other parts of the world than with its direct neighbors. Moreover, linking with countries with similar agroecological and output mixes is all the more interesting when they are technologically more advanced.

Intraregional Technology Spillovers

In recent years, IFPRI has conducted priority-setting exercises on regional agricultural research for each of three SROs,[4] which take into account the potential technology spillovers between countries within each of the subregions. Johnson et al. (2011) discuss the common methodological framework used in these three studies (Box 14.2), as well as the specific strengths and weaknesses of each study. To capture the technology spillover potential across subregions for specific commodities, all three studies use spatial analysis tools to define "development domains" that present similar characteristics in terms of agroecology, climate, population density, and market access, and that can be expected to have similar agricultural development problems and opportunities. These specific domains allow for a more realistic depiction of technology proximity (and, hence, technology spillover potential), as well as a differentiation in innovation approaches.

4 ASARECA (Omamo et al. 2006), CORAF/WECARD (Nin Pratt et al. 2011), and Southern African Development Community (SADC) (Johnson et al. 2014).

BOX 14.2 A common analytical framework

All three regional studies used a common framework, starting with a highly disaggregated spatial analysis based on key biophysical and socioeconomic factors of geographic areas sharing similar characteristics and endowments and, in turn, their degree of agricultural suitability, type of production systems and commodities, and available technology options (Panel A). The resulting distinctive agricultural development domains, which are not limited by political boundaries, provided a measure of the technological proximity of different countries and, hence, the potential for technology spillovers among them. Second, more detailed economic analysis was undertaken using a regional economywide multimarket model (Panel B) and IFPRI's Dynamic Research Evaluation for Management (DREAM) model (Panel C). The DREAM model was typically used to measure the potential magnitude of economic benefits derived from different commodity-based R&D investment options that rely on the distributional pattern of each development domain across countries. The multimarket model was developed to capture economywide implications of the same investments, including the potential benefits from technology spillovers on overall sector growth, incomes, prices, and consumption. Results from the economic analyses were then used to derive alternative rankings of R&D investments based on weighted criteria of a commodity-specific R&D investment's potential to contribute to overall sector growth, generate greater spillover benefits, and provide larger welfare outcomes in terms of regional food security and poverty reduction objectives (Panel D).

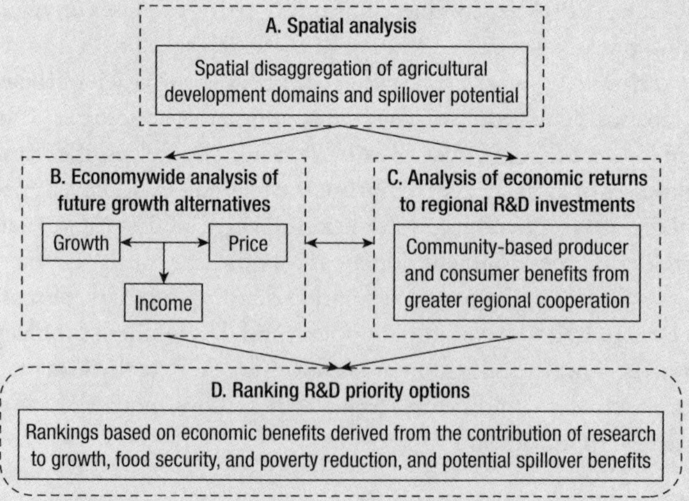

Source: Johnson et al. (2011). Text updated by author.

Notes: IFPRI = International Food Policy Research Institute; R&D = research and development.

In addition to technology proximity, several other factors need to be considered in estimating the technology spillover potential:

1. *The difference in technological advancement between countries,* captured here as yield differences. The technology spillover is expected to run from a country with a high yield to one with a low yield. Moreover, the yield difference needs to be sufficiently large to trigger a spillover.
2. *Whether the recipient country has sufficient absorptive capacity* in the form of adaptive research capacity to capture the full technology spillover potential. The ASARECA and CORAF/WECARD studies assume that all countries in their subregion have the same absorptive capacity, so this factor does not affect their estimates of the potential volume of technology spillover. The SADC study, however, introduces a differentiation in absorptive capacity on the basis of country-specific rates of return for crop and livestock research. Countries with low rates of return on agricultural research investment are assumed to have limited absorptive capacity, and vice versa.
3. *The maximum adoption rate in the recipient country,* which may be affected by such factors as the education of farmers; the quality of the extension services; the availability of credit, land tenure, communications, and market structure; previous exposure to technical change; and so on (Davis, Oram, and Ryan 1987). In all three studies, this factor is ignored, and it is implicitly assumed that the development domains have the same maximum adoption rate across countries.
4. *The time lags in technology development and diffusion*, stylized in Figure 14.2 taken from Davis, Oram, and Ryan (1987), illustrate a process of slow initial adoption, followed by fast adoption and then a return to slower adoption before the adoption ceiling is reached. By differentiating the adoption ceiling across countries, the model also allows the presence of nonadopters to be differentiated. In the country where the research is undertaken, an initial research lag occurs before results become available (research lag A); this is followed by an adoption process that can take another *x* years before the adoption ceiling is reached. If no adaptation of the technology is needed, the adoption process in the recipient country can be assumed to begin at the same time as in the country that conducted the research; hence, the only difference in the adoption process between the country of origin and the recipient country may be the assumed adoption ceiling. However, if adaptive research by the recipient country is needed to adapt the technology to local circumstances, an

FIGURE 14.2 Schematic representation of the ceiling levels of adoption assumed in this study

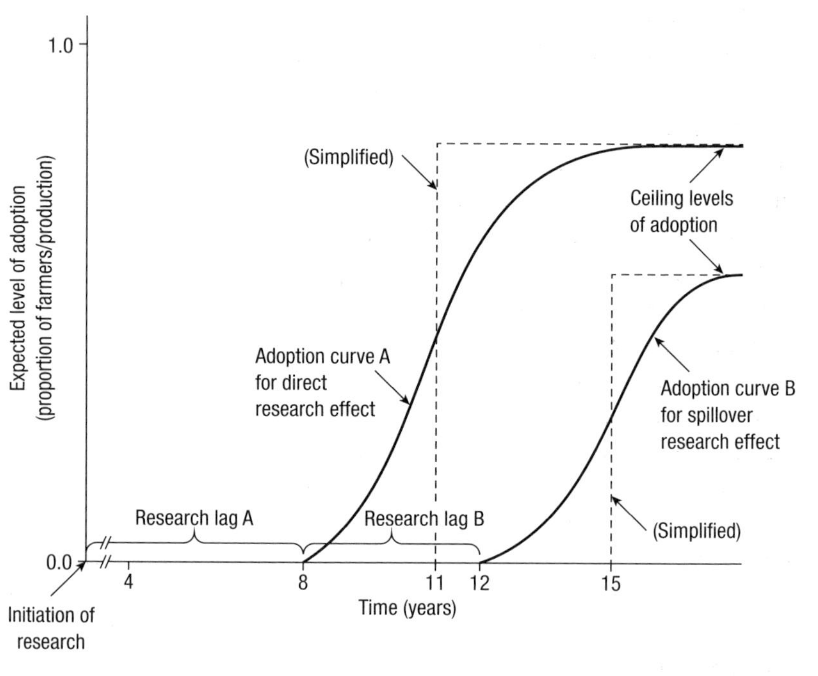

Source: Davis, Oram, and Ryan (1987).

additional lag (research lag B) enters into the picture. Time lags in technology development, adaptation, and adoption are very much technology specific and may result in very different technology spillover patterns ranging from almost complete and immediate to very incomplete and slow. This requires extensive detailed information regarding the technologies being developed for each commodity. In practice, such differentiation has never been attempted at the scale needed for these subregional studies, so a standard technology development, adaptation, and adoption process is usually assumed for all commodities and all technologies, or time lags and adoption ceilings are not considered at all.

By taking technology spillovers into account, the three studies suggest a prioritization of commodities at the subregional level different from when only the national benefits of technology improvements are accounted for. The CORAF/WECARD study does not present the "without spillover" case, but the ASARECA and SADC studies do (Table 14.4):

TABLE 14.4 Ranking of commodities per region based on research-induced benefits, including spillovers

ASARECA benefits				CORAF/WECARD benefits				SADC benefits			
Commodity	National	Spillovers (US$ millions)	Total	Commodity	National	Spillovers (US$ millions)	Total	Commodity	National	Spillovers (US$ millions)	Total
Cassava	5,200	2,581	7,781	Rice	na	na	8,200	Maize	4,870	1,226	6,096
Cow's milk	4,456	2,984	7,440	Cassava	na	na	5,550	Cassava	3,263	264	3,527
Plantains	6,575	659	7,234	Groundnuts	na	na	4,280	Wheat	2,272	71	2,343
Maize	5,659	1,477	7,136	Sorghum	na	na	3,560	Rice	1,722	426	2,148
Beef	3,741	2,409	6,150	Maize	na	na	3,540	Cattle	1,508	294	1,802
Coffee	2,566	1,461	4,027	Yams	na	na	3,390	Chicken and eggs	1,646	78	1,724
Sorghum	1,064	2,059	3,123	Cocoa	na	na	2,375	Beans	877	184	1,061
Vegetables	1,742	956	2,698	Millet	na	na	2,300	(Sweet) potatoes	787	79	866
Dry beans	1,701	626	2,327	Cotton	na	na	2,125	Sheep and goats	797	46	843
Rice	854	1,355	2,209	Bananas	na	na	2,100	Pigs	686	35	721
Mutton/lamb	467	1,399	1,866	Coffee	na	na	460	Sorghum	383	190	573
Groundnuts	553	1,254	1,807	Oil palm	na	na	120	Groundnuts	407	82	489
Potatoes	982	490	1,472					Cotton	234	40	274
Cotton	427	251	678					Sugar	185	11	196
Cashews	396	5	401					Millet	135	26	161
Total	*36,383*	*19,966*	*56,349*	*Total*	na	na	*38,000*	*Total*	*19,772*	*3,052*	*22,824*

Sources: Omamo et al. (2006), Nin Pratt et al. (2011), and Johnson et al. (2014).

Notes: ASARECA = Association for Strengthening Agricultural Research in Eastern and Central Africa; CORAF/WECARD = West and Central African Council for Agricultural Research WECARD and Development; SADC = Southern Africa Development Community; na = not applicable.

1. The intraregional spillover benefits are substantial: 35 percent of the reported research benefits in ASARECA's subregion and 13 percent in SADC's subregion are due to spillovers. Moreover, most (98 percent) of the spillover benefits in the SADC subregion accrue to low-income countries, whereas hardly any accrue to the middle-income countries. However, a substantial part (54 percent) of the technology spillover in the SADC subregion originates from low-income countries, which is somewhat counterintuitive. South Africa, which represents about half the agricultural research expenditure in the SADC subregion (Beintema and Stads 2011) and is technically the most advanced country in that subregion, only generates 19 percent of the technology spillover. This is consistent with the finding of Pardey et al. (2007) that South Africa has relatively little in common with the rest of SSA in terms of agroecology and agricultural production. Given South Africa's large weight in the total, this also explains why spillovers play a much smaller role in the SADC subregion than they do in the ASARECA subregion.

2. Large variations can occur across commodities in terms of the relative importance of the spillover effect. In the case of the ASARECA subregion, the spillover share in benefits ranges from 1 percent for cashews to 75 percent for mutton/lamb. In the SADC subregion, it ranges from 3 percent for wheat to 33 percent for sorghum.

3. The rankings presented in Table 14.4 focus on research benefits, but not all research benefits have the same impact in terms of reducing poverty, improving food security, and so on. This is where IFPRI's economic model (Box 14.2) provides more differentiated information on the expected research benefits, which can be used to give preferred research benefits (for example, those with a higher impact on reducing poverty) an artificially higher weight in the priority-setting process.

4. One of the issues for the SROs is how to interpret the information presented in Table 14.4. The ranking of commodities by total research benefits would be the relevant ranking if an SRO had control over all the agricultural research funding within its subregion. This is not the case, however. Most of the agricultural research funding within a subregion is controlled by national governments. Collectively, they can be expected to prioritize research on the basis of their national benefits, as depicted in the relevant column. This positions the SROs as

complementary actors: they should prioritize their research funding based on the spillover benefits, not the total benefits. Even when a commodity is big within the subregion, the SROs should not get involved (at least this is the logical consequence of the IFPRI approach) when little or no (unintended) spillover benefits accrue.

Technology Spillover and Centers of Excellence

What stands out is that in all three IFPRI studies, the volume of the estimated technology spillover potential is mainly driven by development domain similarities and yield differences. Yield differences also determine the direction of the technology spillover. Other factors in the technology spillover process have been captured with little differentiation (the same values for all commodities) or not at all. The potential spillover effect being captured in these studies is that of *unintended* technology spillovers resulting from the separation of agricultural research into national jurisdictions. What these studies do not capture, however, is the possibility of creating *intended* technology spillovers.

This scenario of creating intended technology spillovers is particularly relevant when development domains are similar, yield differences between countries are low, and the technology exchange is limited (Table 14.5). This is a situation under which all countries do a little bit of the same research (the incidence of research duplication is high), but none of them are really pushing the technological frontier. In such a situation, technology spillovers could be created purposefully by promoting a more differentiated agricultural research landscape at the subregional level by creating centers of research excellence (or specialization or leadership). The idea is that countries would voluntarily assume research leadership for one or more commodities of subregional interest and generate *intended* technology spillovers for the other countries of that subregion. By allocating these leadership roles equally across the subregion

TABLE 14.5 Technology spillover potential

Yield difference	Similarity of development domains	
	Low	High
Low	Low (unintended and intended) technology spillover potential	Low (unintended) technology spillover potential, but a high likelihood of research duplication
High	Low (unintended and intended) technology spillover potential	High (unintended) technology spillover potential

Source: Author.

Note: The combination "high similarity, low yield difference" can also represent a situation of perfect technology spillover.

(which requires the necessary coordination and collaboration), the distribution of costs and benefits among countries of such an approach could be roughly equalized.

The center of excellence approach has been strongly promoted in recent years by the Forum for Agricultural Research in Africa (FARA) and the SROs in the context of the Comprehensive Africa Agriculture Development Programme, which was initiated in 2003 under the New Partnership for Africa's Development to strengthen African agriculture. World Bank funding has backed this approach by providing loans (some $636 million to date)[5] to clusters of countries in each of the subregions, (1) to transform existing national research programs or institutes into centers of research excellence that aim to have subregional impact, and (2) to facilitate the technology uptake by recipient countries through training and joint research activities and standardization of national technology policies. These World Bank, multi-country program loans are sitting together in three subregional agricultural productivity programs, which are coordinated by the three SROs.

IFPRI's measure of technology spillovers only focuses on the "high domain similarity, high yield difference" part of the subregional research agenda, whereas the center of excellence approach focuses on the "high domain similarity, low yield difference (and no technology exchange)" part of the agenda. In a high domain similarity, high yield difference situation, it is quite likely that there are already one or more informal centers of excellence within the subregion. The SROs should identify and support these informal centers of excellence to push the productivity frontier even further (and, hence, maintain the potential volume of spillovers), while countries that are the recipients of technology spillovers should invest (more) in adaptive research and technology transfer.

The center of excellence approach starts with the premise that no centers of excellence are yet in existence and that they need to be created.[6] Moreover, such centers have to link up with research stations and extension services in neighboring countries in order to facilitate technology spillover. However, this aspect has not received the same attention in each agricultural productivity program. In West Africa, far more resources are going to facilitating spillovers—even to (small) countries that do not contribute in the form of a center

5 For more detail, see Table 2.3 in Chapter 2, this volume.

6 In reality, the selection of research programs to become a regional center of excellence was based in part on their past track record and the importance of the commodity to the national economy.

FIGURE 14.3 Decision tree regarding different options of how best to exploit technology spillover potential

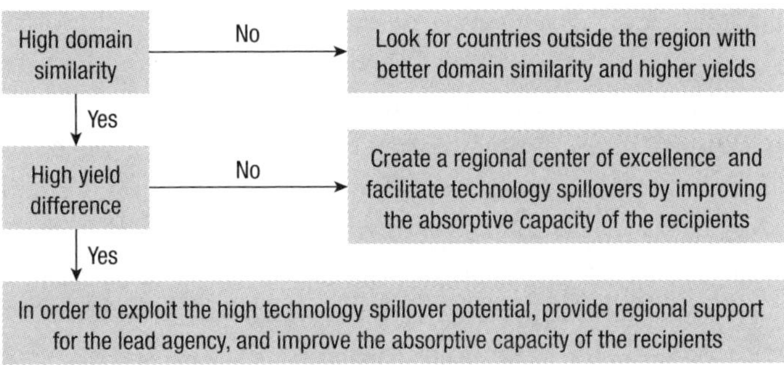

Source: Author.

of excellence of their own—than in East Africa. The stream of "intended" technology spillovers that will be created by such an intervention is not taken into account by the IFPRI studies. They have just ignored this possibility and, hence, portray an incomplete picture of the technology spillover potential. Between these two contrasting situations are many shades of gray. The key issues discussed above are summarized in the form of a decision tree (Figure 14.3).

Benchmarking Crossborder Collaboration Models in Agricultural Research

The political problem facing agricultural research is how to capture the apparent economic benefits of crossborder collaboration in agricultural research. Key to this problem is organizing the political will to address the issue, and finding a way to organize and fund a crossborder agricultural research agenda. To get a better understanding of this issue, this section benchmarks the SSA situation against the United States and the European Union in order to see what can be learned from their experiences.

United States

The overall "architecture" of public agricultural research in the United States is fairly simple, because from the very beginning (the late 19th century) it was designed as a national system with a relatively clear division of federal- and state-level responsibilities. The principal agencies at the federal level are the Agricultural Research Service (ARS), the Economic Research Service (ERS), and the Forestry Service, which all fall under the United States Department of Agriculture (USDA). In addition to its headquarters, ARS also operates four regional agricultural research stations that function as intermediaries for national and state-level agricultural research agendas.

State-level agricultural research has benefited greatly from the federal government initiative of the late 19th century to establish land grant universities in combination with agricultural experiment stations in each of the states. This initiative has resulted in a great deal of uniformity in how agricultural research is organized at the state level throughout the United States. Moreover, this uniformity has been sustained by substantial federal funding (and, in particular, formula funding) going to state-level agricultural research organizations (Table 14.6). The growth in federal funding for state-level agricultural research between 1997 and 2007 mainly came from non-USDA federal resources, such as the National Science Foundation and National Institutes of Health. These agencies operate large, competitive funding schemes that target national research priorities. The schemes stimulate collaboration between research agencies in different states, but it is usually not a binding requirement.

USDA funding for state-level agricultural research is somewhat of a mixed bag, comprising allocations using formula funding, competitive funding schemes, and ad hoc research contracts. USDA's formula funding for state-level agricultural research ($300 million in 2007) primarily targets the local research agenda.[7] In addition, there is a specific funding line for multistate agricultural research projects ($76 million in 2007). The competitive funding schemes for agricultural research managed by USDA (through its newly established National Institute for Food and Agriculture, which replaces the old Cooperative Research, Education, and Extension Service) target multistate, regional, and national research topics. They promote interstate research collaboration, but they do not impose it as a requirement for funding. They channeled some $294 million to state agricultural research in 2007.

The balance between federal- and state-level agricultural research has been quite dynamic over the years. Agricultural research capacity at the state level is now significantly larger than at the federal level, but this has not always been the case. Before 1950, federal agricultural research capacity exceeded the combined state-level capacity. Since 1950, however, state agricultural research capacity has expanded far more rapidly than federal agricultural research capacity (Alston, Christian, and Pardey 1999). This suggests a strong shift toward more localized, applied, and adaptive agricultural research after 1950.

7 Alston, Christian, and Pardey (1999) also point to the redistributive impact of this formula, favoring the poorer and less productive states.

TABLE 14.6 Public funding for agricultural R&D in the United States, 1997 and 2007

Funding source	1997		2007	
	Federally implemented	State implemented	Federally implemented	State implemented
		(US$ millions)		
Federal USDA funding	849	428	1,343	670
Federal non-USDA funding	14	291	40	965
Total federal funding	863	719	1,383	1,635
State funding	None	1,082	None	1,365

Sources: Alston, Christian, and Pardey (1999); King, Toole, and Fuglie (2012).
Note: More background information is available at the National Institute for Food and Agriculture website: http://nifa.usda.gov.

McCunn and Huffman (2000) found strong positive interstate technology spillover effects in their agricultural productivity study across 48 states. They share this finding with several other agricultural productivity studies, but they relate this to the question, "What is an optimal setup for agricultural research in the USA?" They suggest that independent state planning of agricultural research is inefficient, and cooperation across state boundaries, including the establishment of new political jurisdictions for financing public agricultural research, can enhance efficiency. They also suggest that rigid, centralized national planning is inefficient, as technology spillovers tend to concentrate regionally. Huffman and Evenson (2006) compared the impact of "federal formula" versus "federal competitive grant" funding of state agricultural research and found that federal formula funding yielded a higher impact on state-level agricultural productivity than the federal competitive grant funding. They explain this difference by arguing that local (that is, state-level) administrators have better knowledge about local circumstances than do federal administrators located far away and, hence, are in a better position to pick winners.

In more recent years, the long-term decentralization trend in agricultural research capacity in the United States has started to change. The strong growth in federal funding for state-level agricultural research is now coming mainly from national competitive funding schemes and, in particular, from non-USDA competitive funding schemes, which mobilize state-level agricultural research capacity to address national research priorities, representing a move away from local, adaptive research to more upstream applied and basic research.

European Union

The institutional setup and development of agricultural research in the EU is radically different from that of the United States. The EU came into existence

TABLE 14.7 Framework Program and collaborative agricultural research budgets

Framework Program	Time period	Total Framework Program budget	Collaborative research budget	Collaborative agricultural research budget	Collaborative agricultural research per year
			(millions of euros)		
1	1984–1990	3,271	na	na	na
2	1987–1995	5,357	na	na	na
3	1991–1995	6,552	na	na	na
4	1995–1998	13,121	na	na	na
5	1999–2002	14,960	10,843	520	104
6	2003–2006	19,256	14,682	928	186
7	2007–2013	50,806	32,413	1,935	276

Source: European Commission (2013).
Note: na = not applicable.

only in 1957, long after national agricultural research structures had been established across Europe in the late 19th and early 20th centuries. In contrast to the United States, the European agricultural research structure lacks a central design or plan. Moreover, it took the EU another 25 years before research and technology development (including agricultural research) was placed on the EU agenda. The Common Agricultural Policy of the EU, which dates back to the late 1950s, did not do much to promote European agricultural research. This part of the agricultural policy agenda was initially left to the EU Member States.

When research and technology development were added to the common EU agenda in the early 1980s, the idea of creating European (agricultural) research agencies was discussed at length, but was rejected by the Member States on the principle of subsidiarity.[8] Instead, a policy was adopted to mobilize national research capacity to address transborder research problems through crossborder collaboration. This policy has been implemented through a series of Framework Programs for Research and Technological Development. About 60–70 percent of the Framework Program budget goes to competitive funding schemes for collaborative, applied research across countries (Table 14.7).

For the most recent period (2007–2013), the budget reserved by the EU for collaborative food, agriculture, fisheries, and biotechnology research

8 Exceptions are the European Centre for Nuclear Research, the European Space Agency, and the European Molecular Biology Laboratory, but they were created long before research in general was placed on the EU agenda.

yielded €1,935 million, or an average €276 million per year. This collaborative research budget is allocated competitively through calls for proposals. During the first five Framework Programs, the calls for collaborative research proposals were relatively unspecified. As long as proposals fitted the thematic area of the call (usually very broadly defined) and met research quality and cross-country collaboration criteria, the project would be eligible for funding. As of Framework Program 6, however, more effort has gone into identifying a European agricultural research agenda (including broad-based consultations and forecasting exercises) and in making that part of the calls for proposals.

Starting with Framework Program 7, the EU also began to fund basic research projects under its "Ideas Program," managed by the newly established European Research Council (ERC). Under Framework Program 7, €7.5 billion (or 15 percent of the total Framework Program budget) was allocated to this new Ideas Program. In contrast to the Collaborative Program, the research projects funded under the Ideas Program do not require cross-country collaboration and are selected purely on the basis of their scientific excellence. The competition for this funding is high (the approval rate is about 10 percent). It is difficult to identify how much of this investment is of relevance to agriculture. A keyword search for "agriculture" or "agricultural" in the 3,400 research projects in the ERC project database resulted in some 102 project matches. This would suggest that the Ideas Program funds agricultural research to the tune of at least another €32 million per year. It is expected that this line of research funding will increase substantially under Horizon 2020, which is the follow-up to Framework Program 7.

Compared with the national budgets for agricultural research, the European funding for agricultural research is fairly small (Table 14.8): close to 9 percent of the EU's total public agricultural research expenditures. This is in stark contrast with the United States, where 55 percent of public funding

TABLE 14.8 Public funding for agricultural R&D, European Union, 2007

Type of program/funding	EU implemented	Nationally implemented
	(millions of euros)	
Ideas Program (1)	None	32
Collaborative Program (2)	None	276
Total EU funding (1+2)	None	308
National funding	None	3,159

Sources: Data on EU funding are from European Commission (2013); data on national funding are from EUROSTAT (2013).

for agricultural research at the state level comes from the federal government. Moreover, a substantial amount of federal funding ($1,383 million in 2007) is spent on federal agricultural research agencies; the EU has no similar agencies.

The EU considers its rather fragmented research capacity a great disadvantage. One of the key features of the Lisbon Agenda, launched in 2000, is the idea of creating a European Research Area (ERA), which is the research equivalent of the Common Market and aims to improve the integration of the European research base to minimize duplication, encourage excellence, and enhance the contribution of research to economic growth (European Commission 2012). The EU presented a new 10-year economic strategy (that is, the EU 2020 Strategy) in 2010. One of the strategy's seven flagships is the creation of an Innovation Union, which targets a considerably broader set of actors and instruments than does ERA (that is, more emphasis on the actual application and exploitation of knowledge). The idea of an Innovation Union can be seen as a continuation and further strengthening of the policies launched in 2000.

In addition to competitive funding for collaborative agricultural research and fundamental basic research, the EU has introduced various other instruments that aim to stimulate crossborder collaboration and coordination in (agricultural) research:

1. European Cooperation in Science and Technology (COST) is one of the longest-running programs supporting cooperation among scientists and researchers across Europe. Established in 1971, COST facilitates coordination among nationally funded research activities across Europe, focusing on a particular topic or challenge. COST does not fund research (this is left to the participating national governments), but it absorbs the international coordination costs of the research network for a defined period of four years. After this kickoff period, the network members can continue their collaboration, but without assistance. COST's food and agriculture cluster currently supports a portfolio of some 34 networks, and every year, five to seven new networks are established.

2. The objective of the European Research Area Network (ERA-NET) project, which was first launched under Framework Program 6, is to strengthen coordination and collaboration between EU members at the research policy and funding level. Only "program owners" (typically national ministries and regional authorities) and "program managers"

(for example, research councils and funding agencies) are considered eligible partners in an ERA-NET action. ERA-NET actions provide research program owners and managers with a platform from which to explore joint activities, strategy development, and in some instances joint calls for transnational research programs. Opening national calls for research proposals from researchers in other EU countries is also on the agenda.

3. European Technology Platforms (ETPs) were first introduced under Framework Program 6. They aim to bring together relevant stakeholders with various backgrounds (for example, regulatory bodies at various geopolitical levels, industry, public authorities, research institutes and the academic community, the financial world, and civil society) to develop a long-term R&D strategy in areas of interest to Europe. The platforms also have a role in helping to further mobilize private and public R&D investments. The structure of an ETP follows a bottom-up approach in which the stakeholders take the initiative, and the European Commission evaluates and guides the process. The agendas developed by these ETPs guide the calls for proposals under the Collaboration Program.

Africa South of the Sahara

In comparison with the United States and the EU, the development of the institutional structure of agricultural research in SSA also has its peculiarities. In particular, the crossborder dimension of the institutional structure continues to be problematic because of weak political and economic integration at subregional and regional levels. African governments often already have problems raising taxes for national causes, let alone for supranational causes. This void has been filled by donors (Table 14.9) who have backed the research activities of the CGIAR Consortium targeting SSA ($255 million in 2008) and the activities of FARA and the SROs (some $25 million in 2008). The CGIAR centers implement agricultural research primarily through their own research capacities and facilities, and target research problems that have a strong transborder dimension. The advantage of the CGIAR centers is that they do not have to organize crossborder collaboration (with all its inherent political problems) before they can set out to conduct transborder research.[9]

9 Nevertheless, CGIAR is also increasingly trying to coordinate its activities with the other actors in SSA, for example, through the Dublin consultation process.

TABLE 14.9 Public funding for agricultural R&D in Africa south of the Sahara, 2008

Type of program/funding	Africa south of the Sahara	
	Internationally implemented	Nationally implemented
	(US$ millions)	
Supranational agenda, funded by donors	255	25
National agenda, funded by government or donors	None	864[a]

Sources: CGIAR (2009); Beintema and Stads (2011).

Notes: All data are expressed in US dollars to facilitate comparability. [a]This number includes an unknown amount of donor funding. Stads (this volume) reports that the 2011 donor share of funding for the principal NARIs of SSA averaged close to 30 percent, not including Nigeria and South Africa, which together represent about half of all NARS spending in SSA. In relative terms, these two countries receive considerably less donor support; hence, the actual donor share in NARS spending in SSA is probably significantly lower than 30 percent (that is, 15–20 percent). NARI = national agricultural research institute; NARS = national agricultural research system

FARA and the SROs aim to address more or less the same agenda (that is, transborder research problems), but by mobilizing national agricultural research capacity for its implementation. The Framework for African Agricultural Productivity (FARA 2006) proposed a massive increase in investment in this category, from $25 million in 2004 to US00 million per year by 2010. It was expected at that time that FARA and the SROs would manage the allocation of most of these resources through competitive or commissioned agricultural research grant schemes (CARGSs).

A meeting by development partners in Brussels in 2012 reported that funding in the category of "donor-funded, regionally oriented, but nationally implemented" research had reached US$250 million in 2012.[10] However, only a small part of that money is actually being channeled through the CARGSs of the SROs. Despite the establishment of multidonor trust funds for FARA and the SROs, the lack of clear accountability in the form of robust results frameworks has made donors hesitant to channel their money through FARA and the SROs. Hence, the CARGSs have remained small, waiting for more donor funding to materialize. Most of the growth since 2008 in the "donor-funded, regionally oriented, but nationally implemented" research category can be attributed to the subregional agricultural productivity programs that have been funded by the World Bank (at a total volume of $636 million in loans and grants to national governments since 2007; see Table 2.2 in Chapter 2, this volume) and that promote centers of excellence.

10 This figure seems to be rather high, and its source is unclear.

Lessons for Africa South of the Sahara

The idea that crossborder collaboration in (agricultural) research can pay significant dividends is widely accepted. Nevertheless, reaping those dividends is difficult because it requires collaboration across different jurisdictions, whether provinces, states, or nations. Some higher-level authority is seemingly required to make this happen.

The United States stands out as the geographic area where this higher level of authority is most strongly developed in the form of a federal government with major powers and resources, whereas the EU has substantially weaker institutions that are still very much under development (and political debate). In contrast to the EU, the United States has a layer of federal agricultural research agencies that are specifically dedicated to national agricultural research issues. In the case of SSA, higher-level authorities do exist (that is, the various subregional economic unions and the African Union), but the political and economic integration is still very much in its early stages compared with both the United States and the EU.

The capacity and political will to raise local taxes for a supranational agenda are still very much underdeveloped. Nevertheless, SSA has substantial supranational agricultural research capacity in place (particularly in the form of CGIAR centers), but this is funded and organized by the international development community. While for the moment a highly valuable contribution, in the long run African governments will have to assume responsibility for agricultural research (including funding), at not only the national but also the supranational level.

The alternative approach promoted by the three subregional agricultural productivity programs is an interesting institutional innovation, because it attempts to circumvent the funding problem of transborder research by establishing an exchange of research benefits between countries in the form of *intended* technology spillovers. A significant effort is definitely needed for a long time to pull this off. The hope is that once a more differentiated research landscape has been established, it will sustain itself without further financial support by World Bank loans and grants. Whether this will actually happen is doubtful, given experiences with donor support to African NARS causing boom-and-bust cycles in funding (see Chapter 4, this volume). Moreover, the reciprocity on which the model is based can easily be derailed.[11] At the same time, the higher the stakes, the more difficult it will be for countries to pull out.

11 For example, Mali had to pull out of the second phase of the West African Agricultural Productivity Program because of internal political instability and war; CORAF/WECARD has tried to minimize the resulting fallout.

Given it is so much more difficult to mobilize funding for supranational than for national causes, it makes sense to adhere to the subsidiarity principle of keeping implementation (and funding) of government activities at the lowest level possible. In that sense, channeling research funding through the SROs that ultimately ends up funding national research priorities with limited spillover potential should be avoided. These research activities are best funded by national budgets. Interestingly enough, the formula funding that state-level agricultural research entities in the United States receive from the federal government does not seem to adhere to this principle. It funds agricultural research oriented toward local state priorities. However, one argument often used in the United States in favor of the formula funding is that state-level agricultural research creates significant (unintended) spillover benefits for other states. Moreover, formula funding requires a counterpart contribution by state governments. The mechanism is assumed to leverage more state funding into agricultural research, although whether this is actually true is up for debate.

Both the United States and the EU give fairly little attention to the facilitation of technology spillovers, because they are mostly unintended. If, ex ante, it is clear that there will be strong spillovers, researchers are usually advised to solicit funding from a higher-level jurisdiction. Alternatively, and common in the EU, is to have an informal supranational research program that is implemented by a network of national research groups, each taking the lead on a particular aspect of the problem at stake. Under this scenario, each participant funds its own research but benefits from the contributions of others. Such collaboration helps to reduce research duplication at the pretechnology stage but requires upfront coordination. This is another important role that the research program leaders based at the SROs could play.

Conclusion

Although technology spillovers provide the overall economic rationale for crossborder collaboration in agricultural research, it is rare that they are actually quantified ex ante and used to guide the allocation of research funding. In that sense the IFPRI studies are unique. Nevertheless, by focusing only on the potential of *unintended* technology spillovers, these studies have overlooked the potential of *intended* technology spillovers, which can be purposefully created through a more differentiated research landscape (that is, centers of excellence) at the regional level. Although this approach has received extensive support from the World Bank through the subregional agricultural productivity programs in recent years, no attempts have been made to quantify the potential of these *intended* technology spillovers; they are just assumed to exist.

Because SROs only control a small part of their subregion's agricultural research funding, their funding role is indisputably a complementary one and should concentrate on that part of the research agenda that is truly supranational (that is, having wide application across national borders). The advantage of the center of excellence approach is that it circumvents the need for a subregional funding mechanism. The role of the SROs in this approach (as well as in the research network approach) is one of coordination, rather than funding.

Information about technological proximity should help SROs to identify clusters of countries that would benefit from research collaboration on certain commodities or specific challenges. Imposing the same type of collaboration across the whole research agenda and all countries should be avoided. It is more realistic to strive for a patchwork of different coalitions of countries working together on different research topics with different intensities of collaboration. It is definitely not a case of one approach fits all. The high dependence on donor funding makes the overall design of the African agricultural research system—and in particular its supranational component—quite vulnerable. Both stronger national governments and further political and economic integration at the supranational level are needed to create a local funding base for crossborder agricultural research. As the EU example shows, this is not something that can be created overnight; it takes not just years, but decades.

References

Alston, J. 2002. "Spillovers." *Australian Journal of Agricultural and Resources Economics* 46 (3): 315–346.

Alston, J., J. Christian, and P. Pardey. 1999. "Agricultural R&D Investments and Institutions in the United States." Chapter 4 in *Paying for Agricultural Productivity*, edited by J. Alston, P. Pardey, and V. Smith. Baltimore: Johns Hopkins University Press.

Beintema, N., and G. Stads. 2011. *African Agricultural R&D in the New Millennium: Progress for Some, Challenges for Many*. IFPRI Food Policy Report. Washington, DC: International Food Policy Research Institute.

Byerlee, D., and M. Maredia. 1999. "Measures of Technical Efficiency of National and International Wheat Research Systems." In *Global Wheat Improvement System: Prospects for Enhancing Efficiency in the Presence of Spillovers*, edited by M. Maredia and D. Byerlee. Research Report 5. Mexico City: International Maize and Wheat Improvement Center (CIMMYT).

Byerlee, D., and G. Traxler. 2001. "The Role of Technology Spillovers and Economies of Size in the Efficient Design of Agricultural Research Systems." Chapter 9 in *Agricultural Science Policy: Changing Global Agendas*, edited by J. Alston, P. Pardey, and M. Taylor. Baltimore: Johns Hopkins University Press.

CGIAR. 2009. *CGIAR Annual Report 2008*. Washington, DC: CGIAR Secretariat.

Davis, J., P. Oram, and J. Ryan. 1987. *Assessment of Agricultural Research Priorities: An International Perspective*. ACIAR Monograph 4. Canberra, Australia: Australian Centre for International Agricultural Research; Washington, DC: International Food Policy Research Institute.

European Commission. 2012. *High Level Panel on the Socio-Economic Benefits of the European Research Area*. Brussels: DG Research and Innovation, European Commission.

———. 2013. *Development of Community Research Commitments 1984–2013*. Brussels.

EUROSTAT. 2013. "Government Budget Appropriations or Outlays for R&D." Accessed April 21, 2013. http://appsso.eurostat.ec.europa.eu/nui/show.do?dataset=gba_nabsfin07&lang=en.

FARA (Forum for Agricultural Research in Africa). 2006. *Framework for African Agricultural Productivity*. Accra, Ghana.

GFAR (Global Forum on Agricultural Research). 2011. *The GCARD Road Map: Transforming Agricultural Research for Development Systems for Global Impact*. Rome: GFAR Secretariat, Food and Agriculture Organization of the United Nations.

Huffman, W., and R. Evenson. 2006. "Do Formula or Competitive Grant Funds Have Greater Impact on State Agricultural Productivity?" *American Journal of Agricultural Economics* 88: 783–798.

Johnson, M., S. Benin, X. Diao, and L. You. 2011. *Prioritizing Regional Agricultural R&D Investments in Africa: Incorporating R&D Spillovers and Economywide Effects*. Conference Working Paper 15 from the ASTI/IFPRI–FARA Conference "Agricultural R&D: Investing in Africa's Future: Analyzing Trends, Challenges, and Opportunities," Accra, Ghana, December 5–7. Accessed April 2013. www.asti.cgiar.org/pdf/conference/Theme4/Johnson.pdf.

Johnson, M., S. Benin, L. You, X. Diao, P. Chilonda, and A. Kennedy. 2014. *Exploring Strategic Priorities for Regional Agricultural R&D Investments in Southern Africa*. IFPRI Discussion Paper 1318. Washington, DC: International Food Policy Research Institute. Accessed January 2014. www.ifpri.org/sites/default/files/publications/ifpridp01318.pdf.

King, J., A. Toole, and K. Fuglie. 2012. *The Complementary Roles of the Public and Private Sectors in U.S. Agricultural Research and Development*. Economic Brief 19. Washington, DC: United States Department of Agriculture/Economic Research Service.

McCunn, A., and W. Huffman. 2000. "Convergence in Productivity Growth for Agriculture: Implications of Interstate Research Spillovers for Funding Agricultural Research." *American Journal of Agricultural Economics* 82: 370–388.

Nin Pratt, A., M. Johnson, E. Magalhaes, L. You, X. Diao, and J. Chamberlin. 2011. *Yield Gaps and Potential Agricultural Growth in West and Central Africa*. Washington, DC: International Food Policy Research Institute.

Omamo, S., X. Diao, S. Wood, J. Chamberlin, L. You, S. Benin, et al. 2006. *Strategic Priorities for Agricultural Development in Eastern and Central Africa*. Washington, DC: International Food Policy Research Institute.

Pardey, P., J. James, J. Alston, S. Wood, B. Koo, E. Binenbaum, T. Hurley, and P. Glewwe. 2007. *Science, Technology and Skills*. InSTePP Report. St. Paul: University of Minnesota, International Science and Technology Practice and Policy.

Pardey, P., J. Roseboom, and J. Anderson. 1991. "Topical Perspectives on NARSs." Chapter 8 in *Agricultural Research Policy: International Quantitative Perspectives*, edited by P. Pardey, J. Roseboom, and J. Anderson. Cambridge, UK: Cambridge University Press.

Roseboom, J. 2011. *Supranational Collaboration in Agricultural Research in Sub-Saharan Africa*. Conference Working Paper 5 from the ASTI/IFPRI–FARA Conference "Agricultural R&D: Investing in Africa's Future: Analyzing Trends, Challenges, and Opportunities," Accra, Ghana, December 5–7. Accessed July 2013. www.asti.cgiar.org/pdf/conference/Theme4/Roseboom.pdf.

Chapter 15

THE ROLE OF CGIAR IN AGRICULTURAL RESEARCH FOR DEVELOPMENT IN AFRICA SOUTH OF THE SAHARA

Harold Roy-Macauley, Anne-Marie Izac, and Frank Rijsberman

Despite the impressive growth of African agriculture in the past two decades, the pressing need to accelerate agricultural productivity growth is widely recognized. Much of the solution to achieving sustainable food security and nutrition depends on improving production in Africa, without increasing the environmental footprint of agriculture (Chapter 1, this volume). The food gap in Africa south of the Sahara (SSA), currently met through imports, provides a significant potential market for mostly poor and hungry smallholder farmers to expand their output and improve their livelihoods. For this potential to materialize in a situation of increasingly interconnected international markets and in the presence of subsidies in various Organisation for Economic Co-operation and Development countries, farming in the region must become more market oriented and competitive. Improvements in agricultural productivity must come from a farm sector that is rapidly expanding onto more marginal land, with a natural resource base that is often degrading or facing competition from other sectors, and within the context of the adverse impacts of climate change. Moreover, this improved productivity must be achieved, in large part, by farm families operating on less than two hectares of land and with extremely modest physical and human capital.

Turning this situation around constitutes an agenda of transformation that depends on improvements in many dimensions of the agricultural and rural sectors, as has been discussed throughout this volume. A large range of actors are involved in efforts to take advantage of opportunities, achieve major agricultural changes, and create new system-level conditions beyond those of the field or farm. CGIAR has also been evolving over time. Its scientists, for example, used to be held primarily accountable for producing global public goods as measured by publications in high-impact, peer-reviewed journals. Today—although publications continue to be an important indicator of quality—CGIAR scientists are also required to demonstrate how their results trigger development outcomes (ex ante), and to assess the impacts of these outcomes (ex post) (Chapter 12, this volume).

This chapter reviews CGIAR's involvement in agricultural research for development (AR4D) in Africa throughout the four decades of its existence, analyzing its capacity to respond to the region's development challenges. The discussion covers recent CGIAR reform, including potential processes for aligning with African research and development (R&D) to ensure that international public goods make a positive contribution to national development challenges. The chapter further discusses opportunities for CGIAR to strengthen learning mechanisms and develop evolutionary approaches by creating partnerships between CGIAR Research Programs (CRPs) and institutional structures involved in the implementation of national and regional agricultural investment programs. The role and importance of subregional organizations (SROs) in Africa, mandated by regional economic policy institutions, in facilitating this process is also discussed.

CGIAR's Evolving Involvement in Africa

The CGIAR Consortium is a strategic global partnership of organizations and donors involved in agricultural research that was founded in 1971 to provide science-based solutions to low crop productivity, starting with wheat, maize, and rice. Currently, CGIAR is a network of 15 international agricultural research centers employing more than 8,000 scientists and staff operating in more than 100 countries worldwide. The consortium's research in SSA has evolved considerably over time, as have the objectives pursued, the research methods used, and the associated partnerships forged in both the research and the development sectors. Initial attempts to replicate approaches that led to the Green Revolution in Asia did not work, and CGIAR drew lessons from this failure. A short synopsis of the main approaches to research and partnerships implemented during 1971–2009 and since 2010, elucidates lessons and characteristics of the roles CGIAR has played in the region, as well as its comparative advantages in international agricultural research.

CGIAR's Evolving Research Agenda in Africa South of the Sahara

CGIAR was created in response to widespread concerns in the mid-20th century that rapid increases in human populations would continue to lead to widespread famine, which parts of Asia were already experiencing. CGIAR's main goal was focused on reducing hunger by increasing the productivity of staples in small farms; breeding improved cereal varieties was its main priority (Alston, Dehmer, and Pardey 2006). With support from the Rockefeller and Ford foundations, high-yielding, disease-resistant varieties, were developed at the International Rice Research Institute (IRRI) and International Maize and Wheat Improvement Center, which were established in 1960 in the Philippines and in 1963 in Mexico, respectively. These varieties dramatically increased production of staple cereals and sowed the seeds for the Green Revolution in Asia.

India, for example, was turned from a country regularly facing starvation in the 1960s to a net exporter of cereals by the late 1970s. This success was achieved with CGIAR acting as the central node of international breeding networks, including national agricultural research institutes (NARIs), which were selecting locally adapted varieties and facilitating germplasm exchange, through the collective effort of many researchers working on the narrowly focused problem of breeding. Furthermore, the Government of India provided all the institutional and policy support needed for these high-yielding varieties to be adopted and rapidly scaled out.

CGIAR's dominant research approach during the 1970s and most of the 1980s thus focused on the problems of breeding more productive crops, often in the context of irrigated agriculture, and for farmers who had reasonably good access to fertilizer and pest control techniques, inputs, and markets. However, this approach, which was very much inspired by the Green Revolution, was not as successful as hoped in the socioeconomic and biophysical contexts of SSA, where traditional food staples were grown under extremely varying rainfed conditions and with little institutional and policy support. It was not compatible with the large diversity of African agroecological situations and farming systems, often characterized by high and difficult-to-predict production risks.

In addition, the limited success of the Green Revolution model in Africa stemmed from a variety of institutional constraints prevalent among African countries during most of the first three decades of CGIAR's existence. These included inadequate support to research and extension from the state and, consequently, relatively weaker breeding networks in SSA compared with Asia; weaker extension services; and essentially no economic incentives to induce adoption (CGIAR Independent Review Panel 2008). These challenges were further compounded by pronounced market imperfections associated with unfavorable price incentives, ineffective subsidies, and limited credit access—and, consequently, low profitability of input packages and delayed adoption of improved varieties and technologies. Also playing a significant role were weak governance and, thus, political and social instability among the

FIGURE 15.1 CGIAR's evolving scope and mandate

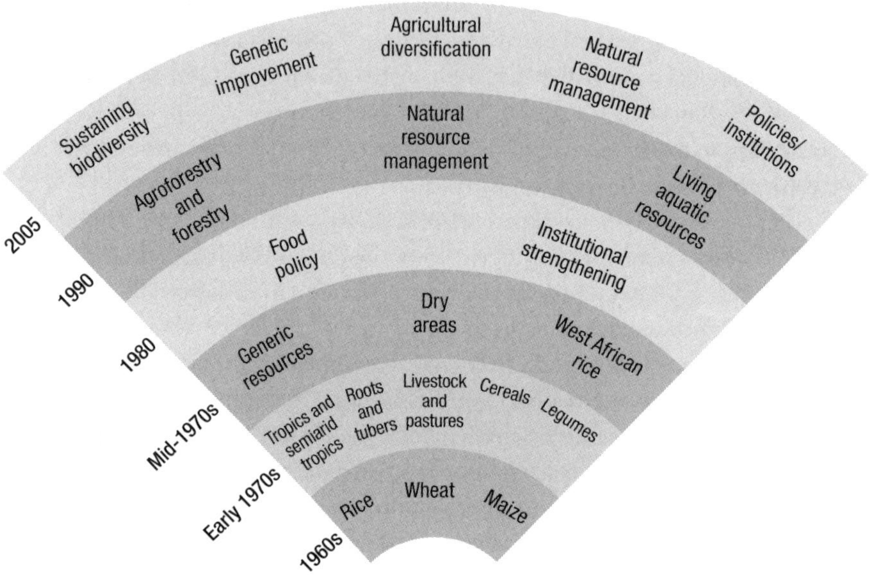

Source: CGIAR Science Council (2006).

newly independent countries, and lower population density with its implications for market access, infrastructure, and service provision.

In the continuing quest to improve the contribution of CGIAR research to agricultural development in Africa, and in view of the very difficult replication of Green Revolution approaches outside of Asia, CGIAR's initial focus evolved over time (Figure 15.1). The focus of research expanded to include other tropical crops and commodities, such as cassava, bananas, plantains, yams, beans, pulses, oil seeds, chickpeas, lentils, sorghum, potatoes, sweet potatoes, millet, livestock, and fish. In parallel with the addition of crops, livestock, and fish was the realization that breeding could not resolve all constraints faced by farmers; improved and more productive germplasm had to be accompanied by improved policies and practices, including natural resource management (NRM), to provide a more enabling environment for the adoption of technologies by farmers (Chapter 12, this volume).

Among the initial four crop-improvement centers, the International Institute of Tropical Agriculture, created in 1967, was headquartered in Africa. This was followed in 1971 by the West Africa Rice Development Association (WARDA), an intergovernmental center now renamed the Africa Rice Center.

Over the years, two other international centers were headquartered in Africa: (1) the International Livestock Research Institute, resulting from the merger of two previously independent centers, the International Laboratory for Research on Animal Diseases (ILRAD) and the International Livestock Center for Africa (ILCA); and (2) the International Center for Research on Agroforestry (ICRAF), since renamed the World Agroforestry Centre. Other centers were created in other continents, and most opened regional offices in various countries in SSA.

By the 1990s, CGIAR comprised 18 centers, some specializing in NRM (forests, trees, soil, and water), and others in policy and capacity building. CGIAR also focused on the research needed to support countries in designing and implementing food, agricultural, and research policies, as well as training to strengthen NARIs and build linkages among them and with other actors in the international agricultural research system (Anderson 1998).

In parallel with the broadening of the scope of CGIAR, research approaches continued to evolve. Following CGIAR's dominant approach of the 1970s of breeding to improve the productivity of key crops, more encompassing approaches to improving agricultural productivity and the welfare of resource-poor farmers were initiated. A principal new approach was that of "farming systems," recognizing that farmers manage landholdings comprising different enterprises and a diverse range of crops, trees, and sometimes animals—a situation prevalent in SSA. This new approach required social scientists and biological scientists to work together—outside of research stations in farmers' fields—to devise not only productivity-enhancing technologies and practices, but also profitable options for resource-poor farmers. An additional new requirement was to work at the level of the entire farming system in order to understand interactions among components of the farmholding and to devise more relevant and robust solutions.

These requirements amounted to a significant scientific cultural change. Using research methods in which data were collected under farmers' conditions and without full control of relevant variables was a challenge. Until then, most scientists had worked under controlled conditions, on research stations and at the plot level. Furthermore, it was not easy for scientists, trained to consider that their discipline provides "the definitive solution to farmers' problems" to accept that other disciplines, in particular the social sciences, can make essential contributions to the design of innovations. The new systems approach embedded in farming-systems research posed various methodological challenges. Achieving success, therefore, required local and national partnerships with other relevant research institutions. The change in the dominant scientific culture that farming-systems research brought about in CGIAR is still ongoing, given that the approach continues to be used and improved by scientists (Collinson 2000).

The integrated natural resource management (INRM) approach, which began emerging in CGIAR centers in the late 1980s and early 1990s, builds on farming-systems research by integrating sustainable management of natural resources as an objective (Chapter 12, this volume), along with productivity and profitability objectives (Izac and Swift 1994; Izac and Sanchez 2001; Clark et al. 2011). This approach requires not only agronomists, breeders, soil scientists, plant pathologists, and integrated pest management experts, for example, to work with social scientists, but also that production and systems ecologists are included and new methodological challenges addressed. This, for instance, includes measuring trade-offs at the system level among (1) productivity increases, (2) sustainability and resilience of resources, and (3) human welfare. Other methodological challenges include working at multiple scales to understand interactions across levels (Izac and Swift 1994) or building interoperational databases across the different types of data collected, including short-term, time-series data from farm surveys and field measurements at landscape/watershed scale and plot level. The results for farmers are very knowledge-intensive options compared with innovations, such as an improved variety. The approaches very often also require that policymakers be involved as stakeholders, so that the policy and institutional environments in which farmers function become more facilitative and supportive of farmers' adoption. CGIAR scientists and their partners continue to address these challenges more effectively.

More recently, the use of participatory action research methods was initiated by CGIAR scientists as a means of increasing farmers' capacity to sustainably adapt their own farming system to constantly changing socioeconomic and environmental circumstances (van de Fliert and Braun 2002; Sayer and Campbell 2003). These methods involve farmers and other stakeholders, from inception in the research for development lifecycle and throughout the different iterative loops. The purpose of this more demand-led approach, based on mutual learning among scientists, farmers, and other key stakeholders, is twofold. First, it ensures that the research undertaken is indeed responding to the actual needs of farmers and key decisionmakers, which in turn results in more rapid and substantial scaling up of results. Second, it brings about more sustainable improvements in farming systems by strengthening farmers' capacity to adapt, to use knowledge-intensive options, and to continue to amend their systems over time (Sayer and Campbell 2003). This approach builds on farming-systems research and INRM approaches. It further requires that scientists who are cognizant of social-learning processes participate, so that appropriate and effective methods of involving and building the capacity of farmers, scientists, and other key stakeholders are implemented as an ongoing part of the participatory action research being undertaken.

CGIAR's research agenda in SSA has thus constantly evolved from a relatively simple focus on achieving significant crop productivity increases, through breeding for a few selected crops in the 1970s, toward a much more encompassing agenda. Just before the beginning of the current reform, the agenda was addressing the challenges of managing complex agroecosystems under conditions of global change. In addition to the initial objective of productivity increases, it required multidisciplinary and interdisciplinary approaches and the assessment of trade-offs among multiple objectives. In parallel with this evolution, the scope of the system enlarged through the creation of new centers specializing in a large range of food crops, livestock, fish, forestry, trees, and policy research. Under such conditions, R&D partnerships had to evolve concomitantly.

CGIAR's Evolving Partnerships in Africa South of the Sahara

The evolution of CGIAR's research agenda in SSA could not have occurred without partnerships at local, national, and international levels. CGIAR's efforts still represent a small component of the overall effort deployed in agricultural research in developing countries. It has thus depended on strategic partnerships to complement its own range of expertise, which is insufficient to successfully address development challenges. While the initial attempt to replicate Asia's Green Revolution in Africa essentially entailed partnering with networks of breeders, since most research took place on CGIAR research stations, the implementation of farming-systems, INRM, and participatory-action research required the involvement of strategic arrays of partners.

Partnerships, in particular with NARIs, became essential to conducting on-farm research from a multidisciplinary perspective; this was especially so, given that the centers implement activities in different countries at the same time, as required by CGIAR's international mandate. The circle of partners grew over time to include research institutions specializing in such areas as the environment, resource conservation, water, soils, and forestry to successfully undertake INRM research. It also included partners working at farm and landscape scales across different countries. With the exception of the very early days when CGIAR partnered with networks of breeders, CGIAR centers do not employ all the needed expertise; developing partnerships has been a successful means of increasing the scope of available expertise to resolve challenges more effectively. Participatory action research required further partnerships with civil society organizations, communities, groups engaged in collective management of resources, farmers' groups, and key decision- and policymakers and development actors, including nongovernmental organizations (NGOs) and development banks. By the time the current CGIAR reform started, the centers had forged partnerships with a range of R&D actors throughout the countries in which they worked. The last external review of the system noted that these partnerships were wide ranging and raised questions about their nature and whether they reflected a strategic approach to partnerships (CGIAR Independent Review Panel 2008).

The quality and efficiency of the partnerships are indeed not easy to circumscribe. The alignment of the partnership models used by the different CGIAR centers with the dynamics of development and African agricultural research is difficult to assess—and the fact that each center has its own approach to partnerships only renders the task more difficult. In the 1990s and 2000s, in response to the interest of policymakers to respond to developmental needs, donors asked CGIAR and research systems in SSA to show the impact of their research outputs (CGIAR Independent Review Panel 2008). In this context, in 1999 the directors general of CGIAR's centers agreed that it was both urgent and important to improve the coordination of research for development activities at all of the centers, as well as among those of their national and subregional partners in West and Central Africa and in East and Southern Africa; WARDA and ICRAF, respectively, were asked to lead these efforts in the two subregions. The overall objective was to create an inventory of CGIAR's activities with its partners and identify ways to improve the alignment of activities, avoid duplications, and fill research gaps. The assumption was that greater alignment of objectives and approaches and more effective use of funds would trigger larger-scale impacts of the work.

Between 1999 and 2002, numerous consultations were conducted involving the SROs, CGIAR scientists, key partner institutions, the NARIs, and representatives of farmers' organizations and NGOs. The main conclusion reached was that improving the quality of the existing collaboration between CGIAR centers and their partners was the best way to integrate research activities effectively across regional partners and maximize the potential for greater impact. The consultations also led to the acknowledgment by both CGIAR scientists and partners of the need for CGIAR to collaborate differently with scientists in the NARIs and with other partners within and beyond the national agricultural research systems (NARSs). Equitable partnerships were widely recognized as more satisfactory, more effective, and more aligned with the needs of partners than the prevailing, more academic type of collaboration. The lack of transparency, particularly in allocating funds to partners, and the "top-down" dimension of partnerships were identified as bottlenecks that had to be addressed. This general recognition and the overall recommendations to the group of directors general were accompanied by concrete proposals for integrated work on joint priorities, joint action plans, and joint fundraising.

Nevertheless, it was not until the release of the final report of the two CGIAR-appointed SSA task forces that concrete action took place (CGIAR Secretariat 2005). Indeed, the Tervuren Consensus, as it was known, determined the need for a single (legal) CGIAR entity for SSA and, as a first step, advocated the development and implementation of integrated medium-term work plans for West and Central Africa and for East and Southern Africa. The Alliance of the International Agricultural Research Centers, created by the 15 CGIAR centers in 2006, thus facilitated the design of an action plan for implementing these recommendations in the two subregions. The Alliance also funded the plans' implementation for a few years, until the commencement of the current reform in 2010, at which time the integrated work plans were discontinued (see Chapter 14, this volume, for further detail).

The impetus to implement the work plans was initially strong, but grew weaker over time—no doubt because institutional obstacles slowed progress, and the political will needed to remove those obstacles was insufficient. Institutional changes in the NARSs also led to increased competition for funding between CGIAR centers and national institutions, lowering scientists' overall interest in integrated research activities. The momentum created by the integrated work plans has not been regained. An external assessment of the impacts of the multiple reforms of CGIAR's research agenda over the past 10 to 15 years would be interesting. One hypothesis is that a series of reforms is often an indication that each reform failed to fully address structural issues—that is, did not go far enough—resulting in calls for further reforms. Another hypothesis is that a minimum of institutional stability is necessary for scientists to successfully develop new sustainable partnerships and global collaborative programs.

In the early 2000s, CGIAR created a new mechanism—challenge programs—to provide even greater incentive for its centers to forge partnerships with external institutions. Five such programs were created, adding to—rather than replacing or displacing—existing programs. Challenge programs involved one or more centers with a variety of external partners. Four programs dealt with issues cutting across continents, such as climate change, food and water management, biofortification, and advanced crop breeding for difficult environments. One was the SSA Challenge Program (SSA-CP), whose main objective was to develop new research approaches to tackle agricultural development problems in Africa. The Forum for Agricultural Research in Africa (FARA) led and managed this program, which involved many of the centers. As part of the current reform, each challenge programs was rolled into the newly created CRPs, so that lessons learned from the SSA-CP could be used by the relevant CRPs. Indeed, some of the research sites and approaches to innovation platforms developed by SSA-CP have been further developed by the Humid Tropics and Dryland Systems CRPs.

Lessons Learned

Among the many lessons that could be drawn from CGIAR's evolving agenda and partnerships, the following five are notable, as they have been used to establish guidance for the centers in the first call for CRP proposals.

THE FIRST LESSON

The generic theory of change CGIAR used in its initial decades—which is still in use in some parts of its current research agenda—is unrealistic and too simplistic. This theory of change assumes (1) that producing new technologies and improved practices will lead to large-scale farmer adoption as long as national-level extension services are in operation, and (2) that producing new scientific evidence will convince policymakers to improve the enabling policy environment for farmers. CGIAR scientists have learned that society will not readily change its behavior in response to new scientific findings and innovations. Today, they are less arrogant about the role of science because they understand that producing new science is only a first step, and that more strategic thinking, time, and effort are needed to ensure that science influences and changes behavior. The evolution of CGIAR's research agenda in SSA exemplifies the search for an effective way of increasing the influence of agricultural science through more integrated and inclusive research approaches and partnerships.

THE SECOND LESSON

Multidisciplinary and interdisciplinary research methods and their associated partnerships, both within CGIAR and with external partners (such as in the Alternatives to Slash and Burn Program), produce relevant and robust results that no monodisciplinary or mono-institutional approach could match (Clark et al. 2011). Such approaches, however, entail high transaction costs for scientists, who need to learn to communicate across different disciplines and to accept the multidimensional nature of the reality—and the associated knowledge base—within which they operate. Hence, CGIAR's interdisciplinary undertakings can still be unwieldy, particularly when biophysical and social scientists need to work together.

THE THIRD LESSON

The governance and management structure of CGIAR programs needs to be aligned with the research for development agenda—not the other way round. This implies that inclusive, transparent, and accountable governance and management mechanisms need to be established to facilitate and support the collaborative research being undertaken. This lesson was further reinforced by the recent external governance and management review of the CRPs (CGIAR Independent Evaluation Arrangement 2014).

THE FOURTH LESSON

It is essential that CGIAR benchmark its work in multiple sites and locations across countries in order to understand the key biophysical and socioeconomic processes that are operating and ultimately determining the performance of innovations in different environments. Indeed, this understanding is key to scaling up options successfully in different environments. To date, most of the scaling up, whether for technological, policy, or institutional innovations, still follows a trial-and-error approach. The need to first understand processes in different environments, and to have sufficient baseline and benchmark data to facilitate this understanding, has only begun to be recognized by the newly created CRPs, principally through the establishment of results-based management.

THE FIFTH LESSON

For CGIAR research to be guided by a theory of change that becomes increasingly more realistic and relevant to the needs of farmers, policymakers, and national partners, scientists must have excellent skills in learning from past mistakes. CGIAR's evolution over time shows that being flexible and adapting to a constantly changing agricultural environment with significant intellectual and scientific humility is essential to a resilient and dynamic research system. This adaptation process takes time, of course, so realistic theories of change cannot be expected to be produced by newly created CRPs; rather, they will emerge over time from this process.

The Reformed CGIAR

To meet the challenges of increased volatility and uncertainty in agricultural markets, the increasingly tangible challenges in agriculture and forestry arising from climate change and natural resources degradation, and the multiple institutional challenges confronting the CGIAR system, CGIAR embarked on a profound reform in 2008. One of the objectives was to fundamentally change CGIAR's business model by focusing on the delivery of development outcomes. The process included improving the engagement of all stakeholders in international agricultural research for development, in order to refocus the efforts of the centers and their partners on major global development challenges. This led to a new vision to reduce poverty and hunger, improve human health and nutrition, and enhance ecosystem resilience, including to climate change, through high-quality international agricultural research and partnerships, and a way of doing business geared toward strengthening research impact and donor harmonization.

The reform aimed to increase the relevance, effectiveness, efficiency, and accountability of CGIAR research to achieve this new vision. The last external review showed that CGIAR research was not driven by the need to resolve development challenges, that its partnerships needed further expanding, and that it was not sufficiently accountable for producing scientific results of demonstrable relevance to development challenges (CGIAR Independent Review Panel 2008). In addition, donors pointed to redundancies across the different research agendas implemented by the centers. The design and creation of CRPs were expected to remedy this situation.

Of the criteria CRPs had to fulfill to be approved, two were of overriding importance. First, all CRPs had to address "big" development challenges related, for example, to food security, rural poverty, or climate change, and to explain what results they proposed to deliver to resolve these challenges. Second, the CRPs had to embody significant new partnerships involving scientists and stakeholders from different CGIAR centers and from a large number of non-CGIAR institutions, with some specialized in research and others in development. This was in recognition that (1) development challenges cut across traditional scientific disciplines and center mandates, requiring scientists from different institutions to collaborate to provide the range of expertise needed; and (2) research alone is insufficient in bringing about sought-after impact, and research practitioners must work alongside stakeholders and development practitioners for their results to generate significant impacts. Accountability was provided by the Consortium Board, created in January 2010 to be responsible for monitoring and facilitating the progress of

CRPs toward delivering their expected results and impacts. The Consortium Board has overall governance and oversight of collective issues regarding CRPs, for the Strategy and Results Framework that underpins their work (CGIAR 2015), and for yearly reporting to donors on the use of CGIAR and bilateral funding by CRPs.

Since 2011, CRPs have been the only mechanism for funding research for development activities from the CGIAR Fund. Hence, as far as the CGIAR Fund is concerned, CRPs have replaced the previously independently conceived research agendas of the 15 CGIAR centers. The Strategy and Results Framework was not approved by donors until the end of 2012, when most CRPs had already been approved. Creating CRPs under such conditions thus had inherent limitations. Despite these limitations, all CRPs have evolved substantially since their inception because of CGIAR's long tradition and experience of learning by doing and readjusting its agenda as the need arises (Lesson 5 from the previous section). This evolution could also be attributed to the fact that both the Consortium Board and donors—organized in a Fund Council as the system's ultimate financial decisionmaking body—provided comments and requests to CRPs to improve various dimensions of their work.

The 15 CRPs collaborate with hundreds of partner organizations, including national and regional agricultural research institutes, civil society organizations, academia, and the private sector. Three of the CRPs focus on farming systems under different agroecological conditions: aquatic agricultural systems, dryland systems, and humid tropical systems. Their objective is to facilitate the sustainable intensification of farming systems in these different zones, targeting small-scale farmers vulnerable to climate and other changes. Another three CRPs address global development challenges concerning the water, soils, land, and ecosystem services nexus; the forest, agroforestry, agriculture continuum, and the role of trees within it; and the mitigation and adaptation of agriculture to climate change for food security. These three CRPs also use a whole-systems perspective with a focus on multiple scales (farming, landscape/watershed, and regional, in addition to global) in designing options to increase systems' resilience and sustainability, and to establish sound adaptation and mitigation strategies.

Seven other CRPs aim to improve the production of various crops and commodities (including livestock and fish), most often by integrating breeding into a value-chain approach. Among them they cover the main cereals grown in developing countries, as well as roots, tubers, bananas, and a range of grain legumes. One CRP specializes in policies, institutions, and markets and on the science–policy interface, providing methodological support to the large number of other CRPs that also engage in work on policy and value chains from the perspective of their own commodity/NRM focus. Finally, one CRP specializes in improving nutrition and health through agriculture and also provides methodological support to the CRPs undertaking nutrition work from their particular perspectives.

All CRPs focus on developing, testing, evaluating, and facilitating the scaling up of the agricultural knowledge, technologies, institutional arrangements, and policies they produce. Capacity building is accomplished through degree and nondegree training and joint research. Collaborative and joint activities provide the opportunity to share knowledge and the methods used in producing it. These activities are delivering outputs, which include (1) new knowledge on research methods for complex problems in agricultural development; (2) improved varieties and breeding lines of major food crops, livestock, fish, and indigenous fruit trees; (3) new knowledge, methods, practices, and tools for the efficient management of integrated crop, livestock, and natural resources, as well as diseases and pest management; and (4) new knowledge and implementation methods on appropriate policies and institutional arrangements to provide a more favorable environment for small-scale farmers to adopt innovations.

The portfolio of 15 CRPs, after just a few years of operation,[1] is well positioned to deliver significant results and impacts regarding food security and rural poverty reduction from the perspective of cereals, roots, tubers, bananas, native fruits, and grain legumes. This research has a long CGIAR history and is scientifically robust in terms not only of crop breeding, but also of (1) climate-change research to mitigate adverse changes and support agricultural systems and crops in adapting; (2) scaling up work, in particular through value-chain approaches; (3) policy and methods development; and (4) research on the agriculture–tree nexus. Gender research is clearly being mainstreamed and contributes to an increased focus on the circumstances of the poorest and most vulnerable farmers.

The CRPs are gathering momentum in the delivery of results to improve nutrition and health, for example, by scaling up biofortified crops, improving the nutritional status of the poorest and most vulnerable, and developing methods and policy implications and recommendations. This work is proceeding rapidly but needs further expansion, for example, concerning (1) the extremely underresearched human health dimension of food systems and CGIAR's contributions to these systems and (2) the implications of the

1 The first CRPs were created in January 2011, and the last one was created in January 2013.

development challenge of providing more balanced, diversified, and nutritious diets for the overall balance of CGIAR investment.

Regarding sustainable NRM in agriculture, the portfolio has the potential to deliver more results on decreasing the environmental footprint of agriculture. Work on these issues is relatively limited to the three CRPs specializing in specific aspects of NRM (for example, trees and forests, water and land management, and climate change). Shrinking the environmental footprint of agriculture is, however, a prerequisite for strengthening agricultural resilience and sustainability and, thus, needs to be integrated more effectively into the work undertaken by the commodity-improvement CRPs. At the portfolio level, more attention is needed on ecosystem services and their contributions to resilience, sustainability, and system productivity, and to in situ biodiversity management as a potential pathway to balancing productivity, profitability, and resilience under specific circumstances.

The CRPs have begun to strengthen their approaches to value chains, innovation platforms and networks, and gender research. They can now draw lessons from their collective experiences to progress more rapidly on these issues. Ongoing tasks include monitoring progress toward the delivery of results (which is the responsibility of the Consortium Board), and developing a much more robust culture of conducting research to evaluate impacts, in particular NRM research. Traditional monitoring and impact assessment methods are suited to assessing the relatively straightforward progress toward, and consequences of, the adoption of improved crop varieties. For instance, the more complex world of multistakeholder platforms, value-chain development, or landscape assessments of the multiple effects of climate-smart agricultural practices, or of sustainable intensification of entire agroecosystems, requires new approaches. In general, and at this juncture, the ongoing CGIAR reform is seen to facilitate the contribution of the CGIAR system to achieving the bigger picture depicted in the United Nations' post-2015 Sustainable Development Goals and, more especially for Africa, by the Comprehensive Africa Agriculture Development Programme (CAADP).

The reformed CGIAR is well positioned to bring together scientists and stakeholders from various (internal and external) disciplines, experiences, and institutions to implement this bigger-picture, transdisciplinary approach (which is discussed further in the next section in the context of opportunities for Africa). While it was considered a more effective and rapid approach to solving agriculture's complex development challenges, CGIAR did not seize the opportunity in the first call for CRP proposals. As of mid-2015, however, CGIAR was introducing additional institutional changes that should encourage the implementation of transdisciplinarity. For the call for new CRP proposals that will be implemented in 2017, these changes include (1) integrating the CRPs as a unified portfolio, rather than just a set of individual programs; (2) introducing a preproposal phase to allow concepts to be evaluated, aligned with the existing portfolio, and refined through high-level, constructive guidance before being developed into full-blown proposals; and (3) defining a "site integration" mechanism to improve county-level coordination and collaboration at selected locations.

Whether CGIAR scientists will ultimately see the value and relevance of these processes and embrace novel transdisciplinary approaches is difficult to predict. Given that barriers to the adoption of the new approaches currently exist, formally and systematically integrating them into the call for proposals would hasten the process. Indeed, it is currently financially rewarding for CRPs to make accurate, quantifiable predictions of their results a few years ahead of time, but transdisciplinary research approaches do not lend themselves well to such predictions. For truly novel approaches to be implemented widely within CRPs, such institutional barriers will need to be removed (for example, by donors).

New Opportunities for Africa

In Africa, institutional frameworks for agricultural research have shifted significantly, and new platforms, coordinating bodies, and processes have been established to increase the relevance and impact of research on development and poverty reduction. The ongoing CAADP process and recent redesign of the CRP portfolio present important opportunities for CGIAR to collaborate closely with African R&D systems to support agricultural transformation in Africa. Components of the CAADP agenda, including regional, subregional, and national agricultural and food security investment plans, have been developed with established goals, targets, and priorities. The Framework for African Agricultural Productivity, developed by agricultural stakeholders in Africa under the leadership of FARA, has encouraged SROs and their partners to broaden their focus by looking into practical, new knowledge on innovation processes; strengthening the capacity of collaborative programs to deliver; fostering leadership for pro-poor innovations; and facilitating effective communication and knowledge sharing among stakeholders in innovation processes (Chapter 2, this volume).

The establishment of national and regional agricultural investment plans (NAIPs and RAIPs) and flagship subregional agricultural productivity programs in East, West, and Southern Africa (EAAPP, WAAPP, and APPSA) provide the opportunity for enhanced partnerships between CRPs and SROs.[2] They are also leveraging strong partnerships with NARIs outside Africa, especially the Brazilian Agricultural Research Corporation (Embrapa), and donor support. These initiatives are currently considered by the SROs as models for sustainable investments in agricultural research to create country-level impact and are emerging, next to CGIAR, as critical providers of solutions for agricultural productivity, NRM, and development policy advice.

Linkages with CRPs will provide more opportunities for CGIAR to increase the effectiveness of its work through more equitable national-level partnerships; adapt to the dynamics of AR4D in SSA at the national level; help countries meet their needs; and facilitate country-level impact, drawing lessons from all the countries in which the CRPs work. Nevertheless, coordination among CRPs, SROs, and NARSs—while seemingly less difficult than before—still lacks mechanisms to facilitate effective and coherent interactions; hence, the looming issue is establishing the necessary institutional architecture to enable these interactions. With CAADP now into its second decade of implementation, and the RAIP and CRP processes moving into their second phases, the opportunity exists to improve the institutional alignment and linkages among NARSs, SROs, and CGIAR.

Recognition of the need to do so emerged from discussions during the Dublin process, initiated in 2011. The purpose was to identify where the greatest value addition could be generated to support the next decade of CAADP's implementation regarding science, technology, and innovation. This process provided a platform for the African Union Commission (AUC), FARA, the SROs, World Bank, United States Agency for International Development, EU, and CGIAR to engage in discussions that explored filling gaps in research investments, avoiding duplication of efforts, matching investments to priorities, and defining ways to enhance the linkages between NAIPs/RAIPs and CRPs to support CAADP's ongoing momentum. Discussions also focused on deepening the alignment of Africa's agricultural research, extension, and education programs and of collaborations with those of CGIAR and its partners to facilitate the process of African agricultural

2 The NAIPs and RAIPs are being coordinated by regional economic communities, especially in West and Central Africa through the Economic Community of West African States and Economic Community of Central African States. The agricultural productivity programs have been developed with World Bank–funded national loans and are based on the principle of regional integration and the creation of spillover effects (Chapter 14, this volume); their implementation is being coordinated by the SROs.

transformation. Various work streams were designed and executed via a partnership among AUC, NEPAD Agency (the planning and coordinating agency of the New Partnership for Africa's Development), FARA, the SROs, CGIAR, and development partners.

In January 2013 a memorandum of understanding (MOU) between AUC and CGIAR formalized the process, outlining six key activities:

1. developing an effective plan to align CRPs with the research programs of African institutions and CAADP investment plans;
2. developing a Science and Technological Agenda for African Agriculture;
3. developing a joint plan to support regional and subregional research activities to increase the efficiency of research investment Africa-wide;
4. developing joint African and CGIAR technology platforms (that is, subregionally based partnerships) to assist countries in identifying, accessing, and using the latest knowledge and technology to support the priority commodities and value chains specified in national CAADP investment plans;
5. providing demand-driven, client-oriented technical support in the design of national and regional medium- and long-term plans under CAADP; and
6. establishing a process for jointly sharing information and developing a knowledge base to underpin the dissemination of best practices in institutional development, policy development, and capacity building for agricultural research.

The Dublin process, which has gained a high profile in Africa, within CGIAR, and among development partners, is characterized by joint planning and priority setting and greater clarity about each institution's role in addressing specific agricultural issues. In addition, the process further integrates CGIAR programs with the CAADP agenda in terms of planning processes, capacity building, and implementing investment plans. The July 2015 meeting of the steering committee implementing the AUC–CGIAR agreement reviewed progress to date, recognized that numerous achievements had been made but little value had been captured, and recommended ways to improve the implementation of key activities in line with the Post-Malabo Implementation Strategy and Road Map (African Union 2015). CGIAR will need to put in place a strong monitoring, evaluation, and learning system to effectively and efficiently capture the achievements being made through the implementation of the MOU.

CGIAR's past and current experience in implementing systemwide programs, challenge programs, and CRPs has shown that—beyond interdisciplinary research (and naturally along with reductionist, monodisciplinary research)—transdisciplinary research is needed to successfully address new and complex development challenges in agriculture. Transdisciplinarity is an approach to solving complex problems that cuts across disciplinary boundaries and integrates knowledge, tools, and ways of thinking from a large array of relevant sciences and from the perspective of key stakeholders (National Research Council 2014). This approach facilitates the design of a comprehensive framework that addresses scientific and societal challenges as they interface with multiple fields of expertise. Transdisciplinary approaches, for example, can generate a more rigorous and systematic understanding of the interconnected issues that characterize a challenge in different environments, such as the dynamics of biological, social, economic, and ecological factors in improving the productivity of a given farming system, while decreasing its environmental footprint (National Research Council 2014). Such understanding is often well developed concerning biophysical factors, but not very robust when it comes to interconnections among biophysical and socioeconomic factors. By merging scientific and other diverse expertise, including such stakeholders as farmers, transdisciplinarity stimulates innovation, but it requires an open and inclusive culture, a common set of concepts and metrics, and a shared set of institutional and research goals (National Research Council 2014). As mentioned in the previous section, additional changes being introduced within CGIAR as of mid-2015 should help CGIAR institute greater transdisciplinarity.

CGIAR will progress more rapidly toward the fulfillment of its mission in the region by working more closely with the African R&D community through the Dublin process. Indeed, these partnerships and new scientific approaches will enable existing gaps to be addressed. The innovation platforms that support transdisciplinary approaches are expected to generate an understanding of the interconnections needed for innovations to be scaled up effectively—that is, beyond current trial-and-error approaches (Chapter 13, this volume). The current paucity of information about the long-term (positive and negative) social, economic, and ecological consequences of CRP innovations can also be addressed through innovation platforms and transdisciplinary approaches. These consequences are directly related to agriculture's long-term environmental footprint, and CGIAR must integrate this perspective throughout its future work with its African partners on improving agricultural and food systems in SSA.

The establishment of centralized programs by the SROs, strongly aligned with those of NARSs, and the creation of strong coordinating and facilitating tools have proved to be successful. More than ever, the SROs are demonstrating their effectiveness and efficiency in coordinating and facilitating agricultural research at the subregional level. They are establishing centers of excellence within NARIs to consolidate the implementation of national and subregional agricultural research based on a model that embraces the emergence of a critical mass of agricultural research stakeholders within the context of the innovation process (Chapter 2, this volume). While site integration may be new to CGIAR, the SROs developed site integration by establishing innovation platforms involving different categories of stakeholders working together on a common constraint. Most of the time, the sites targeted for the establishment of innovation platforms were already hosting government-funded projects. Under such conditions, upscaling of results is automatic. The establishment of country dialogues between NARIs and CRPs, facilitated by the SROs, will definitely render site integration a demand-driven process for CGIAR. Seen this way, site integration is something CGIAR should capitalize on to ensure alignment, linkages, upscaling, and effective fostering of transdisciplinary approaches.

Given the complexity of innovation and research processes, these centers of excellence could position themselves as catalyzing and linking agents within regional networks. They could help to identify successful experiences in multiple countries, link innovators with sources of scientific and technical information in distant locations, use action research to help adapt foreign experiences to local conditions, and promote the emergence of global research networks. Such a system would leverage stronger partnerships with advanced research institutions at the international level, including CGIAR, and promote a clear role for CGIAR within this new architecture, with NARIs acting as the central nodes of a system of decentralized experimentation with centralized learning. The performance of such arrangements should be evaluated not only via traditional research indicators, such as peer-reviewed publications, but also in terms of contributions to farmers' welfare, agricultural productivity, and the sustainable use of natural resources in agriculture at the national level. The SROs are, indeed, central coordinating mechanisms to rationalize joint planning and priority setting between CGIAR and NARSs; the CGIAR system just needs to recognize the SROs' role and empower them to do just that.

The SROs certainly do not have a funding base for this role, but the question remains as to whether the CGIAR system wants to or will give this great consideration, within its present funding windows.

* * * * *

It should be noted that, as of December 2015, the reform of the CGIAR system is still ongoing and will be until donors approve whatever the transition team proposes as the new institutional umbrella. Donors recently agreed to reducing the number of CRPs from 15 to 12, and more changes are afoot. As a result, CGIAR scientists now have to contend with constant changes, not only in the context of ongoing planning, but also in the context of short-term institutional and budget-related modifications. The transaction costs of such a constantly changing institutional environment cannot be over-emphasized and constitute a very high risk for the sustainability of the CGIAR system.

References

African Union. 2015. *Implementation Strategy and Roadmap to Achieve the 2025 Vision on CAADP: Operationalizing the 2014 Malabo Declaration on Accelerated African Agricultural Growth and Transformation for Shared Prosperity and Improved Livelihood.* Addis Ababa, Ethiopia. Accessed August 2016. http://pages.au.int/caadp/documents/implementation-strategy-and-roadmap-achieve-2025-vision-caadp.

Alston, J., S. Dehmer, and P. Pardey. 2006. "International Initiatives in Agricultural R&D: The Changing Fortunes of the CGIAR." In *Agricultural R&D in the Developing World: Too Little, Too Late?* Edited by P. Pardey, J. Alston, and R. Piggott. Washington, DC: International Food Policy Research Institute.

Anderson, J. 1998. "Selected Policy Issues in International Agricultural Research: On Striving for International Public Goods in an Era of Donor Fatigue." *World Development* 26 (6): 1149–1162.

CGIAR. 2015. *CGIAR Strategy and Results Framework 2016–2030: Redefining How CGIAR Does Business until 2030.* Montpellier, France: CGIAR Consortium.

CGIAR Independent Evaluation Arrangement. 2014. *Review of CGIAR Research Programs Governance and Management, Final Report.* Rome.

CGIAR Independent Review Panel. 2008. *Bringing Together the Best of Science and the Best of Development.* Independent Review of the CGIAR System. Report to the Executive Council. Washington, DC.

CGIAR Science Council. 2006. *Positioning the CGIAR in the Global Research for Development Continuum.* Rome: Science Council Secretariat.

CGIAR Secretariat. 2005. *Report of the CGIAR Sub-Saharan Africa Task Forces: The Tervuren Consensus.* Accessed July 2015. www.cgiar.org/www-archive/www.cgiar.org/exco/exco8/exco8_ssa_tf_report%20.pdf.

Clark, W., T. Tomich, M. van Noordwijk, G. Guston, D. Catacutan, N. Dickson, and E. McNie. 2011. *Boundary Work for Sustainable Development: Natural Resources Management Research at the CGIAR.* Proceedings of the National Academy of Sciences of the USA. Washington, DC: National Academies Press. Accessed September 2014. www.pnas.org/content/early/2011/08/11/0900231108.short.

Collinson, M. 2000. "Introduction." In *A History of Farming Systems Research,* edited by M. Collinson. Wallingford, UK: CABI Publishing.

Izac, A., and P. Sanchez. 2001. "Towards a Natural Resources Management Paradigm for International Agriculture: The Example of Agroforestry Research. *Agricultural Systems* 69: 5–25.

Izac, A., and M. Swift. 1994. "On Agricultural Sustainability and Its Measurement in Small-Scale Farming Systems in Sub-Saharan Africa." *Ecological Economics* 11: 105–125.

National Research Council. 2014. *Convergence: Facilitating Transdisciplinary Integration of Life Sciences, Physical Sciences, Engineering and Beyond.* Washington, DC: National Academies Press.

Sayer, J., and B. Campbell. 2003. "Research to Integrate Productivity Enhancement, Environmental Protection and Human Development." In *Integrated Natural Resources Management: Linking Productivity, the Environment and Development,* edited by B. Campbell and J. Sayer. Wallingford, UK: CABI Publishing.

van de Fliert, E., and A. Braun. 2002. "Conceptualizing Integrative, Farmer Participatory Research for Sustainable Agriculture: From Opportunities to Impact." *Agriculture and Human Values* 19: 25–38.

PART 6

Synthesis

Chapter 16

UNLOCKING AFRICA'S AGRICULTURAL POTENTIAL

Johannes Roseboom, Nienke Beintema, John Lynam, and Ousmane Badiane

The preceding chapters of this volume have covered the spectrum of challenges facing Africa south of the Sahara (SSA) in its efforts to develop an effective and efficient agricultural research and development (R&D) system at national, subregional, and regional levels. This chapter endeavors to synthesize the key messages of the individual chapters in order to present an overview of the actions required to unlock the inherent potential of agricultural R&D in the quest for faster growth and more broadly shared development outcomes. The discussion begins with a summary of why resolving the issue of underinvestment in agricultural R&D is fundamental to advancing the region's technical progress and, hence, raising agricultural productivity. Next, the case is made for the essential need to develop rural innovation capacity to motivate the adoption of new technologies by farmers and increase farm productivity. Finally, the chapter presents the key strategies needed to address current limitations and inefficiencies in agricultural R&D financing, human resources, organization and management, and systems-level structuring.

The Case for Increasing Investment in African Agricultural R&D

Having languished for decades, often below the rate of population growth, the rate of agricultural growth in SSA has both accelerated and spread more broadly across the region in recent years, although with significant national and subregional differences (Chapter 1, this volume). Not all countries—especially the smaller ones—partook of this acceleration, and for many it was insufficient, ruling out any hope of achieving the Millennium Development Goal target of halving the incidence of extreme poverty by 2020. Much of the agricultural growth recorded in Africa stems from expanded use of resources, some of which—like land—are finite; only a small amount of growth can be attributed to total factor productivity (TFP) growth (Chapter 3, this volume).

On the other hand, accelerated growth in agricultural gross domestic product (AgGDP) in recent years can be correlated with modest TFP growth after several decades of stagnation or decline. This is definitely good news, but the productivity of African agriculture remains very much lower than in the rest of the developing world, and the gap is increasing rather than decreasing (Chapter 3, this volume).

This low productivity undermines the competitiveness of African agriculture not only in the global market, but also in the home market. Africa is becoming increasingly food-import dependent, which most countries can ill afford (except, perhaps, for a few mineral-rich countries). Moreover, agriculture's contribution to the overall economy in many African countries remains substantially below what could be expected for countries at the same stage of economic development elsewhere in the world (Chapter 1, this volume). Looking to a future of rising world prices and production costs, African countries will pay a double penalty—in terms of the higher cost of imports and the lost opportunity to increase agricultural export earnings—if they fail to invest sufficiently in raising agricultural productivity. Population growth and urbanization would only exacerbate this reality, further increasing Africa's dependence on food imports and compounding food insecurity and poverty.

The imperative of productivity growth, and the case for investment in agricultural R&D to achieve it, derives from the reality that the current trajectory of growth based on increased resource use cannot be sustained. Smallholders remain the principal farming system responsible for the greater share of overall agricultural production, and increased productivity is essential for both higher incomes and increased market participation. Higher rural incomes also reduce the incentive for rural dwellers to migrate to urban areas, a trend that exacerbates pressures on urban infrastructure and employment opportunities. In short, a productive agricultural sector is essential for balanced and equitable growth in African economies.

Technical change is a necessary contributor to productivity growth, either through spill-ins or investment in domestic research capacity (Chapter 14, this volume). In seeking spillovers of agricultural technologies, low- and middle-income countries have relatively little to gain from high-income countries, because the commodities they produce and the agroecologies within which they produce them are too different. This is unfortunate, especially compared with other types of technologies, where such gains can be massive (consider information and communications technologies, for example). More often than not, this means that Africa's low- and middle-income countries have to rely either on their own agricultural R&D to develop and adapt the technologies they need, or on research capacity, such as CGIAR or regional research networks. SSA's large countries make a better return on their agricultural research investment than do the majority of small countries, which suggests that economies of scale are a key factor (Chapter 3, this volume). This has implications for how best to organize agricultural research (discussed further below).

Private investment in agricultural R&D in SSA is still very small and concentrated in a few select niches (such as hybrid seeds and export-oriented value chains) and in a few countries (most important, South Africa) that have private companies big enough to have their own research facilities and programs (Chapter 7, this volume). The most prominent private investors in agricultural R&D are multinationals, whose markets expand across national borders and, hence, whose technology platforms and staff can be shared across multiple countries. Factors that stimulate private investment in agricultural R&D include size and growth of agricultural input markets, limited government intervention in such markets, improved protection of intellectual property rights, and a stimulating R&D environment. For the medium-term future, however, most of the investment in agricultural R&D in Africa will have to rely mainly on government funding (derived from general tax revenues or through donor contributions).

As evidenced in various chapters in this book, underinvestment in agricultural R&D by governments in SSA has been widespread and persistent for decades. Only a few African countries have met the minimum investment target of 1 percent of AgGDP set by the New Partnership for Africa's Development (NEPAD) and United Nations. In fact, the region's average agricultural R&D intensity ratio has steadily declined in recent years (Chapter 4, this volume), which indicates that regional agricultural R&D spending has not kept pace with growth in agricultural output. Raising Africa's investment in agricultural research to 1 percent of AgGDP implies a major expansion of Africa's agricultural research capacity.

When AgGDP grows by 5 percent per year, agricultural research expenditures should increase by 5 percent as well, just to stay on par. Under the current scenario (whereby AgGDP continues to grow by 5 percent per year), agricultural research budgets need to grow at 10 percent per year for 15 years, in real terms, in order to double the research intensity ratio from 0.5 to 1.0 percent. This growth level may push the boundaries of what is feasible in terms of government budgets, the supply of qualified staff, management capacity, infrastructure expansion, and so on. Nevertheless, it illustrates the type of acceleration needed to realize this very crucial target, as has repeatedly been expressed in various high-level policy documents.

The Comprehensive Africa Agriculture Development Programme (CAADP) has set out to increase government support to agriculture, including agricultural R&D, to at least 10 percent of national budgets; however, only a few countries have achieved this target to date. Moreover, agricultural R&D does not score very high among reported government investment priorities. Investments in activities that generate diffuse and uncertain benefits far into the future are not very attractive to politicians who want to deliver quick and concrete results of benefit to their specific constituencies (Chapter 5, this volume). Politicians and policymakers need more education on the importance of sustained investments in agricultural R&D.

In summary, more investment in agricultural R&D is required to raise Africa's agricultural productivity and competitiveness and, in turn, to increase rural incomes and reduce poverty, reduce food-import dependence, increase food security, and put a stop to environmental degradation. Failing to do so will jeopardize the sustainability of the current recovery process and significantly reduce the prospects for equitable and balanced growth in African economies.

Developing Rural Innovation Capacity

Increasing productivity requires both the new techniques and knowledge arising from agricultural R&D and improvements in rural innovation capacity (Chapter 13, this volume). Innovation capacity in this context refers to the inherent capacity of farmers to innovate, as well as the institutional environment that facilitates such innovation. Understanding how to build the capacity for farmers to innovate has become a focus of CGIAR research in development work. Leeuwis et al. (2014, 5) have identified these capabilities as the capacity to

- continuously identify and prioritize problems and opportunities in a dynamic systems environment;
- take risks, experiment with social and technical options, and assess the trade-offs that arise from these;
- mobilize resources and form effective support coalitions around promising options and visions for the future;
- link with others in order to access, share, and process relevant information and knowledge in support of the above; and
- collaborate and coordinate with others during the above, and achieve effective concerted action.

Farmer innovation capacity, however, must be complemented by a conducive institutional environment that promotes market integration in particular. In the past, lack of access to markets and of associated support institutions has constrained the uptake of improved technologies. Nonetheless, such capacity has improved rapidly over the past decade or so following the structural adjustment and market liberalization of the mid-1990s. Input and output markets have improved significantly, although access to appropriately priced fertilizer and improved seed remains a particular constraint on productivity increases.

Several other major developments are helping raise farmer innovation capacity. For instance, rural banking is expanding, significantly aided by the capacity for cash transfers using cellular phones. The increased use of cellular phones in rural areas has also spawned improved access to information related to prices, market buyers, the quality control of inputs, and extension. Crop and livestock insurance is also being piloted based on expanded access to weather data. Access to many of these services, particularly credit and agricultural extension, are being provided through the enhanced formation of farmers' groups, which is significantly reducing the transaction costs involved while improving social capital.

There are obvious synergies between farmers' enhanced innovation capacity and improved access to input and output markets, credit, insurance, and information, and numerous organizational innovations are facilitating these linkages. One example is enhanced interaction between farmers and service delivery agents, which is occurring through the decentralization of public services, such as extension and veterinary services. Experimentation with innovation platforms is facilitating organizational linkages between markets, service delivery agents, and farmers, most often within the context of value chains. Such organizational innovation is key to the strategic objective of commercializing smallholder agriculture, which is often central to national CAADP investment plans. Organizational innovation—which most often builds on the formation of farmers' groups and the encouragement of collective action—underlies efficient markets and effective service delivery by reducing the transaction costs inherent in a smallholder agrarian structure, especially where transport infrastructure in rural areas is still underdeveloped. Over the past decade and a half, rural African economies have significantly enhanced their rural innovation capacity, which in turn has been responsible for the increases in smallholder agricultural productivity and in agricultural growth rates.

Enhanced rural innovation capacity creates much stronger and more dynamic demand for new technologies and management practices. The scope of the national research agenda to intensify African smallholder agriculture is extraordinarily wide, given highly diverse farming systems. The research system will need to respond in terms of (1) integrating higher-value crop and livestock activities into smallholder systems; (2) increasing the productivity of staple food crops; (3) integrating soil and water management techniques; and (4) enhancing crop, tree, and livestock management (including exploiting complementarities among the three). Simple reliance on improved agricultural inputs, particularly fertilizer and high-yielding varieties—while necessary—will not be sufficient to achieve sustained increases in overall farm productivity and TFP.

Given the complexity of smallholder farming systems and the spatial heterogeneity of these systems, managing the scope of the research agenda by national agricultural research institutes (NARIs) is especially complicated, particularly with limited budgets and human resource capacity. Regional approaches and links to international research networks will be essential. The organizational challenge of improving connectivity within an agricultural innovation system, while managing regional and international linkages, adds to the overall administrative and management challenges facing African national agricultural research systems (NARSs), as is discussed in the next section.

Increasing the Effectiveness and Efficiency of Agricultural R&D

Financial Resources

Organizations under financial stress often spend an extremely high share of their budgets on salaries. Not only is this inefficient, it can also seriously undermine the viability of research programs and negatively affect staff morale when research cannot be properly implemented due to lack of facilities, services, and equipment—from basic office and laboratory space, to the necessary agricultural inputs and vehicles, to computer equipment and software, and even to such fundamentals as water and electricity.

The relatively low level of investment in agricultural R&D in Africa is further constrained by the fact that it is also highly volatile, largely because of the patterns of donor funding, which often lead to "boom-and-bust" spending cycles. A certain level of volatility may be expected at the institute level, perhaps the result of large investments in training or infrastructure, but when it comes to the day-to-day operation of research programs or the maintainance of fundamental research infrastructure, volatile spending patterns are extremely counterproductive because they interfere with the planning, conduct, and efficacy of research. It is therefore important that governments provide stable and sustainable levels of funding, not just to secure researcher salaries, but also to enable necessary operating and capital expenditures (Chapter 4, this volume).

Generating revenues internally—either through the sale of goods and services (such as improved seed, laboratory tests, contract-based research, or income from intellectual property rights) or by winning national or international competitive research grants—remains a modest source of funding for most NARIs. Governments can be overly optimistic about the share of funding that can be generated internally, and the conditions and timeframe required to make it possible; for example, research partnerships take time to develop, developing winning research proposals requires skills and experience, and the framework for securing intellectual property rights may be totally absent (Chapter 4, this volume). Hence, governments cannot rely too much on NARSs being able to internally mobilize funding sufficiently and rapidly enough to meet the financial needs of their R&D systems.

Donor funding for agriculture and agricultural research has risen in recent years, after several decades of neglect. Such funding is being derived not only from traditional donors and multilateral agencies, but also from new donors—such as Brazil, China, India, and Saudi Arabia—and new philanthropic agencies—most notably, the Bill & Melinda Gates Foundation (Chapter 6, this volume). Funding by donors always comes with strings attached, including ideas on how best to tackle Africa's agricultural development challenges. Governments that rely too heavily on external funding for agricultural research risk having their research agenda diverted from national priorities. Such ideas are usually strongly influenced by donors' own experiences, such as China's achievements in rural development, India's Green Revolution, and Brazil's success in the Cerrado. Recipients of donor funding need to make sure that such ideas are in line with their own agricultural development strategies. Rather than relying too much on donors and development banks to fund critical research areas, governments need to more clearly identify their own long-term national priorities and design relevant, focused, and coherent agricultural R&D programs accordingly. Donor and development bank funding needs to be closely aligned with these national priorities, and consistency and complementarities among donor programs need to be ensured (Chapters 4 and 6, this volume).

CAADP's aim of raising government spending on agriculture to at least 10 percent of the government budget can help to secure additional funding for agricultural research, but it is not a guarantee. Many African governments seem to prioritize other agricultural spending options higher than agricultural R&D. More advocacy is definitely needed within the agricultural sector itself to raise national investment in agricultural research (Chapter 1, this volume).

Human Resources

Lack of sufficient and appropriately trained and experienced human resource capacity, particularly in terms of staff with relevant MSc- and PhD-level qualifications, still places a major constraint on the quality and volume of agricultural R&D outputs in Africa. Fundamental to building strong human resource capacity in agricultural research is the development of comprehensive recruitment, training, and succession plans to fill existing and anticipated staffing gaps, and establish proper career paths for researchers. Countries and institutions with uncompetitive salary and benefits packages need to take steps to redress these barriers. The ability to build human resource capacity, however, depends on the longer-term financial and institutional capacity to do so, and on the supporting or limiting factors inherent in the policy environment. The new donor-funded regional capacity-building initiatives could play an important role in rebuilding the region's cadre of agricultural researchers, but they will need to be upscaled to other countries and sectors (Chapter 8, this volume).

For the past few decades, student enrollment at African universities has grown rapidly, facilitated by the expansion of existing universities (including the upgrading of diploma-granting colleges to universities) and the establishment of new public and private universities. This growth took place under usually severe government budget constraints, forcing (both public and private) universities to seek part or all of their funding privately (student fees, consultancies, and so on). African universities are often overwhelmed by the number of graduate students to be trained. At the same time, many African universities are trying to move up the educational ladder and, hence, are increasingly offering specialized postgraduate programs leading to an MSc or PhD degree.

In the case of African faculties of agriculture, the number of students enrolled in most of these postgraduate programs is still quite small, which makes them relatively expensive and inefficient because they tend to lack critical mass. Moock (Chapter 10, this volume) argues that cross-country collaboration among faculties of agriculture is needed so that the necessary level of postgraduate program specialization and quality assurance can be achieved effectively and at a viable cost. This also involves a more innovative learning approach with students attending different providers based on their desired specializations, studying outside their home country, and taking advantage of the expanding capacity in distance learning. Strategic networks of African faculties of agriculture are already emerging to bolster university-based postgraduate training and research.

In most African countries, there is a short supply of MSc- and PhD-level agricultural scientists, which is a particular challenge for agricultural research organizations and faculties of agriculture, as they are confronted by a wave of retirement of postgraduate agricultural scientists trained overseas in the 1970s and 1980s. Opportunities for overseas training in agricultural sciences fell sharply after 1990 as a result of (1) a contraction in donor funding; (2) doubts about the effectiveness of overseas training—that is, lack of relevance to African conditions and low return rate; and (3) an assumption that local faculties of agriculture would be able to supply MSc- and PhD-level agricultural specialists with relevant skills, of a decent quality, and in sufficiently large numbers. This assumption has proved incorrect; hence, developing postgraduate programs of sufficient quality that address the needs of African agriculture and ensure appropriate research skills and methods remains a central challenge (Chapter 9, this volume).

Organization and Management

Agricultural research entities, like all organizations, need clear strategies to translate their mission and vision statements into concrete objectives and activities. Staying focused on core objectives can be difficult in the face of—often overwhelming—day-to-day problems. A strategy needs to be followed up with a concrete plan that prioritizes activities and their implementation over time, explicitly allocating the resources required and determining the necessary trade-offs between the costs and benefits of research. A number of economic tools are available, and NARIs should increase the use of these tools during their planning phases. This in turn requires investment in better databases, data collection, and analytical capacity (Chapter 11, this volume).

Monitoring and evaluation (M&E) complements the planning process by providing feedback on the implementation of planned activities, their results and outputs, and ultimately their outcomes and impacts. M&E tools are expected to provide accountability to government and funding agencies that money was well spent; enable implementers to learn from experience in terms of identifying problems, flexibly adjusting activities, and readjusting goals in real time; and reduce risk in decisionmaking. With some positive exceptions, such as the Kenya Agricultural Research Institute, weak M&E is widespread among agricultural research organizations in Africa; however, this mostly stems from a lack of proper planning to begin with. NARIs need to improve both their planning and their M&E efforts, which ultimately should go hand in hand. This is an obvious area where targeted training and experience are essential (Chapter 12, this volume).

While agricultural research organizations have traditionally focused on developing knowledge and technologies, they are now also being held accountable for the application of such knowledge and technologies. This accountability is forcing organizations to closely address how farmers innovate, what deters them from adopting, and what can be done to eliminate these barriers. Many factors other than just the technologies themselves will need to be put in place to ensure that innovations reach farmers and, hence, have impact. This demands a far more holistic approach to agricultural innovation and requires that researchers interact far more intensively across disciplines and among other stakeholders, including farmers. This in turn requires a massive retooling and mind shift of agricultural researchers, but also giving them the right training, incentives, support, resources, and flexibility to do so (Chapter 13, this volume).

Systems-Level Issues

Importanst systems-level developments need to be addressed, at both national and supranational levels, for African NARSs to be sufficiently effective and efficient. National issues include the following:

1. **Coordination among agricultural research agencies, particularly when their agendas align or overlap.** As NARSs evolve, they tend to become more complex, which calls for improved coordination mechanisms, often constrained to date by lack of resources, commitment, and goals. Nevertheless, a number of countries (Ethiopia, Kenya, and Tanzania, for example) are in the process of restructuring public agricultural research to streamline and coordinate their NARSs. Furthermore, agricultural research capacity at universities has become comparatively stronger over the years, which has added greater diversity and

complexity to African NARSs. Joint research planning and projects between faculties of agriculture and NARIs are recommended both to avoid duplication of effort and to maximize use of resources and research staff.

2. **The introduction of new funding modalities, such as competitive mechanisms and private contract-based research or public–private cofinancing of research.** This requires the necessary institutional arrangements and flexibility to mobilize multisource funding, as well as a more entrepreneurial attitude (Chapters 4 and 7, this volume).[1] An added benefit is that such funding instruments can be used to improve coordination and alignment among agencies.

3. **Improved linkages with farmers and private and nonprofit sectors to ensure more effective agricultural R&D.** NARIs provide necessary—but insufficient—inputs into the rural innovation process. To ensure the adoption of their research outputs, NARIs must interact with an expanding private sector and an increasing array of farmers' and civil society organizations that are important partners in identifying problems and validating technologies. These partners can also play a crucial role in lobbying government for support to agricultural R&D (Chapter 2, this volume). Through market liberalization and democratization, agricultural research has an opportunity to move from a state of near isolation to developing essential organizational linkages for rural innovation and smallholder development (Chapter 13, this volume).

4. **Improved linkages between agricultural research and extension providers.** The success of investment in agricultural research is heavily dependent not only on the quality of the research, but also on the strength of the links between research and extension providers. In addition, it is important that these two functions are structured in such a way that they do not compete with each other, especially for the same scarce resources, and that practical channels and incentives for interaction are fostered (Chapter 13, this volume).

---

1 Government grants also now come with more strings attached as governments are adopting results-based funding allocation; hence, new funding methods are often also used to cement linkages among the various entities involved in conducting agricultural research (such as research institutes and higher-education or private actors).

5. **The importance of convincing policymakers of the significant impact of agricultural R&D on national development goals** (Chapter 11, this volume). Policymakers need information about agricultural challenges and opportunities and the alternative solutions and options that may be available. Agricultural research can play a crucial and active role in informing the policy dialogue with such information. One of CAADP's key contributions is stimulating agricultural policy dialogue in participating countries, but this is an opportunity that agricultural research organizations have only partially seized. More can and should be done in this area.

Supranational-level systems issues have received significant attention over the past two decades, resulting in the establishment of the Forum for Agricultural Research in Africa (FARA), the subregional organizations (SROs), and more recently the subregional agricultural productivity programs funded through bilateral loans to the participating countries by the World Bank (Chapters 2 and 15, this volume). Both institutional developments aim to strengthen crossborder collaboration in agricultural research motivated by the potential efficiencies of joint resource use in the development of new technologies of mutual benefit to multiple countries.

Critical bottlenecks in addressing the supranational agricultural research agenda are the limited political and economic integration of Africa and the almost complete lack of African funds to finance joint initiatives. All supranational initiatives, particularly those implemented by FARA and the SROs, are heavily dependent on donor funding. This cannot continue indefinitely, so it is essential to start strategizing about regional funding. The "center of excellence" approach promoted by the subregional agricultural productivity programs avoids this problem by attempting to generate reciprocal spillovers of benefits among participating countries. Time will tell whether this approach will work and whether it can be sustainable in the absence of donor support.

In addition to these regional and subregional initiatives, CGIAR has continued to play a pivotal role in addressing African agricultural research issues at the supranational level. The coexistence of these various initiatives requires coordination to ensure efficient use of resources, effective coverage of topics, and alignment with CAADP priorities (Chapter 15, this volume). The Science Agenda for Agriculture in Africa, launched in 2014, is a step in this direction.

Conclusion

Agriculture in SSA is at a prospective tipping point. Agricultural growth has increased in the past decade, probably in response to the reforms of a decade before. This growth path, however, relies on the unsustainable tactic of increasing the use of finite resources. Shifting to a growth path based on increased productivity—as in the rest of the developing world—is essential if Africa is to increase rural incomes and compete in both domestic and international markets. The yield gap in African agriculture is significant; scenarios on feeding the world into the future highlight the need to increase Africa's agricultural production. Shifting to this growth trajectory will require building on evolving improvements in market efficiency, the expanded capabilities of cellular phones to deliver a range of services, the improved responsiveness of public services to decentralization, and the expansion of farmers' organizations—all of which, in effect, amounts to deepening rural innovation capacity. An essential component of innovation, however, is a continuous supply of improved agricultural technologies and management practices stemming from an effective and efficient agricultural research system.

The design of the agricultural R&D system in Africa must incorporate the small-country problem, the wide scope of research needs, and the heterogeneity in agroecological and socioeconomic conditions. These factors particularly affect the efficiency of agricultural research, especially in the context of limited government budgets and reliance on highly variable donor aid flows. Heterogeneity requires a significant adaptive research capacity that can most effectively be provided by NARIs. Moreover, economies of scale and scope in agricultural research point to regional approaches, with focused research programs engendering the necessary critical mass.

The basic "architecture" of such an agricultural R&D system in SSA is essentially in place—namely, CGIAR and other global research partners working with the SROs, in turn supporting the wide network of NARIs across the continent. Nevertheless, this so-called architecture has not yet coalesced into a fully *interactive* and *integrated* system with clear divisions of labor and effective subsidiarity. A significant reason for this is the underinvestment in national systems, together with maintenance of a wide scope of research programs, which reduces their effectiveness. National agricultural investment plans under CAADP have not solved this problem because agricultural R&D has not been given sufficient priority in such plans. A few "large" African countries do invest in agricultural R&D at a scale that allows them to spearhead new technologies; while they are critical in subregional programs in providing centers of excellence, they do not replace the capacity needed in the "small" country programs to adapt such new technologies to local conditions. Increasing national agricultural R&D investment remains a critical prerequisite for achieving balanced agricultural growth in Africa.

References

Leeuwis, C., M. Schut A. Waters-Bayer, R. Mur, K. Atta-Krah, and B. Douthwaite. 2014. *Capacity to Innovate from a System CGIAR Research Program Perspective.* Program Brief AAS-2014-29. Penang, Malaysia: CGIAR Research Program on Aquatic Agricultural Systems.

Contributors

Jeffrey Alwang (alwangj@vt.edu) is a professor in the Department of Agricultural and Applied Economics at Virginia Tech, United States. Jeffrey earned a PhD in agricultural economics from Cornell University, United States. His research focuses on analysis of effects of agricultural and other rural-focused policy on the poor in developing countries.

Ousmane Badiane (o.badiane@cgiar.org) is Director for Africa at the International Food Policy Research Institute (IFPRI), where he oversees the Institute's regional offices for West and Central Africa, based in Dakar, Senegal, and Eastern and Southern Africa, based in Addis Ababa, Ethiopia. Ousmane earned a PhD in agricultural economics from the University of Kiel, Germany. His research focuses on structural transformation, agricultural policymaking processes, and other agricultural and economic development issues in African countries.

Nienke Beintema (n.beintema@cgiar.org) is head of Agricultural Science and Technology Indicators (ASTI), which is led by IFPRI. Nienke earned an MSc in economics from the University of Groningen, The Netherlands. Her research focuses on international indicators of and trends in human, financial, and institutional capacity and development in national agricultural research systems in developing countries.

Samuel Benin (s.benin@cgiar.org) is a research fellow in IFPRI's Development Strategy and Governance Division. Samuel received a PhD in econometrics, natural resource economics, and development economics from the University of California–Davis, United States. His research focuses on the impact of public spending and on strategies that can better connect public investments and policies with development goals.

Derek Byerlee (dbyerlee@gmail.com) is an adjunct professor in the School of Foreign Relations at Georgetown University, Washington, DC. He earned a PhD in agricultural economics from Oregon State University, United States. His research focuses on agribusiness, land use, and agricultural science policy globally.

Julia Collins (j.collins@cgiar.org) is a research analyst in IFPRI's West and Central Africa Office. Julia earned an MSc in agricultural and applied economics from the University of Wisconsin–Madison, United States. Her research focuses on agricultural development issues in African countries.

Adipala Ekwamu (e.adipala@ruforum.org) is the Executive Secretary of the Regional Universities Forum for Capacity Building in Agriculture (RUFORUM), a network of 55 member universities in Africa. Adipala earned a PhD in plant pathology from Ohio State University, United States. His research focuses on plant pathology, crop improvement, networking and advocacy for higher education institutions, and engaging with policy- and decision-makers in efforts to create, institutional mechanisms, frameworks, and an enabling environment for favorable postgraduate training in Africa.

Howard Elliott (howardelliott@shaw.ca) is an independent consultant in the area of agricultural research policy and systems. At the time of contributing to this work he was a consultant to major regional, international, and private organizations, including the Food and Agricultural Organization of the United Nations, World Bank, Forum for Agricultural Research in Africa, Association for Strengthening Agricultural Research in Eastern and Central Africa, and Bill & Melinda Gates Foundation. Howard earned a PhD in economics from Princeton University, United States. His research focuses on university and research system development and regional collaboration.

Kathleen Flaherty (k.flaherty@cgiar.org) is senior research analyst with ASTI. Kathleen received an MSc in development studies from University College Dublin, Ireland. Her research focuses on agricultural research investment and capacity in low- and middle-income countries.

Keith Fuglie (kfuglie@ers.usda.gov) is an economist with the Economic Research Service, U.S. Department of Agriculture. Keith received a PhD in applied economics from the University of Minnesota, United States. His research focuses on the economics of technical change, international agricultural productivity, and science policy.

Anne-Marie Izac (a.izac@outlook.com) is an independent scientist and chair of the Independent Steering Committee of the CGIAR Research Program on Forestry, Agroforestry, and Trees. At the time of contributing to this work she was the chief science officer for the CGIAR Consortium. Anne-Marie earned a PhD in natural resources and environmental economics from the University of Western Australia. Her research focuses on the interface between economics and ecology in agriculture, specifically to increase resilience, productivity, and the design of new approaches to address complex systemic issues under global change.

John Lynam (johnklynam@gmail.com) is chair of the Board of Trustees at the World Agroforestry Centre. At the time of contributing to this work he was an independent consultant based in Nairobi, Kenya. John earned a PhD in applied economics at the Food Research Institute of Stanford University, United States. His research focuses on smallholder-led, agricultural development in Africa south of the Sahara.

Linden McBride (lem247@cornell.edu) is a PhD candidate in the Charles H Dyson School of Applied Economics and Management at Cornell University, United States. At the time of contributing to this work she was a research assistant with IFPRI's Development Strategy and Governance Division. Linden's research focuses on agricultural development and poverty targeting.

Joseph Methu (j.methu@asareca.org) is head of Partnerships and Capacity Development with the Association for Strengthening Agricultural Research in Eastern and Central Africa (ASARECA). Joseph earned a PhD in dairy cattle nutrition from the University of Reading, United Kingdom. His research focuses on animal production and agricultural innovation.

Tewodaj Mogues (t.mogues@cgiar.org) is a senior research fellow in IFPRI's Development Strategy and Governance Division. Tewodaj earned a PhD in agricultural and applied economics from the University of Wisconsin–Madison, United States. Her research focuses on the returns to and determinants of public expenditures in agriculture and rural development, and the governance of public service delivery.

Joyce Lewinger Moock (joycemoock@gmail.com) is a board director of the African Economic Research Consortium and of the International Network for Higher Education in Africa. At the time of contributing to this work she served on the International Advisory Panel of RUFORUM and was a member of the Patrons Advisory Group for the Consortium for Advanced Research Training in Africa. She is a former associate vice president for the Rockefeller Foundation and, since retirement, has served as a consultant to several foundations. Joyce earned a PhD in anthropology from Columbia University, New York. Her research focuses on capacity development in Africa.

Latha Nagarajan (lnagarajan@ifdc.org) is a senior economist at the International Fertilizer Development Center. At the time of contributing to this work she was a research associate in the department of food and resource economics at Rutgers University, United States. Latha earned a PhD in applied economics from the University of Minnesota, United States. Her research primarily focuses on private-sector R&D in agriculture, seed systems, technology adoption, and agricultural input markets.

Paul Nampala (p.nampala@ruforum) is the Grants Manager of RUFORUM. Paul earned a PhD in pest management and entomology from Makerere University, Uganda. His research focuses on crop-pest interactions, integrated pest management in farming systems, evidence-based policy analysis, multistakeholder engagements in the research for development continuum, and human-capital development.

George Norton (gnorton@vt.edu) is professor of Agricultural and Applied Economics at Virginia Tech. He received a PhD in agricultural economics from the University of Minnesota, United States. His research focuses on agricultural research evaluation, integrated pest management, and international agricultural development.

Moses Osiru (m.osiru@ruforum.org) is Deputy Executive Secretary of RUFORUM. Moses earned a PhD in plant pathology from Makerere University, Uganda, and has worked with CGIAR in Uganda, Malawi, and Mali as a regional Plant Pathologist. His current work focuses on higher agricultural education in Africa and strengthening the developmental focus of African universities.

Prabhu Pingali (plp39@cornell.edu) is a professor in the Charles H. Dyson School of Applied Economics and Management in the College of Agriculture and Life Sciences at Cornell University, United States, with a joint appointment as a professor in the Division of Nutritional Sciences, and Director of the Tata-Cornell Agriculture and Nutrition Initiative. Prabhu received a PhD in economics from North Carolina State University, United States. His research focuses on food policy.

Carl Pray (pray@aesop.rutgers.edu) is a distinguished professor in the Department of Agricultural, Food and Resource Economics at the School for Environmental and Biological Sciences of Rutgers, the State University of New Jersey, United States. Carl earned a PhD in economic history from the University of Pennsylvania, United States. His research focuses on agricultural science and technology policy in China, South Asia, Africa, and Latin America.

Nicholas Rada (nrada@ers.usda.gov) is a research economist with the Food Security and Development Branch in the Market and Trade Economics Division of the Economic Research Service, U.S. Department of Agriculture. Nicholas earned a PhD in agricultural and resource economics from Oregon State University, United States. His research evaluates the policies and factors accelerating long-term agricultural productivity growth in the rapidly emerging countries of Brazil, China, India, Indonesia, and Russia, and in less food-secure countries of Africa south of the Sahara.

Frank Rijsberman is the Chief Executive Officer of the CGIAR Consortium, which is the largest agri-food research partnership in the world, comprising 15 research centers and employing 10,000 staff in over 70 countries. Frank earned a multidisciplinary PhD in water resources planning and management, and civil engineering from Colorado State University, United States. He leads the implementation of CGIAR's vision through a portfolio of impact-focused research programs dedicated to increasing food and nutrition security, reducing rural poverty, and protecting the environment.

Johannes Roseboom (j.roseboom@planet.nl) runs his own innovation policy consultancy and works for a wide range of international clients. Johannes earned a PhD in social sciences from Wageningen University, The Netherlands. His research focuses on agricultural research investment, its impact pathways, and how to organize and manage agricultural research to secure optimal economic and social impact.

Harold Roy-Macauley (h.roy-macauley@cgiar) is director general of the Africa Rice Center, one of the 15 Centers of the CGIAR System Organization. At the time of contributing to this work, he was executive director of the West and Central African Council for Agricultural Research and Development. Harold earned a PhD in tropical plant biology from the University Denis Diderot, France. His research focuses on plant biodiversity improvement and agricultural development.

David Spielman (d.spielman@cgiar.org) is a senior research fellow in IFPRI's Environment and Production Technology Division. David earned a PhD in economics from American University, United States. His research focuses on agricultural science, technology, and innovation policy; seed systems and input markets; intellectual property rights; and agricultural research and extension.

Gert-Jan Stads (g.stads@cgiar.org) is ASTI's senior program manager. He earned an MSc in economics from the University of Maastricht, The Netherlands. His research focuses on international indicators of and trends in human, financial, and institutional capacity and development in national agricultural research systems in low- and middle-income countries.

Michael Waithaka (m.waithaka@asareca.org) is the manager of ASARECA's Market, Market Linkages and Trade theme. He holds a PhD in agriculture from Humboldt University, Germany. His research focuses on harmonizing and rationalizing regulations, standards, and procedures in key food-commodity sectors, and options for smallholder agricultural growth in face of climate change.

Fatima Zaidi (f.zaidi@cgiar.org) is a research analyst in IFPRI's Environment and Production Technology Division. Fatima holds an MSc in public administration from the School of International and Public Affairs at Columbia University, United States. Her research focuses on agricultural science, technology, and innovation; governance and input markets; and agricultural research and development.

Index

Page numbers for entries occurring in boxes are followed by a *b;* those for entries in figures, by an *f;* those for entries in notes, by an *n;* and those for entries in tables, by a *t.*